Chemistry of Winemaking

A. Dinsmoor Webb, *Editor*

A symposium sponsored by
the Division of Agricultural
and Food Chemistry at the
165th Meeting of the
American Chemical Society,
Dallas, Texas,
April 12–13, 1973.

ADVANCES IN CHEMISTRY SERIES 137

AMERICAN CHEMICAL SOCIETY
WASHINGTON, D. C.   1974

Library of Congress ⊂ℙ Data

Chemistry of winemaking.
(Advances in chemistry series, 137)

"A symposium sponsored by the Division of Agricultural and Food Chemistry at the 165th meeting of the
American Chemical Society, Dallas, Texas, April 12–13,
1973."

Includes bibliographical references.

1. Wine and wine making—Congresses.
I. Webb, Albert Dinsmoor, 1917- ed. II. American
Chemical Society. Division of Agricultural and Food
Chemistry. III. Series.

QD1.A355 no. 137 [TP544] 540'8s [663'.2'00154]
74-19186
ISBN 0-8412-0208-7   ADCSAJ 137 1-311 (1974)

Advances in Chemistry Series

Robert F. Gould, *Editor*

FOREWORD

Advances in Chemistry Series was founded in 1949 by the American Chemical Society as an outlet for symposia and collections of data in special areas of topical interest that could not be accommodated in the Society's journals. It provides a medium for symposia that would otherwise be fragmented, their papers distributed among several journals or not published at all. Papers are refereed critically according to ACS editorial standards and receive the careful attention and processing characteristic of ACS publications. Papers published in Advances in Chemistry Series are original contributions not published elsewhere in whole or major part and include reports of research as well as reviews since symposia may embrace both types of presentation.

CONTENTS

Preface vii

1. Winemaking as a Biotechnological Sequence 1
   F. Drawert

2. Chemistry of Grapes and Other Fruits as the Raw Materials
   Involved in Winemaking 11
   James F. Gallander

3. The Chemistry of Red Wine Color 50
   Pascal Ribéreau-Gayon

4. Chemistry of Winemaking from Native American Grape Varieties 88
   A. C. Rice

5. Chemistry of Wine Stabilization: A Review 116
   George Thoukis

6. The Present Status of Methods for Wine Analysis and Possible
   Future Trends 134
   Maynard A. Amerine

7. Malo–Lactic Fermentation and Winemaking 151
   Ralph E. Kunkee

8. The Enzymology of Malo–Lactic Fermentation 171
   Richard Morenzoni

9. Analytical Fractionation of the Phenolic Substances of Grapes and
   Wine and Some Practical Uses of Such Analyses 184
   Vernon L. Singleton

10. Wine Quality Control and Evaluation 212
    Richard G. Peterson

11. Chemical Aspects of Distilling Wines into Brandy 232
    James F. Guymon

12. Some Aspects of the Wooden Container as a Factor in Wine
    Maturation 254
    Vernon L. Singleton

13. The Chemistry of Home Winemaking 278
    A. Dinsmoor Webb

Index 307

CONTENTS

PREFACE

Man has been fascinated with fermentation since early civilization. Both the bubbling produced by the liberated carbon dioxide and the seemingly miraculous change from a sweet to a euphoric drink reinforced the fascination. Thus, very early many basic principles underlying present technology of beer and wine production were worked out by trial and error. Concomitantly, regional and individual specialized preference or tastes developed. Today's regional preferences for oakwood flavor in certain wines and for retsina or rosin flavor in others undoubtedly arose because oak timber was the most suitable wood from which to build barrels and, where there were no oak trees, it was necessary to seal the seams of animal skins with pitch to make them wine tight. From today's perspective it seems incredible that such a large and diversified practical art could have developed without scientific knowledge of the fundamentals involved. Nevertheless, it is only within the past century that chemists and biochemists have been able to work out the complicated steps involved in the glycolytic pathway.

Although Lavoisier studied some of the changes of fermentation as early as 1789, and Gay-Lussac in 1810 established the basic stoichiometry for conversion of hexose sugars to ethanol and carbon dioxide, it was only with the work of Louis Pasteur starting in 1848 and culminating in 1861 that it was demonstrated unequivocally that yeast cells were strictly necessary for fermentation. Pasteur demonstrated, in addition, that a number of the by-products of fermentation resulted from yeast activity. In the 50–60 years following Pasteur's pioneering discovery, numerous researchers have contributed the exquisite detail to the complicated pathway from hexoses to ethanol and carbon dioxide. With disclosure of these details came the knowledge of how to control and guide the winemaking process itself.

The wine industry in the United States has not seen even growth; it has been characterized by periods of rapid expansion interspersed with times of little growth. Our nationwide experiment designed to prohibit production, transport, and sale of alcoholic beverages effectively destroyed the commercial wineries and fine varietal vineyards of this country. As a side effect, however, it introduced many people to the mysteries of making wine at home and the reconstitution of fine varietal grape vineyards with higher yielding sorts that would withstand transcontinental transport. Today both the grape growing and winemaking industries are in a period of feverish expansion, and there is tremendous public interest in grapes and wines.

A. DINSMOOR WEBB

Davis, California
July 1973

Winemaking as a Biotechnological Sequence

F. DRAWERT

Institut für Lebensmitteltechnologie und Analytische Chemie der Technischen Universität Munchen, D-8050 Freising-Weihenstephan, München, Germany

The chemistry of winemaking is characterized by intensive metabolism of the ingredients, beginning with the grape and ending with wine. We therefore view the chemistry of winemaking as a biotechnological sequence. This sequence originates with the grape, proceeds to the grape destemming, crushing, and pressing technology and, finally, is decisively influenced by the fermentation. The ingredients of the grapes, their dependence on variety, climate, vineyard management (fertilization), and biochemical processes during the crushing and pressing operations, are important to the biosynthesis of aroma substances.

The chemistry of winemaking is still in the formative stages, and most of the chemical problems are analytical in nature. Wine contains hundreds of components (*1, 2, 3*) of widespread concentrations (ng-grams/l.). It is not clear what sensorial contribution each of these components makes to the taste and aroma of a particular wine. Neither is it clear how grape variety influences aroma and taste, even though the influence is obviously great. The varietal influences are pronounced in wine made from such grapes as Concord, Muscat, Burgundy, and Riesling, but analysis indicates little difference in such gross measures as alcohol content, extract composition, sugar content, acidity, and ash content. This implies that the sensorial characteristics of wine are derived from specific components. With modern analytical techniques, such as gas–liquid chromatography and mass spectrometry, it is possible to identify the specific components. In the past few years several groups (*1, 2, 3*) have made inventories of several hundred aroma components in various wines. However, experience and critical evaluation of the inventories suggest that the inventories are still insufficient to characterize wine adequately or to describe winemaking chemistry. We believe that wine production and winemaking chemistry are best understood by considering the specific components to be end products of a biotechnological sequence (2), beginning with the vine and ending with bottled wine.

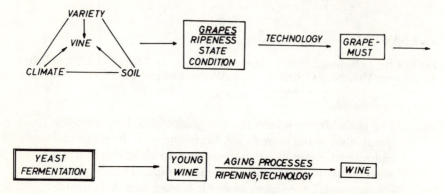

Figure 1.   The biological-technological sequence of wine

Results and Discussion

The Biotechnological Sequence. Figure 1 shows the biotechnological sequence for winemaking, beginning with the vine and including the effects of climate, soil, and grape variety. During the sequence, grape components are modified by ripening and by the physical condition of the berries. The technology involved in crushing may also greatly influence the components of the grape must. This influence is of particular importance (4, 5, 6) in the subsequent yeast fermentation caused by the extraordinary reduction potential and the ability of the yeasts to synthesize various aroma components depending on the composition of the

Table I.   Dependence of the Content

	Fertilizer			Nitrate,	Amino Acids, mg/liter					
Variety	N	K	P	N_2O_5	Total Amino Acids	His	Lys	Arg	NH_3	Asp
Aris[b]	0,5	2,0	2,0	3,5	960	+	+	480	29	39
	2,0	2,0	2,0		1598	20	+	801	39	22
	4,0	2,0	2,0		2651	15	+	853	82	57
	6,0	2,0	2,0	29,0	2972	36	+	828	87	70
Riesling	0,5	2,0	2,0	1,5	800	16	+	174	10	49
	2,0	2,0	2,0	2,0	1064	23	+	314	20	36
	4,0	2,0	2,0	4,4	2445	48	13	810	53	63

[a] Ser + Glu–NH₂ + Asp–NH₂.

substrate. Finally we have the aging processes. These depend, for example, on the reduction potential and/or the oxygen content and are very influential on the final character of the wine.

The components of the grape vary according to variety during the first stages of the sequence. This is shown by analytical data on sugar acids, amino acids, proteins, phenolic compounds, coloring matter, and aroma compounds in the wine. We should, however, be extremely careful in interpreting specific differences between samples of the same types of wines because the differences are only quantitative. The differences may change in magnitude depending on the climate, soil, and particularly on the kind of fertilization. Little is known, for example, about the influence of nitrogen fertilizer on the nitrogen content of grapes or on the nitrogen content of free amino acids.

Table I shows remarkably increasing amounts of free amino acids occurring in the grapes if the amount of fertilizer applied (ammonium nitrate) is increased. The grapes come from Aris (a new hybrid) and Riesling vines (7). The vines were grown in culture vessels according to Mitscherlich's method. The applied doses of potassium sulfate and calcium hydrogen phosphate were kept constant throughout the experiments. During the yeast fermentation, the amino acids enter the yeast metabolism with a definable biochemical valence with respect to the fermentation side products which are aroma substances. We recently succeeded in differentiating vine varieties (Figure 2) by using the distribution patterns of grape proteins which we obtained from disc electrophoresis and isoelectrical focusing in polyacrylamid gels (8, 9).

If we dye the enzymes distributed in the gels together with the proteins according to substrate specificity, we obtain the so-called zymograms for phenolase, peroxidase, esterase, and malatdehydrogenase (9). The resulting zymograms exhibit heterogenity of the isoenzymes, a prob-

of Free Amino Acids on Fertilization

Amino Acids, mg/liter

Thr	Ser[a]	Glu	Pro	Gly	Ala	Val	Met	Ileu	Leu	Phe
26	128	222	+	+	63	19	+	12	20	22
38	204	203	+	+	107	27	15	25	35	62
44	470	216	480	7,5	252	27	21	31	50	66
88	495	268	590	4,5	318	46	—	31	45	67
30	65	132	123	+	86	22	21	26	33	31
49	78	—	136	7,5	136	40	30	31	42	48
123	305	370	179	22,5	454	45	24	23	32	49

[b] New wine variety.

lem in its own right. This is noticed, for example, when purified poly-phenol oxidase (isolated from grapes)is allowed to react on catechol-containing base systems. When measuring the oxygen consumption with a Warburg apparatus, we found oxygen consumption to vary with the addition of certain amino acids (10) (Table II). Therefore, the following reactions are possible:

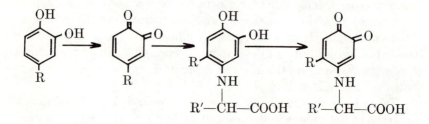

Grapes release numerous hydrocarbons in the adjacent gas volume. We have found alkanes, alkenes, and aromatic hydrocarbons (11). Accord-ingly, we extracted the skin waxes of the grapes with n-pentane and investigated the extracts by using a combination GLC–mass spectrometer. The total amounts of the wax extracts contained about 0.5% paraffins, 89% of which were saturated hydrocarbons ranging from C_{13} to C_{31}.

Table II. Consumption of Oxygen after Adding Amino Acids to the System Phenolase–Catechol (1,2-Dihydroxybenzene)

Phenolase + Catechol + Amino Acid	µl O_2/60 min
—	140
glycine	270
valine	230
methionine	230
threonine	220
arginine-HCL	200
histidine-HCL	170
gluthathione	110
aspartic acid	90
glutamic acid	70

About 11% were unsaturated hydrocarbons ranging from C_{14} to C_{29}. The main components were the odd-numbered, saturated hydrocarbons with carbon numbers 23, 25, and 27. The even-numbered hydrocarbons are found in lesser amounts and have a maximum at C_{24} (12) (Table III).

Considering the biotechnical sequence, it is possible to explain the presence of hydrocarbons in wine. To characterize a further step in the biotechnical sequence which bears a vital influence on the aroma events, an example from the many biochemical and chemical processes which take place in connection with the crushing and pressing of the grape berries is made. During crushing and at the moment of destruction

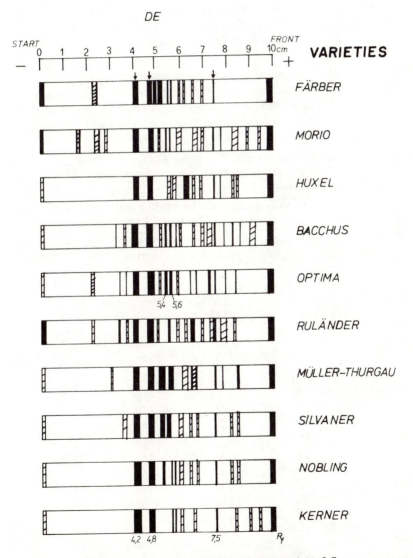

Figure 2. Protein pherograms (disc electrophoresis) of different grape varieties (harvest 1971). The separation distance was redrawn to a 10 cm scale. Scale: Rf-values, Arrows: Main bands common to all varieties.

of the cell structure, enzymatic–hydrolytic and enzymatic–oxidative processes occur with amazingly high speed (*13, 14, 15*). Depending on pH and temperature, the first ones cause a more or less rapid cleavage of natural esters; the effects of the latter are especially severe and are causing new synthesis of aroma substances.

To determine these enzymatic–oxidative processes quantitatively on a laboratory scale, we homogenized, for example, grape berries or vine leaves without inhibiting the enzymes in a phosphate buffer. Within 10 min, increased amounts of hexanal were apparent (as indicated in Table IV, column b); they increased from 90 to 410 μg/100 grams, and *trans*-2-hexen-1-al increased 130 to 13600 μg/100 grams. We proved in numerous, thorough experiments that these enzymatic–oxidative processes occur if oxygen (in air) had access to the substrate and if the enzymes had not been inhibited previously. Hexanal and hexenals are essential components of the grass taste and odor; this is why grape juices made

Table III. Hydrocarbons in the Waxlayer of Grapes (Riesling)
3.924 mg/10 kg
n-Alkanes: 3481 μg = *ca.* 88.8%
n-Alkenes: 443 μg = *ca.* 11.2%

Number of Carbon Atoms	n-*Alkanes*		n-*Alkenes*	
	μg/10 kg	%	μg/10 kg	%
10	tr[a]	—	tr	—
11	tr	—	—	—
12	tr	—	6	0.2
13	4	0.1	—	—
14	6	0.15	7	0.2
15	8	0.2	—	—
16	12	0.3	2	0.05
17	15	0.4	—	—
18	15	0.4	3	0.05
19	21	0.6	5	0.1
20	29	0.7	9	0.2
21	195	5.0	9	0.2
22	90	2.3	80	2.0
23	725	18.5	49	1.2
24	112	2.8	115	3.0
25	1090	27.8	52	1.3
26	72	1.8	51	1.3
27	550	14.0	30	0.8
28	30	0.8	15	0.4
29	312	8.0	10	0.2
30	15	0.4	tr	—
31	180	4.6	tr	—
32	tr	—	tr	—

[a] Traces.

Table IV. Formation of C₆-aldehydes and C₆-alcohols (μg/100 grams) in Grapes and in Vine Leaves

	Grape, Inhibition of Enzymes[a]		Leaf, Inhibition of Enzymes[a]	
	Instant	After 10 Min	Instant	After 10 Min
Hexanal	15	300	90	410
cis-3-Hexene-1-al	10	60	—	300
trans-3-Hexene-1-al	2	10	—	100
trans-2-Hexene-1-al	10	1700	130	13600
[c]	5	20	10	850
1-Hexanol	—	160	5	30
cis-3-Hexene-1-ol	5	500	150	2400
trans-2-Hexene-1-ol	3	170	170	260

[a] Sbl. 2-19-58—newly cultivated vine variety: European x American.
[b] Vine leaves of the same variety.
[c] Not identified for certain.

from relatively unripe pickings taste and smell so unpleasantly grassy. We reported extensively on the formation of C₆-aldehydes and -alcohols from grapes of various vine varieties and from apples (*13*).

By inserting radioactively labeled compounds and applying radio gas chromatography, we know that hexanal is formed from the enzymatic cleavage of linolenic acid; the hexenals are likewise formed by enzymatic cleavage of linolenic acid. These reactions may be briefly characterized and formulated on both of the reaction formulas of Figures 3 and 4. The aldehydes formed by cleavage can be reduced to their respective alcohols, enzymatically and during the yeast fermentation. These alcohols are also flavor substances. The behavior of the phenolic compounds during the crushing and mashing processes of the grapes should also be mentioned in this context. They pose, if taken from a purely analytical point of view, an additional problem of their own.

The next step of the biotechnical sequence, yeast fermentation, is of the utmost importance to the chemistry of winemaking as well as to the formation of flavor substances. We have investigated this previously using ¹⁴C-tagged compounds (*16*). Amino acids, for example, enter the yeast fermentation with a quasi biochemical valence with regard to the formation of metabolic side products like alcohols and esters. In that respect, the composition of the fermentation substrate, the grape must, is highly important to the formation of aroma substances by yeasts.

Materials and Methods

Fertilization experiments were made in Mitscherlich culture vessels and contained about 23 kg of sand and peat compost. There were two plants per pot and they were watered so that water and dissolved fertilizer dripped through the soil. The drip-water was collected for reuse. Five-tenths nitrogen means 0.5 gram N per pot.

The amino acids were analyzed from the press juices of the berries by an amino acid analyzer with ion-exchange columns as previously reported (17, 18). We described in detail the methods of disc electrophoresis and the isoelectrical focusing in polyacrylamid gels in another publication (8).

The polyphenol oxidase was obtained from extracts of the pressed grape juice sediment by acetone precipitation, purified on Sephadex G 25 and DEAE-Sephadex A 25, and further concentrated by ultrafiltration. Model experiments were done in a Warburg apparatus (10). The grape skin waxes were extracted with pentane. The resulting extracts were analyzed by a GLC (Varian Aerograph, model 1201) which was coupled via a helium separator according to Watson and Bieman (19) to a mass-

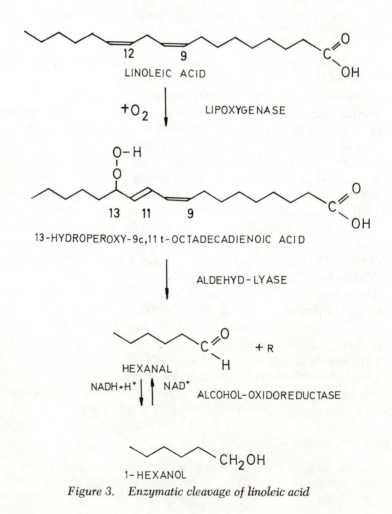

Figure 3.   Enzymatic cleavage of linoleic acid

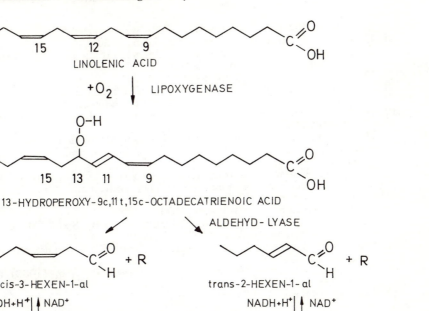

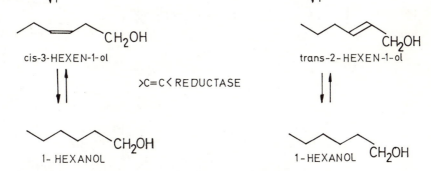

Figure 4.   Enzymatic cleavage of linolenic acid

spectrometer (Varian-MAT, Bremen, W. Ger., model CH7). The separation column was 2.5 m stainless steel, 1/16-inch diameter, filled with 10 wt % FFAP on Chromosorb (80/100 mesh). Carrier gas flow rate was 5 ml N_2/min . Temperature was 100°–270°C (2°/min). We have reported in detail the isolation and separation of C_6-aldehydes and -alcohols elsewhere (*13*).

Literature Cited

1. Webb, A. D., *Advan. Appl. Microbiol.* (1972) **15**, 75.
2. Drawert, F., Rapp, A., *Vitis* (1966) **5**, 351–376.
3. Drawert, F., Rapp, A., *Chromatographia* (1968) **1**, 446–457.

4. Drawert, F., Rapp, A., *Vitis* (1964) **4**, 262–268.
5. Drawert, F., Rapp, A., Ulrich, W., *Vitis* (1965) **5**, 20–23.
6. Drawert, F., Tressl, R., "Abstracts of Papers," 164th National Meeting, ACS, Aug.-Sept. 1972, AGFD 066.
7. Drawert, F., Alleweldt, G., unpublished data.
8. Drawert, F., Görg, A., *Chromatographia* (1972) **5**, 268–274.
9. Drawert, F., Görg, A., *Z. Lebensm. Unters. Forsch.*, in press.
10. Gebbing, H., "Über pflanzliche Polyphenoloxidasen," Doctoral Dissertation, University of Karlsruhe, 1968.
11. Drawert, F., "Moderne physikalisch–chemische Methoden in der Erforschung der Aromastoffe des Weines," in "Chemie und landwirtschaftliche Produktion," Wien, 1971.
12. Speck, M., "Enzymatisch–oxidative Bildung von Aldehyden und Aldehydcarbonsäuren beim Zerstören pflanzlicher Zellverbande und über die Oberflächenwachse von Bananen und Trauben," Doctoral Dissertation, Technische Universität München, 1971.
13. Drawert, F., Tressl, R., Heimann, W., Emberger, R., Speck, M., *Chem. Mikrobiol. Technol. Lebens.* (1973) **2**, 10–22.
14. Drawert, F., Heimann, W., Emberger, R., Tressl, R., *Naturwissenschaften* (1965) **52**, 304–305.
15. Drawert, F., Heimann, W., Emberger, R., Tressl, R., *Ann.* (1966) **694**, 200–208.
16. Drawert, F., Rapp, A., Ullemeyer, W., *Vitis* (1967) **6**, 177–197.
17. Drawert, F., *Vitis* (1963) **4**, 49–56.
18. Drawert, F., Heimann, W., Rolle, K., *Z. Lebensm. Unters. Forsch.* (1970) **144**(4), 237–245.
19. Watson, I. T., Bieman, K., *Anal. Chem.* (1964) **36**, 1135.

RECEIVED September 20, 1973.

Chemistry of Grapes and Other Fruits as the Raw Materials Involved in Winemaking

JAMES F. GALLANDER

Department of Horticulture, Ohio Agricultural Research and Development Center, Wooster, Ohio 44691

The chemical composition of raw fruits is significant in winemaking and wine quality. The chemistry of grapes is particularly important to winemaking, but the composition of other fruits such as apples, pears, cherries, and blackberries is also discussed. Sugars in fruits and their importance in winemaking are discussed. Since wine acidity is essential to biological stability and tartness, information concerning organic acids is presented for several fruits. The occurrence and significance to winemaking of other essential constituents of fruits such as vitamins, enzymes, pectins, minerals, aromatic volatiles, and nitrogenous and phenolic substances are also discussed.

With any processed food, the quality of the raw material is fundamentally important in the production and marketing of a successful product. This is true of wine, but defining the raw product quality in terms of the chemical composition is rather complicated. Not only do the chemical components of the raw product contribute directly to the sensory properties of wines, many are required in the fermentation process. During the chemical and biological processes of fermentation, the transformation of organic substrates such as sugars and nitrogenous substances yields many organic compounds that are essential to wines and their quality. Furthermore, the amount and type of products produced by alcoholic fermentation are greatly influenced by vinification practices. These steps of winemaking include harvesting, yeast selection, sulfur dioxide application, fermentation temperature, aging, and many others. Although certain guidelines have been established within each step, they are often varied depending on the desired wine type. Certainly the wine industry is endowed with many wine types, and new wines are constantly being developed. Further, the quality of wines is quite variable and is frequently related to the winemaker's art in the execution of many winemaking steps. These facts serve to illustrate the complexity of wines and winemaking and suggest the further complications introduced by variability of the raw product. Since wines are complex products and are known for their sensory sophistication, knowledge of the chemical composition of the raw material is especially critical in the production of a successful wine.

Considerable research has been conducted on the chemical composition of fruits commonly used in wines. The variability of fruit composition is well documented especially for grapes. Factors known to influence the chemical composition include variety, maturity, soil, rootstock, climatic conditions, crop yield, and post-harvest handling. Since most wines are derived from grapes, this paper will emphasize the chemical composition of grapes and its significance to winemaking along with a general account of changes in the chemical composition of grapes and other fruits. Although the subject has been partially covered in several texts (1, 2, 3), this paper will review and provide additional information of the chemistry of fruits with reference to winemaking.

Water

Water is the most important quantitative constituent of fruits and has an obvious influence on the yield of juice. Although this relationship is an important economic factor in wine production, there appears to be a lack of basic information concerning the per cent of water in fruits and its effect on juice yields. This is probably because juice yield depends on several factors such as variety, maturity, fruit condition, climatic conditions, type of crushers and presses, and method of fermentation. Amerine and Joslyn (2) reported that the average juice yield from grapes varies between 160 and 190 gallons per ton and is associated with a number of factors including the wine type, i.e., white or red wine. Occasionally, wines such as Sauternes of France are made from grapes infected with *Botrytis cinerea*. This mold shrivels the berry, thus moisture is lost and the relative sugar content is increased. These wine types are usually costly because the volume of juice is markedly reduced by the *Botrytis* infestation.

Since the moisture content of raw fruits varies widely, it is difficult to express typical values for water contents for various fruits. Amerine and Joslyn (2) stated that water in grapes ranges from 70 to 85%. The moisture content of several tree fruits grown in British Columbia was reported by Strachan et al. (4). The approximate moisture contents included: apple, 85.1%; pear, 83.9%; peach, 86.9%; and cherry, 78.4%.

The water contents of several soft fruits such as strawberry, 90.2%; raspberry, 85.0%; and grape, 77.0%, were obtained by Zubeckis (5). These values are only general averages because the moisture contents are influenced by many known factors.

Sugars

The major sugars of fruits which are normally used for making wine are glucose, fructose, and sucrose. The two reducing sugars predominate in grapes and occur in approximately equal amounts, and sugar content is one of the most important measures of grape maturity. Harvesting at the proper stage of maturity is essential for obtaining maximum wine quality. In California, sugar content (usually expressed in degrees Brix) together with total acidity is used to determine harvest maturity in grapes. Amerine and Winkler (6) demonstrated that the Brix/acid ratio is useful in classifying grape varieties for wine within certain regions. The same authors reported guidelines for table wine grapes as pH 3.0 to 3.35, acidity 0.65 to 0.90%, and Brix 19 to 23° (7). Studies have shown that the Brix/acid ratio for certain grape varieties and regions can be used to predict optimum wine quality (8, 9). Sugars are also fundamentally important in that the most essential wine component, ethyl alcohol, is derived from the fermentation of sugars. For many years, the alcohol content has been used to classify, tax, and evaluate wines. Amerine *et al.* (10) have stated that ethyl alcohol is one of the few wine constituents possessing odor, taste, and feel. The maximum theoretical amount of ethyl alcohol that can be derived through fermenting a 6-carbon sugar such as glucose is 51.1% of the original sugar content. However, in practice this alcohol yield is unobtainable for a number of factors including variability in the amount and type of sugar present in the must. Within a certain range, the amount of sugar occurring in the must is directly related to the final alcohol content of the wine. However, grape variety, season, wine type, and region all have an effect on the amount of alcohol predicted from a Brix reading of the must (11). Also, the optimum fermentation rate for grape juice occurs in the range 15–20° Brix (12). Above 25° Brix the must fermentation is retarded, and a sugar concentration higher than 70% will inhibit wine yeast activity (3). Gray (13) reported that the fermentation efficiency (alcohol produced per gram of sugar fermented) was diminished as the glucose concentration exceeded 5%. Since most fruits contain a high per cent of fermentable sugars (glucose and fructose) or easily hydrolyzed sugars such as sucrose, the type of carbohydrate is generally not critical in the alcoholic fermentation. In addition to ethyl alcohol, other wine components are derived from the fermentation and are important to the sensory properties of wines. These constituents include glycerol, acetic and lactic acids, higher alcohols, and acetaldehyde. Although they are produced from sugars during fermentation, their amounts in wines are generally dependent on a number of vinification practices such as fermentation temperature, must aeration, yeast strain selection, and sulfur dioxide application. Sugars also will influence sweetness in wines. The degree of sweetness (sweet to dry) is an essential sensory property of wines and is often scored in a critical examination of wines (10, 14, 15). The wine industry demonstrates the value of sweetness by controlling the sugar content in their wines. Judging the sweetness preference of pink wines, 12 experienced wine tasters were nearly divided in their preference for dry vs. sweet wines (16). The average threshold values for glucose and fructose in water were reported as 0.42 and 0.14 gram/100 ml, respectively (17). In a study concerning the influence of various wine constituents on sucrose thresholds, Berg et al. (18) found that ethyl alcohol enhanced the sweetness of sucrose while tannin increased the minimum detectable level. Sugars, particularly residual reducing sugars, have been suggested as a source of energy for lactic acid bacteria. Melamed (19) reported that malo–lactic fermentation tended to decrease the glucose and fructose content in wines. This secondary fermentation, usually following the alcoholic fermentation, has been observed in many wine regions and is generally promoted in high acidity wines.

Sugars identified in berries of Vitis vinifera L. var. Thompson Seedless included stachyose, manninotriose, raffinose, melibiose, maltose, sucrose, galactose, glucose, and fructose (20). Within this list it is generally known that glucose and fructose are the most important sugars in grapes on a quantitative basis. These sugars have been studied and reviewed extensively (21, 22, 23). Most investigators have found that glucose predominated in unripe grapes, and the glucose/fructose ratio is near 1.0 at maturity. Amerine and Thoukis (22) reported that ratios of these two sugars varied widely among several varieties at maturity from 0.72 (White Reisling) to 1.20 (Gamay Beajolais). Concentrations and relative amounts of glucose and fructose varied with variety and maturity (24). The glucose/fructose ratios for eight varieties were near 1.0 ± 0.1 at maturity, but the ratios decreased rapidly as the grapes became overly mature. Later, Kliewer (23) reported glucose and fructose concentrations in 78 virus-free grapes of early and late harvested fruits. He found that the ratios of 50 wine varieties varied between 0.74 and 1.05 with mean values of 0.94 and 0.85 for early and late harvested grapes, respectively. In every variety the glucose/fructose ratios remained the same or decreased with late harvested fruits. Kliewer (25) studied the glucose/fructose ratios of fruits of 26 species of Vitis and showed that the ratios ranged from 0.47 to 1.12. At maturity, only two species contained more glucose than fructose. Lott and Barrett (26) showed that the fructose/glucose ratios of 39 grape clones varied from 0.81 to 1.49, one clone was less than 1.00.

Carroll *et al.* (27) and Johnson and Carroll (28) studied the glucose and fructose concentrations in varieties of *Vitis rotundifolia*. They found that the glucose/fructose ratios were near 1.00 at maturity. The average fructose content of 12 muscadine varieties was 5.51% while the glucose content averaged 5.16% (27). Fructose and glucose were the primary sugars in six wine varieties which included two standard American varieties, *V. labrusca*, and four French hybrids (Table I). The range of these sugars in mature fruits varied widely (glucose 8.50 to 5.60 grams/100 ml, and fructose 10.60 to 6.40 grams/100 ml). The glucose/fructose ratios were relatively low compared with values obtained by other workers, with an average of 0.88.

Table I. Total Soluble Solids, Glucose, Fructose and Glucose-Fructose Ratio of Several Wine Varieties Grown in Ohio

Variety	Total Soluble Solids, °Brix	α-D-Glucose, grams/100 ml	β-D-Glucose, grams/100 ml	D-Fructose, grams/100 ml	Glucose/Fructose Ratio
Delaware	19.7	4.00	4.50	10.60	0.90
Catawba	16.9	2.75	3.80	7.70	0.85
Aurora	15.8	3.00	3.85	7.75	0.88
Villard blanc	13.6	2.50	3.10	6.80	0.82
Chancellor	15.9	2.60	3.35	6.40	0.93
DeChaunac	18.3	3.70	5.00	9.00	0.97

Carbohydrates other than glucose and fructose are usually found in trace amounts in grapes of *V. vinifera*. Kliewer (29) found the concentrations of sucrose and raffinose in eight *vinifera* varieties. At maturity, these sugars varied from 0.019 to 0.18% (sucrose) and from 0.015 to 0.34% (raffinose). Earlier, Amerine and Root (30) reported that sucrose was below 0.10% for three *vinifera* grapes. However, sucrose may be found in rather significant amounts in grapes of other commercially important species, *V. labrusca* and *V. rotundifolia* for example. Caldwell (31) reported that most varieties of *V. labrusca* and standard American hybrids contained sucrose. The maximum amount of sucrose was 5.57% and occurred in a *V. labrusca* variety. Sucrose was very erratic among the 66 varieties of this five-year study; it was not present in any variety for all five years and tended to be indicative of immaturity. Gallander (32) found up to 6.06% sucrose in the juice of 18 grape varieties grown in Ohio. Sucrose was not detected in the standard eastern varieties, Catawba, Concord, and Delaware. Carroll *et al.* (27) reported sucrose levels varied widely in 12 varieties of *V. rotunifolia.* In the popular variety Scuppernong, no sucrose was found, but the Hunt variety contained 5.20% sucrose.

Sugar contents in terms of reducing sugars after inversion has been reported for several fruits (5, 33, 34). Starchan *et al.* (4) determined the amounts of reducing sugars and sucrose in many tree fruits. For 86 apple varieties the average reducing sugar content was 8.37% while sucrose was 3.06%. Results of numerous determinations by Caldwell (35) of 82 American-grown French cider apples and other apples showed that the reducing sugar and sucrose contents varied widely among the varieties. Lee *et al.* (36) and Lott (37) found that fructose was the dominant sugar in apples, with sucrose second in quantity, and glucose in a relatively low concentration. Widdowson and McCance (38) reported that dessert apples contained 6.80% fructose, 3.62% sucrose, and 1.72% glucose. In pears, fructose generally exceeds glucose, and sucrose is the least abundant of the three sugars (38, 39). However, Lee *et al.* (36) found that the sucrose concentration in three pear varieties was greater than glucose. They reported that fructose averaged 6.77%, sucrose 1.61%, and glucose 0.95%. In contrast to apples and pears, the sugar content of cherries is mostly reducing sugars with very small amounts of sucrose (4, 36). An average sugar content of 13.6% for cherries was reported by Strachan *et al.* (4). The sucrose content ranged from 0 to 0.6% in the same cherry varieties. The peach is one of the few fruits in which the amount of sucrose exceeds the reducing sugar content. Strachan *et al.* (4) found that the average sugar content of several peach varieties was 8.96%, and sucrose comprised 70% of the total sugar content. Generally, the major soluble components of soft fruits are reducing sugars, with sucrose in low amounts (36, 40). Zubeckis (5) determined the total sugar content (invert sugar) for strawberries (4.74%) and raspberries (6.00%).

Carbohydrates other than glucose, fructose, and sucrose have been found in the above common fruits. However, these sugars are usually found in trace amounts and are quite variable. Trace sugars in these fruits have included mannose, galactose, arabinose, xylose, maltose, and sorbitol (36, 39, 41).

Organic Acids

Although their amounts are low in wines, organic acids are important constituents which affect the sensory properties of wines, particularly tartness (10). Berg *et al.* (17) reported both the threshold and minimum concentration differences for a number of acids found in grape wines. In a comprehensive study and review of acids, Amerine *et al.* (42)

determined the order of tartness for several wine acids, reporting malic, lactic, citric, and tartaric acids in decreasing order. The authors concluded that wine tartness was related to pH and titratable acidity.

Organic acids, particularly tartaric and malic acids, have a significant influence on wine pH. The importance of pH to wine color, spoilage, and stability has been investigated extensively. Red wines with high acidity (low pH) had more brightness than low-acid wines (*43*). In addition, the color of pink wines is generally improved by selecting low pH grapes (*44*). Since fixed organic acids buffer wines at a relatively low pH (3.0 to 4.0), a favorable condition exists for yeasts but not for spoilage bacteria. Table wines of pH 3.4 or below have more resistance to spoilage than wines above this value (*2*). Bacterial spoilage may occur at high pH values, and a pH above 4.0 is always hazardous. The growth of lactic acid bacteria in fermenting must, sometimes beneficial in high acidity wines, is also influenced by the pH, low values inhibiting malo–lactic fermentation (*45*).

Sulfur dioxide is used widely in the wine industry as a sanitizing agent and antioxidant (*46, 47, 48*). In the pH range of 3.0 to 5.0, a slight decrease in pH increases the antiseptic action of sulfur dioxide (*49, 50, 51*).

Organic acids and pH are also important factors which influence tartrate stability. An excess of potassium bitartrate occurs in newly fermented wines and is routinely removed by chilling or ion exchange. These steps are essential in providing a commercial wine with a high degree of clarity without sediment. The importance of tartaric acid and pH was established by using these wine components to calculate the potassium bitartrate concentration–product value (*52*). This value is recommended as a guideline test for determining tartrate stability in wines (*53*). Another wine disorder, iron clouds, is affected by the acidity and pH of the finished product. Amerine and Joslyn (*2*) stated that clouding of white wines by ferric phosphate *casse* occurs only in the pH range from 2.9 to 3.6.

The major organic acids of grapes are tartaric and malic. Malic acid is common to many fruits, but tartaric acid is rarely present in other fruits. Amerine and Winkler (*54*) reported that these two major acids constitute well over 90% of the total acid content of *vinifera* grapes. They also found that malic acid decreased more rapidly than tartaric acid during ripening, and tartaric acid was the predominant acid. Similar results were obtained by Amerine (*55*) for 12 *vinifera* varieties. The malic acid content decreased during ripening from 14 to 45% at maturity. Since the tartrate/malate ratios vary widely with variety, the *vinifera* wine varieties were classified by their low and high malate contents by Amerine (*56*). In an extensive study of 28 table and 50 wine varieties

Table II.   Total Titratable Acidity, pH, Total Tartrates, Total Malates,

Variety	Year	Total Titr. Acidity, grams Tartaric Acid/100 ml	pH
Concord	1971	0.80	3.37
	1972	0.64	3.38
Catawba	1971	0.94	3.08
	1972	0.85	3.16
Chancellor	1971	0.99	3.26
	1972	0.81	3.16
DeChaunac	1971	1.19	3.06
	1972	0.86	3.20
Vidal 256	1971	1.02	3.06
	1972	0.92	3.08
Villard blanc	1971	1.03	2.93
	1972	0.99	3.08

of *Vitis vinifera*, Kliewer *et al.* (57) found that concentrations and relative amounts of these two major acids and their salts varied widely with variety and maturity. The total titratable acidity of early and late harvested wine varieties was 0.75 and 0.58, respectively. The varieties also were classified into high, moderately high, intermediate, and low malate types based on the tartrate/malate ratios. Studies have also shown that high temperatures and light intensities during ripening tend to decrease the total acidity and malates in grapes (58, 59). An interesting result by Kliewer (58) indicated that the commercially important grape species of the United States, *Vitis vinifera*, *V. labrusca*, and *V. rotundifolia*, and *V. Spp.* (from Afghanistan) were the only species of 26 species which had a greater percentage of their total acidity from tartrates rather than malates. Recently, concentrations of tartaric and malic acids were reported for mature *vinifera* grapes grown in Arizona (60). The malic acid concentrations were lower than tartaric acid in all grapes except for the variety Robin.

Generally, grapes grown in eastern United States were relatively high in total acidity (31, 61, 62). High acidity values were also obtained in recent studies of evaluating grape varieties and selections for the eastern wine industry, United States and Canada (63, 64, 65). In Ohio, the total acidity expressed as tartaric acid for 41 grape varieties and selections at maturity ranged from 0.46 to 1.50% and averaged 0.96% (66). Zubeckis (5) reported similar results in Canada with an average of 1.01% total acid expressed as tartaric acid and ranging from 0.43 to 1.57%. Although there appears to be a lack of information concerning organic acids in varieties of *V. labrusca*, tartaric acid is considered the dominant acid. Studies have indicated that the major acid in the Concord

and Tartrate-Malate Ratio of Several Wine Varieties Grown in Ohio

Total Tartrates, grams Tartaric Acid/100 ml	Total Malates, grams Malic Acid/100 ml	Tartrate/Malate Ratio
0.57	0.41	1.39
0.37	0.21	1.76
0.65	0.48	1.35
0.43	0.33	1.30
0.67	0.69	0.97
0.66	0.34	1.94
0.71	0.84	0.85
0.53	0.37	1.43
0.74	0.30	2.47
0.63	0.37	1.70
0.79	0.30	2.63
0.63	0.27	2.33

variety of V. *labrusca* was tartaric acid (*67, 68, 69*). Also, malic acid was the second major acid in Concord. These results were in contrast to those reported by Amon and Markakis (*70*) who found that malic acid was the dominant acid and tartaric acid was slightly lower in concentration.

The concentration of total tartrates exceeded total malates in two *labrusca* varieties, Concord and Catawba, for two years (Table II). These varieties were harvested at maturity, and the tartrate/malate ratios were nearly the same for each season. According to Kliewer *et al.* (*57*), and based on their tartrate/malate ratios, these *labrusca* varieties are classified as moderately high malate varieties. Generally, the ratios of the other varieties (French hybrids) varied widely between variety and season. These ratios ranged from 0.85 to 2.63 with Villard Blanc being a low malate variety. In contrast to most varieties of V. *vinifera* and V. *labrusca*, high levels of malate compared with tartrate were found in several V. *rotundifolia* varieties (*27*). The average malate content was 0.50% while the tartrate concentration averaged 0.26%.

Several organic acids other than tartaric and malic acids have been identified in grapes. Kliewer (*20*) found the following acids in V. *vinifera*: malic, tartaric, citric, isocitric, ascorbic, *cis*-aconitic, oxalic, glycolic, glyoxylic, succinic, lactic, glutaric, fumaric, pyrrolidonecarboxylic, α-ketoglutaric, pyruvic, oxalactic, galacturonic, glucuronic, shikimic, quinic, chlorogenic, and caffeic. Many of these acids were also identified and determined quantitatively in the *labrusca* variety Concord (*70*).

The major organic acids in fruits other than grape are usually citric and malic acids (Table III). However, quinic acid, in addition to citric and malic acids, was a principal acid in peaches (*71, 72*). The main acid

Table III.   Organic Acids

Per Cent Acids in Fruits as grams/100 ml or 100 grams

Fruit	Citric	Malic	Quinic	Succinic
Apple	tr–0.10	0.10–1.00	0.05–0.24	tr
Pear	<0.39	0.14–0.88	0.05–0.62	<0.01
Blackberry	tr	0.57–0.82	tr	tr
Black raspberry	1.89	0.08		
Peach	0.03–0.25	0.34–0.60	0.09–0.38	tr
Strawberry	0.61	0.11		
Cherry	major			

constituent of apple, pear, and blackberry was identified as malic acid, with lesser amounts of citric, quinic, and succinic acids. Citric acid was reported to be the predominant acid in black raspberry, strawberry, and cherry. Other organic acids have been reported and are generally found in trace amounts (Table III).

Nitrogenous Substances

Nitrogen compounds are essential for the growth and development of yeasts in the fermentation process, and they greatly influence wine quality. Certain of these compounds, particularly amino acids, may be stimulatory for lactic bacteria which are responsible for malo–lactic fermentation (45, 73, 74, 75). The utilization of nitrogen compounds by yeasts during fermentation has an important influence on the keeping quality of wines. If these compounds are not completely utilized during fermentation, refermentation and bacterial spoilage may occur in the finished wines.

Ammonia and ammonium salts are of particular importance in satisfactory development and reproduction of yeasts during alcoholic fermentation. Tressler *et al.* (76) found that adding ammonium salts stimulated the fermentative ability of yeasts in sparkling ciders. To successfully ferment other fruit wines, addition of extra nitrogen in the form of urea or ammonium phosphate was recommended by Yang (77) and Fuleki (78). However, grape musts usually contain sufficient quantities of nitrogenous substances to support an active fermentation. In fact, the inherent ammonia content of grape musts is recognized as one of the most important nutritional factors in wine fermentation and has been used as an indicator of fermentation rates (8, 79, 80).

The ammonia nitrogen of grapes is quite variable. Studies have shown that the ammonia content of musts is associated with maturity,

Reported for Several Fruits

Other Acids	References
citramalic, galacturonic, chlorogenic, shikimic, *p*-coumarylquinic, lactic, phosphoric, caffeic	*72, 249, 250, 251*
galacturonic, chlorogenic, lactic, shikimic, tartaric, phosphoric	*252, 253, 254*
isocitric, lacto-isocitric, oxalic, shikimic, galacturonic	*72, 255, 256*
galacturonic	*72, 256*
galacturonic, phosphoric	*71, 257*
galacturonic	*72, 256*
	72

variety, rootstock, location, and season (*8, 81, 82*). The ammonia values of grape juices from California varied between 24 and 309 mg/liter with an average of 123 mg/liter (*82*). The juices were also analyzed for total nitrogen and contained an average of 849 mg/liter. In France, Lafon-Lafourcade and Peynaud (*83*) found that musts contained an average of 70 mg ammonia/liter with a total nitrogen content of 454 mg/liter. Results of a study by Lafon-Lafourcade and Guimberteau (*84*) indicated a decrease in ammonia nitrogen during maturation for the varieties Cabernet Sauvignon and Merlot at two locations. They reported a range of ammonia content of 45 to 89 mg/liter (average 63 mg/liter) for the grapes at maturity. The average total nitrogen content for both varieties was 350 mg/liter. In Ohio, the ammonia contents of grape juices from nine varieties of *V. labrusca* and French hybrids were determined for two seasons (*85*). The values ranged from 15 to 86 mg/liter, with an average of 34 mg/liter, and tended to be less than those reported in the other studies. This difference may be associated with maturity, variety, season, climatic conditions, and known factors.

Information concerning the ammonia content of fruits other than grapes is limited. Burroughs (*86*) reported the ammonia content of eight sweet ciders averaged 0.2 mg/liter. This value was lower than the ammonia contents of juices obtained in Ohio (*85*). The average ammonia content of juices from the apple varieties Red Delicious and Golden Delicious was 4.5 mg/liter. This preliminary investigation also determined the ammonia content of other fruit juices. The ammonia values (mg/liter) for the fruits included strawberry, 21.4; blackberry, 122.4; and cherry, 156.3. Ammonia in peach and pear juices was not detectable in this study. Since this investigation was preliminary and many factors are known to affect the ammonia content of fruits, an extensive study seems necessary before typical values can be established for these fruits.

Unlike ammonia values, total nitrogen contents have been established for many fruits (77, 87). Apples and pears are relatively low in total nitrogen as compared with other common fruits used in winemaking. Yang (77) reported the average nitrogen content of apples and pears of 0.01 and 0.03%, respectively. The total nitrogen content of cherries and blackberries was approximately 0.08%, and it was slightly less than the 0.10% in vinifera grapes. Burroughs (87) reported the total nitrogen content of several fruits, including strawberries, which ranged from 150 to 300 mg/100 grams.

Certain amino acids are good sources of nitrogen and are readily assimilated by yeasts during fermentation (88). Castor (89) reported that 11 amino acids were largely removed from musts during yeast fermentation.

The formation of higher alcohols is associated in part with amino acids (90). Higher alcohols or fusel oils may occur as taste components in wines. Taste thresholds of isoamyl alcohol ranged from 100 to 900 ppm (average 300 ppm) in dry white wines for seven panelists (91).

An extensive review of analytical methods used in determining amino acids in grapes and wine was reported by Ough and Bustos (92). This paper also includes a comprehensive review of the amino acid composition of musts and wine. In addition, Amerine (21), Amerine and Joslyn (2), and Amerine et al. (3) reviewed amino acids in grapes and wine.

The literature indicates that most studies pertaining to amino acids have been devoted to grape varieties of the species V. vinifera. In general, investigations have shown that grape varieties of the species V. vinifera and V. labrusca are qualitatively similar in amino acid composition. However, the concentrations of amino acids vary over a wide range, and the varietal effect is significant (83, 93, 94, 95, 96, 97, 98, 99, 100).

The amounts of amino acids in grapes as reported by several investigators in the United States are shown in Table IV. Most of the reported results were obtained from studies concerning the amino acid composition of vinifera grapes grown in California. The remaining results were from studies conducted in Ohio on the amino acid contents of V. labrusca grapes and French hybrids. Although the amount of each amino acid varies over a wide range, several are found in relatively large concentrations in grapes. These amino acids include alanine, arginine, glutamic acid, proline, and serine. In vinifera grapes, proline and arginine were reported to be the predominant amino acids, with glutamic acid and alanine next in importance (99, 100). In addition, Ough (101) reported that proline accounted for 5–43% of the total nitrogen in the juice from vinifera grapes, depending on variety and maturity. Studies of grapes with V. labrusca parentage indicated that alanine was the major amino acid (*98, 99, 100*). Kliewer (*100*) suggested that V. *labrusca* may possess the gene responsible for the large amount of alanine. Both investigators also reported large amounts of arginine, glutamic acid, and serine in labrusca-type grapes. Recently, glutamine was found in a surprisingly high concentration (416 mg/100 ml) in the *labrusca* variety Concord (*102*). Since this study was preliminary, and some artifact may be involved in the analytical method, this high level of glutamine is uncertain at this time.

Changes in amino acid composition in grapes are influenced by maturity, season, location, and environmental conditions as reported by several workers (*8, 84, 96, 101, 103, 104, 105*).

A limited survey of amino acids occurring in some fruits is presented in Table IV. In general, the major amino acids in apples were reported to be asparagine, aspartic acid, and glutamic acid. Moderate amounts of serine, α-alanine, γ-aminobutyric acid, valine, isoleucine, and 4-hydroxymethylproline were found in most apple varieties (*86*). Amino acid patterns and levels in pears were similar to those in apples except that pears usually contained large amounts of proline. In addition, pears contained a relative abundance of serine and alanine and less asparagine, aspartic acid, and glutamic acid. The occurrence of a high level of 4-hydroxymethylproline was also found in pears (*86*). This amount and its presence in apples and pears may be unique to these fruits (*106*). Also, the occurrence of 1-aminocyclopropane-1-carboxylic acid was reported by Burroughs (*107*) in apple and pear juices.

In cherries, alanine and proline were the predominate amino acids (Table IV). Other amino acids in cherries included moderate amounts of γ-aminobutyric acid, asparagine, cysteine, glutamine, isoleucine, serine, tyrosine, and valine. Silber *et al.* (*108*) observed a high level of pipeolic acid which may be the primary amino acid in cherries. Substantial amounts of alanine, asparagine, glutamic acid, and glutamine were reported in strawberries, asparagine and glutamine usually found in the largest amounts. Proline is absent in strawberries. Blackberries were reported to contain large amounts of asparagine and glutamic acid with moderate levels of glutamine, serine, and alanine. Silber *et al.* (*108*) found medium amounts of alanine, aspartic acid, γ-aminobutyric acid, valine, and leucine in peaches.

Commercial wines are commonly tested for protein stability. Wine proteins, upon denaturation by heat or cold, may cause cloudiness and unsightly deposits after bottling. In addition, proteins may combine with iron and copper salts to form flocculate material in bottled wines. The reaction and absorption of proteins on bentonite is an effective means of removing protein from wines (*109, 110, 111*). Therefore, fining wines

Table IV.   Amino Acid Content

Amino Acid	Range, mg/100 ml Grape
Alanine	5.0–167.1 (7,8,11,12,13,17)[a]
γ-Aminobutyric acid	2.8–131.1 (8,11,12,13,17)
Arginine	7.0–236.0 (3,4,5,6,7,8,11,12,13,14,17)
Asparagine	<4.8 (2,8,17)
Aspartic acid	2.2–30.6 (3,4,6,7,8,11,12,13)
Cysteine	
Cystine	0.2–2.1 (3,4,6,7,17)
Glutamic acid	5.4–182.3 (3,4,5,6,7,8,11,12,13,17)
Glutamine	<415.9 (8,17)
Glycine	0.3–3.1 (3,4,6,7,8,17)
Histidine	<14.3 (3,4,6,7,8,17)
Isoleucine	0.2–10.0 (3,4,6,7,8,17)
Leucine	0.3–8.5 (3,4,6,7,8,17)
Lysine	0.3–4.4 (3,4,6,7,17)
Methionine	0.3–1.8 (3,4,6,7,8,17)
Ornithine	tr (17)
Phenylalanine	2.9–7.8 (3,4,6,8,17)
Proline	0.9–349.0 (5,7,8,11,12,13,14,15,17)

Reported for Grapes and Other Fruits

Apple	Pear	Cherry	Strawberry	Blackberry
Relative Amounts[b] of Amino Acids				
++ (1,9,10,16,17)	+++ (1)	+++ (16,17)	+++ (2,16,17,18)	++ (2)
++ (1,9,10,16,17)	++ (1)	++ (16,17)	++ (2,16,17)	++ (2)
++ (1,9)	+ (1)		++ (17)	+ (2)
+++ (1,9,10,16,17)	++ (1)	++ (16,17)	+++ (2,16,17,18)	+++ (2)
+++ (1,9,10,16,17)	+++ (1)	+ (16)	++ (2,16,18)	++ (2)
++ (2,17)		+ (17)	+ (2,17)	
++ (17)		+ (17)	+ (17)	
+++ (1,9,10,16,17)	+++ (1)	++ (16,17)	+++ 2,16,17,18	+++ (2)
++ (1,10,16)	+ (1)	++ (16)	+++ (2,16,18)	++ (2)
++ (1,9,10,17)	+ (1)	+ (17)	+ (2,17)	
++ (1,9)	++ (1)	++ (17)	++ (17,18)	
++ (1,10)		++ (17)	+ (2,17,18)	+ (2)
+ (1)	+ (1)	+ (17)	++ (17)	
		+ (17)	+ (17)	
		+ (17)	++ (17)	
+ (1,9)	+ (1)	+ (17)		
++ (1,9,10,16)	+++ (1)	+++ (16,17)		+ (2)

Table IV.

Amino Acid	Range, mg/100 ml Grape
Serine	5.0–55.0 (5,7,8,11,12,13,17)
Theronine	4.2–42.4 (5,7,8,11,12,13,17)
Trytophan	2.2–6.5 (3,4,6)
Tyrosine	1.0–4.6 (3,4,6,8)
Valine	3.6–9.4 (3,4,6,7,8,17)
Pipeolic acid	tr (17)

[a] References:
1—Burroughs (86)
2—Burroughs (87)
3—Castor (93)
4—Castor (89)
5—Castor and Archer (258)
6—Castor and Archer (88)
7—Gallander et al. (98)
8—Gallander (102)
9—Hulme and Arthington (259)
[b] +++ major, ++ minor, + trace.

with bentonite is a widely used method for obtaining protein stability (112, 113).

Bayly and Berg (114) studied wine proteins and their effect on wine stability. They found that the determination of total protein was not reliable in predicting wine stability. They reported juice proteins of 11 samples ranging between 20 and 260 mg/liter with an average of 121 mg/liter. Koch and Sajak (115) found soluble proteins in the juice of six grape varieties, but the nature of the proteins was not identical for each variety. Recently, Beelman (116) reported the protein content in cold- and hot-pressed Concord grape juice was 92 and 80 mg/liter, respectively. During the first two weeks of fermentation, he observed a rapid increase in protein content to approximately 280 mg/liter. Twenty weeks later the protein content decreased to about 215 mg/liter. This value was considerably greater than the protein content of the original juice samples. In this regard, Bayly and Berg (114) found an occasional increase in total protein content of California musts following fermentation. They stated that this increase was attributable to peptides released by yeasts during fermentation.

Other fruits generally have low levels of protein. An extensive review of fruit proteins has been reported by Hansen (117). Also, a

Continued

Relative Amounts[b] of Amino Acids

Apple	Pear	Cherry	Strawberry	Blackberry
++ (1,9,10)	+++ (1)	++ (16,17)	++ (2,16,18)	++ (2)
++ (1,10)	+ (1)		+ (2,18)	+ (2)
+ (9)				
		++ (17)		+ (2)
+ (1,9,10)		++ (17)	+ (2,18)	+ (2)
+ (1)		+ (17)		

10—Hulme and Arthington (260)
11—Kliewer (105)
12—Kliewer (99)
13—Kliewer (100)
14—Kliewer and Ough (261)

15—Ough (101)
16—Rockland (262)
17—Siber et al. (108)
18—Tinsley and Bockian (263)
[b] +++ major, ++ minor, + trace.

review by Hulme (39) on apples and pears includes a section of nitrogenous substances including proteins. The protein contents of apples, pears, blackberries, black raspberries, and strawberries varied from 0.02 to 0.56% (118, 119, 120). Within this list of fruits, apples and pears were lowest in protein content.

Phenolic Substances

Phenolic substances significantly affect color, appearance, taste, and body of wines. Complete accounts of phenolic compounds in grapes and their influence on wine quality have been reported by several workers (2, 3, 21, 121). An excellent contribution to the grape and wine phenolic literature was written by Singleton and Esau (122). This book includes over 1200 references and covers the importance of phenolic substances in grapes and wine. Several reviews pertaining to the significance of phenolic compounds in foods have also included some discussions on phenolics in grapes or wine (123, 124, 125). A comprehensive bibliography with brief abstracts regarding grape and wine anthocyanin research was provided by Mattick and Weirs (126). Also, an excellent summary of knowledge regarding grape anthocyanins was reported by Webb

(*127*). Recently, an extensive review of fruit phenolics was given by Van Buren (*128*). The available literature indicates that a tremendous amount of research effort has been applied to phenolic components of fruits, particularly grapes and their important role in winemaking and quality.

Color is one of the most important quality attributes of wine. This character is routinely scored during the sensory examination of wines (*10*). The color of red wines is primarily derived from fruit anthocyanins through heat extraction or fermenting-on-the-skins. The amount of color in grapes and other fruits is affected by variety, maturity, season, location, climatic conditions, and other known variables. Winkler and Amerine (*43*) reported that grapes from warm seasons and regions, other factors being equal, tended to have less color. They also reported that the anthocyanin content of Zinfandel grapes increased with ripening to 26.2° Brix. However, Clore *et al.* (*129*) found a decrease in the anthocyanin content of grapes at a late harvest date. An interesting result by Singleton (*130*) showed that the tendency of browning in juice was greatest at the onset of anthocyanin synthesis in red grapes. Further, the color stability and browning of red wines, particularly in aging wines, is associated with anthocyanin pigments. Amerine and Winkler (*131*) reported that the color loss during wine aging depends on the variety of grape and the season. More recently, a study of New York State wines found that certain anthocyanins, diglucosides, were more stable to decolorization than the corresponding monoglucosides (*132*). However, the diglucosides were found more susceptible to browning.

Vinifera grapes and their wines contain few or no diglucosides (*2, 122*). This is contrary to *V. labrusca* and some direct-producer varieties (French hybrids) which contain diglucosides in relatively large amounts (*133, 134, 135*). Therefore, a test for diglucosides is used to distinguish pure *vinifera* wines and *labrusca* or other wines containing diglucosides. In contrast, a method for the detecting adulteration in Concord grape juice (*V. labrusca*) by *V. vinfera* juice has been studied in the United States (*136*).

Tannins, another group of phenolic substances, are known for their sensory effects on wine. In addition to their influence on the body of wines, tannins produce an astringent taste. Amerine and Joslyn (*2*) reported that the usual range of tannin content for white wines is from 0.01 to 0.04% and from 0.10 to 0.20% for red table wines. The threshold for tannin was reported as 0.10 gram/100 ml for a white table wine and 0.15 in a red table wine (*137*). Other studies have reported that tannin affects the detection levels for sweetness and tartness in wines (*18, 138, 139*). In addition to their influence on taste, tannins are responsible, in part, for the oxidative browning in wines (*3*). This is especially critical in white wines since browning is a serious defect.

The rate of browning depends also on the amount of other phenolic substances such as catechins and leucoanthocyanins present. These phenolics are associated with variety, maturity, and region in which the grapes are grown (*130, 140, 141, 142*). Berg and Akiyoshi (*143*) reported that variety was the predominant factor that influences the rate of browning. Also, other factors such as sulfur dioxide, iron, copper, citric acid, and oxygen affected browning in wines.

The importance of phenolic compounds to potassium bitartrate precipitation in wines was postulated by Pilone and Berg (*144*). They reported that the removal of pigments and other polyphenols decreased the tartrate-holding capacity of red wines. Later, Balakian and Berg (*145*) found that a grape-skin extract containing pigments and tannins increased the solubility of potassium bitartrate in both decolorized wines and model solutions. Evidence was reported to indicate that the anthocyanins were responsible for complexing the tartrate anion and not potassium.

Studies have shown that many phenolic substances of grapes and wine may possess some antibacterial properties (*146, 147, 148*). For an extensive review and discussion pertaining to the bactericidal effect of grape and wine phenolics, *see* Ref. *122*.

A brief listing of some important anthocyanins reported in these grape species, *V. vinifera, V. labrusca,* and *V. rotundifolia,* is presented in Table V. Excellent reviews of anthocyanins in *vinifera* and non-*vinifera* grapes are reported by Ribéreau-Gayon (*121*), Webb (*127*), Singleton and Esau (*122*), and Amerine and Joslyn (*2*). In general, *V. vinifera* grapes principally contain monoglucosides and only trace amounts of diglucosides. This is in contrast to the anthocyanins in *V. labrusca* and *V. rotundifolia* and several French hybrids in which diglucosides are characteristic.

Several workers have surveyed pigments in grapes and established pigment patterns (*128, 149, 150*). Somers (*151*) reported that the pigments of six *vinifera* grapes consisted of anthocyanins, acylated anthocyanins, and tannin pigments. A regional difference in the pigment composition was also noted in this study. Somaatmadja and Powers (*152*) reported that Cabernet Sauvignon grapes contained six anthocyanins, all monoglucosides. The *V. vinifera* variety Flame Tokay was studied by Akiyoshi *et al.* (*153*) who found that cyanidin-3-monoglucoside was the major anthocyanin. Albach *et al.* (*154*) reported on the anthocyanin pigments of Tinta Pinheira, *V. vinifera,* and found several 3-monoglucosides acylated with *p*-coumaric acid. The principal anthocyanin pigment for the hybrid grapes, Rubired and Royalty, was mal-

Table V.   Anthocyanins Reported for Grapes and Other Fruits

Fruit	Anthocyanins[a]	References
Grape		
Vitis rotundifolia	cy, del, mv, peon, pet diglucosides	121, 264
Vitus labrusca	cy, del, mv, peon, pet-3-glucosides and 3,5-diglucosides, and acylated monoglucosides and diglucosides, p-coumaric and ester of del-3-glucoside, p-coumaric acid ester of cy-3-glucoside	121,133,134
Vitis vinifera	cy, del, mv, peon, pet-3-glucosides; traces 3,5-diglucosides and some acylated monoglucosides	121,265
Apple	cy-3-galactoside, cy-3-glucoside, cy-3-arabinoside, cy-3-xyloside, and acylated derivatives of all four glucosides, cy-7-arabinoside	266,267
Pear	cy-3-galactoside, cy-3-arabinoside, cy-7-arabinoside	267,268
Cherry (sour)	cy-3-glucosylrhamosylglucoside, cy-3-gentiobioside, cy-3-rutinoside, cy-3-rhamnoglucoside	269,270
Blackberry	cy-3-glucoside, cy-3-rutinoside	271
Strawberry	pg-3-glucoside, cy-3-glucoside	272,273

[a] Abbreviations used: cy = cyanidin, del = delphinidin, mv = malvidin, peon = peonidin, pet = petunidin, pg = pelargonidin.

vidin-3,5-diglucoside (155, 156). Recently, Liao and Luh (157) reported that malvidin-3-monoglucoside and malvidin-3-monoglucoside acylated with p-coumaric acid were the major anthocyanins in Tinto-cao grapes.

In addition to the many studies concerning anthocyanins, a considerable amount of information is available on other phenolic substances in grapes. A complete review of phenolic substances in grapes and their significance is included in an excellent book by Singleton and Esau (122).

Singleton (130) reported that the total extractable phenols per berry increased during the ripening period. The total phenolics were quite variable by variety and ranged from 2 to 10 mg/gram. The relative amounts of the phenolic substances in 12 grape varieties were estimated by Singleton et al. (141). Their results indicated that the pulp of the grapes contained lower levels of phenolics as compared with juice and seeds. The phenolic composition of the grapes included the following flavanols: catechin, epicatechin, and epicatechin gallate. Flavanols or tannins identified by Hennig and Burkhardt (158) were d-catechin, l-epicatechin and l-epigallocatechin.

The principal source of leucoanthocyanins, astringent substances in grapes, was reported to be the seeds (*159*). These workers also found a relatively large amount of leucoanthocyanins in stems, and grape varieties differed in levels of these phenolic substances. Somaatmadja et al. (*148*) identified four leucoanthocyanins in extracts from Cabernet Sauvignon grapes. These phenolics included two leucocyanidins, one leucodelphinidin, and one unknown. In a purified leucoanthocyanin preparation from the variety French Colombard, Ito and Joslyn (*160*) extracted and identified the following phenolics: *d*-catechin, *l*-epicatechin, *l*-epicatechin gallate, protocatechuic acid, phloroglucinol, kaempferol, and quercetin. Ribéreau-Gayon (*161*) noted that the flavanols kaempferol, quercetin, and myricetin were present in the form of monoglucosides and the monoglucuronoside of quercetin. Grapes also contain hydroxy cinnamic acids and other phenolic compounds such as caffeic, shikimic, ellagic, chlorogenic, gallic, and *p*-coumaric acids (*2*).

In addition to the detection and identification of anthocyanins in fruits (Table V), other phenolic substances have been extensively studied in the past several years. An excellent review of fruit phenolics was recently provided by Van Buren (*128*). Also, phenolics were included in the review of the biochemistry of apples and pears by Hulme (*39*).

The phenolics (+)catechin and (−)epicatechin are common flavanols in several fruits (*128*). Apples and pears contain other phenolic compounds such as quinic, shikimic, chlorogenic, and caffeic acids (*39*). Durkee and Poapst (*162*) reported that the two major phenolic constituents of core tissues and seeds of McIntosh apples were chlorogenic acid and phloridzin. After hydrolysis of extracts from core tissues, the identified phenolics were: phloretin, caffeic acid, *p*-coumaric acid, phloretic acid, and trace amounts of ferulic acid. Studies have shown that apple leucoanthocyanins yield catechin, epicatechin, cyanidin, and pelargonidin after hydrolysis (*163, 164*). Van Buren et al. (*164*) also reported that a purified leucoanthocyanin from apples was either a dimer or oligomer containing (−)epicatechin, and 5,7,3′,4′-flavin-3,4-diol.

Two catechins, (+)catechin and (−)epicatechin, and two leucoanthocyanins were isolated from Bartlett pears by Nortje (*165*). Durkee et al. (*166*) identified arbutin (*p*-hydroxphenyl-β-D-glucoside) in mature and immature pears. Several flavanol glycosides such as quercetin 3-glucoside, isorhamnetin-3-glucoside, isorhamnetin-3-rutinoside, isorhamnetin-3-galactoside, and quercetin-7-xyloside were identified by Duggan (*167*).

Recently, nine flavonol glycosides including 3-β-monoglucosides and 3-β-glucuronides of quercetin and kaempferol were detected and purified from strawberry fruit (*168*). A summary of phenolic compounds in soft fruits is presented by Green (*40*).

Inorganic Constituents

Inorganic components of grapes and other fruits significantly influence the production and quality of wines and are important for human nutrition. During the intermediate reactions of the fermentation process, potassium, manganese, magnesium, iron, and copper are essential for the synthesis of alcohol (2). This is particularly true for phosphorus, an essential element for yeasts. In general, fruits supply sufficient quantities of inorganic elements for the fermentation process in wines, but ammonium phosphate is frequently added to certain fruit musts to aid fermentation (77).

In addition to their importance in the chemical reactions of alcoholic fermentation, some inorganic components have a significant effect on the stability of wines. Problems frequently are associated with excesses rather than deficiencies of certain inorganic constituents. Generally, excessive levels produce undesirable effects by altering the color, appearance, or taste of the wines (48). An excessive level of inorganic constituents in wines may arise from many sources such as the inherent content in musts, winery equipment, cellars, and vineyard materials.

Newly fermented wines are usually supersaturated with potassium bitartrate. Wineries routinely remove the excess potassium bitartrate in wines by refrigeration or ion exchange procedures. These steps are necessary to obtain a wine free of tartrate deposits after bottling. Calcium may also combine with tartrates which contribute to the deposits in wines. Generally, the stabilization practices for potassium bitartrate are sufficient to remove calcium tartrate from wines.

Iron and copper in wines may form complexes with other components to produce deposits or clouds in white wines. Iron clouds generally occur at a pH range from 2.9 to 3.6 and are often controlled by adding citric acid to the wines (2). Copper clouds appear in wines when high levels of copper and sulfur dioxide exist and are a combination of sediments, protein–tannin, copper–protein, and copper–sulfur complexes (169). Further, the browning rate of white wines increases in the presence of copper and iron (143). The results of this study indicate that iron increased the browning rate more than copper.

Since significant levels of other inorganic constituents usually occur from sources other than the fruit themselves, a discussion of these components is not included in this article. An excellent review of inorganic constituents of wines has been provided by Amerine (48).

Many factors influence quantity and type of inorganic constituents in fruits. Some of the significant factors are: maturity, variety, season, location, production practices, and climatic conditions (170). Since these factors are of critical importance to plant nutrition, growth, development, and product quality, the available literature is rather extensive for most fruits, particularly grapes.

The changes in potassium, calcium, and sodium during maturation of the grape variety Semillon were reported by Amerine (*56*). The calcium content remained nearly constant while the potassium and sodium contents increased with maturity. The amounts of potassium, calcium, and sodium found at maturity were 1240, 20, and 120 mg/liter, respectively. Puissant (*171*) also found that the potassium content of several varieties increased with maturation and varied between varieties and locations. In this study, the calcium and magnesium levels in the grapes decreased significantly as the season progressed. The amounts of calcium and magnesium for all varieties at maturity were about 208 and 188 mg/liter, respectively. Robinson *et al.* (*62*) found that the potassium content (K_2O) of 152 juice samples of Concord type grapes ranged from 105 to 335 mg/100 grams with an average of 225 mg/100 grams. Recently, Mattick *et al.* (*69*) reported that the potassium content in Concord grape juices ranged from 0.118 to 0.257%. Studies during two seasons in Ohio indicated that the potassium contents of musts from eight grape varieties of *V. labrusca* and French hybrids ranged from 860 to 1920 mg/liter with an average of 1380 mg/liter (*102*).

The phosphate content of grape musts ranged from 0.02 to 0.05% (*3*). Garino-Canina (*172*) reported that phosphorus content of grapes increased during maturation and ranged from 150 to 300 mg/liter at maturity. The average phosphorus content in 12 grape varieties in Canada was 27.0 mg/100 grams (*173*). These workers also reported that the copper and iron content in grapes averaged 0.17 and 0.46 mg/100 grams, respectively. Copper levels in must from other regions varied from 0.22 to 0.83 mg/100 ml (*48*). In California, grape musts contained iron ranging from 1.5 to 25.5 mg/liter (*174, 175*). The iron content of the various parts of the grape berry was different and was influenced by the iron level in the soil (*176*). These data showed that the iron content of the whole fruit ranged from 0.34 to 1.68 mg/100 grams.

The literature contains considerable data on inorganic constituents in fruit tissues. However, most of the data pertain to apples and grapes, and relatively little information is available on other fruits. A summary of typical averages of mineral contents of some common fruits is presented in Table VI.

Vitamins

In addition to their importance to human nutrition, vitamins are essential microbial growth factors. The significance of vitamins in winemaking is attributed to their influence on the fermentation process.

Table VI.   Inorganic

Amounts in mg/100 grams Fresh Weight

Fruit	K	Ca	Fe	Mg
Apple	118	12.13	0.27	4.20
	138	5.29		5.89
	140	4.17		5.07
Pear	111	16.15	0.22	5.25
	130	8.00	0.30	
	108	6.50	0.17	5.76
Cherry (sour)	169	21.24	0.44	6.23
	191	22.00	0.40	
	273	16.60	0.34	9.86
Blackberry	170	32.00	0.90	
	208	63.30	0.85	29.50
Strawberry	197	26.14	0.44	9.49
	164	21.00	1.00	
	161	22.00	0.71	11.70

Vitamins are required for satisfactory development or function of most yeasts. Wickerham (177) devised a complete yeast medium which included eight vitamins: biotin, pantothenic acid, inositol, niacin, p-aminobenzoic acid, pyridoxine, thiamine, and riboflavin. The concentrations of these growth factors varied widely with inositol in the greatest concentration and biotin in trace amounts. Many of these vitamins are considered major growth factors for yeast multiplication and development, as noted in several studies and reviews (178, 179, 180, 181, 182). Generally, the benefit of adding vitamins to musts and wines has not been established as a normal winery practice. This lack of response is because vitamins occur naturally in sufficient quantities in grapes and are produced by yeasts themselves (3).

A study of the B vitamins was reported by Morgan et al. (183). Perlman and Morgan (184) studied the stability of added B vitamins in grape juices and wines in California. Most of the vitamins were retained in the wines except riboflavin which was largely destroyed in wines bottled in clear glass. Castor (185) reported the changes in B vitamin content during the fermentation of wines. The riboflavin content was greater in young wines than in the grape musts, but wine aging for five years decreased the content by 36%. He further reported that p-aminobenzoic acid content increased during fermentation but that the amount returned to its original must level after fermentation. Generally, the other vitamins decreased during fermentation. A similar study concerning B vitamins in grapes, must, and wines was performed by Hall et al. (186). Thiamine and pantothenic acid decreased rapidly during fer-

Constituents for Several Fruits

    Amounts in mg/100 grams Fresh Weight

Cu	P	Mn	References
0.05	11.65	0.06	*173*
	15.52		*274*
	13.95	0.06	*275*
0.06	14.62	0.11	*173*
	11.00		*276*
0.12	9.58		*277*
0.08	19.40	0.13	*173*
	19.00		*276*
0.08	17.40		*277*
	19.00		*276*
0.12	23.80		*277*
0.05	26.66	0.24	*173*
	21.00		*276*
0.13	23.00		*277*

mentation, and the latter was nearly eliminated in sulfited wines. The riboflavin and folic acid contents of the musts were considerably less than amounts found in the grapes. In musts treated with thiamine, Lafon-Lafourcade *et al.* (*187*) found that wines contained less pyruvic and α-ketoglutaric acids. Also, the wines had less fixed sulfur dioxide and volatile acidity. Ough and Kunkee (*80*) studied the biotin content of grape juice and its effect on the fermentation rate. They found that biotin content together with total nitrogen could be used to predict yeast activity and fermentation rate. The presence or absence of vitamins in musts may have a profound effect on the formation of higher alcohols in wines (*2*). Furthermore, since vitamins influence the growth of bacteria, they may affect the induction or occurrence of malo–lactic fermentation in wines.

A review of the vitamin composition of grapes and musts has been reported in recent texts (*2, 3*). Also, vitamins in grapes were reported by Amerine (*21*) in an extensive review of organic components of wines.

Ascorbic acid level in grapes is relatively low compared with other fruits. Zubeckis (*5*) in Canada reported that ascorbic acid in fresh grapes varied from 1.1 to 11.7 mg/100 ml juice, except for the variety Veeport. This Canadian variety contained a high content of ascorbic acid (33.8 mg/100 ml juice). More recently, Ournac (*188*) determined the ascorbic acid content of seven grape juices and found a range from 2.0 to 6.0 mg/100 ml.

Early studies in the United States on the B vitamins in grapes and juices were conducted by Morgan *et al.* (*183*) and Perlman and Morgan

(*184*). Smith and Olmo (*189*) reported pantothenic acid and riboflavin contents in juice of diploid and tetraploid grapes. They found that pantothenic acid contents were significantly different between varieties and tended to be higher in the juice of inter-specific hybrids of *labrusca* × *vinifra* than in *vinifera* varieties. Later, Castor (*185*) determined the amounts of B vitamins in several musts of California grapes. The riboflavin content varied from 6.3 to 25 μg/100 ml juice and pantothenic acid ranged from 51 to 138 μg/ml. Also, Hall *et al.* (*186*) in California found nine to 16 μg riboflavin/ml in various musts, and the pantothenic acid content varied from 50 to 100 μg/100 ml. Matthews (*190*) reported the content of B vitamins in Swiss grape juices from four varieties. The thiamine content of these juices ranged from a trace to 25.0 μg/100 ml. Generally, these thiamine levels were less than those observed in musts of California by Perlman and Morgan (*184*) and Hall *et al.* (*186*). Thiamine ranged from 17 to 50 μg/100 ml juice in six grape varieties in France by Ournac and Flanzy (*191*).

The pyridoxine content of two grape varieties, Merlot and Cabernet Sauvignon, varied with season, location, and maturation (*192*). They reported the free pyridoxine content ranged from 0.16 to 0.53 mg/liter for 17 must analyses and averaged 0.32 mg/liter. These values were similar to Castor's (*185*) report of 0.48 mg/liter but were generally lower than the 0.60–1.06 mg/liter reported by Hall *et al.* (*186*). Castor (*185*) also reported the nicotinic acid in must samples of five *vinifera* varieties ranging from 79 to 375 μg/100 ml. In addition, Hall *et al.* (*186*) and Matthews (*190*) found high levels of nicotinic acid in grapes and musts.

Another vitamin constituent, inositol, was found in relatively high levels in grapes. Castor (*185*) showed that musts contained about 390 mg/100 ml. Similar inositol amounts were obtained by Matthews (*190*) with juices containing a range from 340 to 482 mg/100 ml. Peynaud and Lafourcade (*193*) studied inositol levels in two Bordeaux grape varieties during maturation. Ripening increased the inositol content which ranged from 330 to 640 mg/100 ml at maturity. Significant differences in inositol content were obtained between varieties and seasons.

Choline, *p*-aminobenzoic acid, folic acid, and biotin were rather low in grapes (*185, 190*). Lafon-Lafourcade and Peynaud (*194*) reported that musts of Bordeaux grapes averaged 47 μg *p*-aminobenzoic acid and 33 μg choline/liter. The folic acid content of California musts ranged from 1.1 to 3.9 μg/liter (*186*). Ough and Kunkee (*80*) stated that the biotin content of juices from red grapes was greater than from white fruit with amounts ranging from 1.2 to 2.7 μg/liter.

Since most fruits are generally good sources of ascorbic acid (vitamin C) and β-carotene (vitamin A), these vitamins have been extensively studied in many fruits. A review of vitamins in fruits was made by Mapson (*195*). The content of ascorbic acid in various fruits in relation to variety varied over a wide range (*4, 5, 196, 197*). This wide range was also found in fruits grown at the Research Center, Wooster, Ohio (Table VII). Although ascorbic acid in fruits is influenced by several factors such as maturity, location, and climatic conditions, the concentrations of ascorbic acid were rather typical for these fruits. The average ascorbic acid values for cherries and pears were reported by Strachan *et al.* (*4*) to be 9.4 and 3.9 mg/100 grams, respectively. These workers also stated the results of an extensive study on the carotene content of several fruits. Peaches were a good source of carotene with an average of 788 μg/100 mg. Additional data concerning the carotene content of apple and pear fruits were included in an excellent review by Hulme (*39*). This publication also states the content of some B vitamins in both fruits by other authors.

Table VII. Ascorbic Acid Content of Several Fruits Grown in Ohio

| | | | Ascorbic Acid Content in mg/100 grams | |
Fruit	*Number of Years*	*Number of Varieties*	*Range*	*Average*
Apple	1	9	1.6–5.4	4.1
Blackberry	3	6	12.5–24.1	18.9
Black raspberry	2	16	6.1–20.7	11.3
Peach	3	26	2.3–17.8	7.3
Strawberry	3	29	30.0–74.2	48.9

Pectins

Pectic substances in raw fruit juices include protopectin, a water-insoluble material, and soluble pectins. Pectin is a general term and is often used to include hydrolysis products such as pectinic and pectic acids. Most of the pectins in fruits are lost as a precipitate during fermentation. This precipitate consists of colloidal material and tends to reduce the rate of natural clarification. The protective colloidal action of pectins causes small particles to remain suspended and results in cloudy juice or wine. This material is often difficult to remove by filtration without prior treatment of the wines. One common method for fining, a process in which various agents are added to wines to help promote clarification, is the addition of pectic enzymes to musts and wines. The hydrolytic action by these enzymes results in a loss of the stabilizing properties of the pectins. This action allows the suspended particles to settle, thus helping to clarify the wines. These enzymes are also used in crushed grapes to increase juice yields by improving the pressing properties of the pomace. However, results from use of pectic enzymes in California under controlled conditions were variable (198). Generally, the commercial use of pectic enzymes is governed by certain conditions such as grape variety and production techniques.

Also, pectins in musts may contribute to the methyl alcohol content of wines (199). This undesirable wine constituent is probably derived from hydrolysis of naturally occurring pectins. The pectin changes occurring in cider fermentation have been extensively studied by Pollard and Kieser (200).

Generally, reports indicate that *labrusca* grapes contain a greater amount of pectin than *vinifera* varieties (2). Flanzy and Loisel (199) determined the total pectin content (pectinic acid, pectic acid, and methanol) in four grape musts of V. *vinifera*. They reported a range from 452 to 913 mg/liter. These amounts were lower than pectin contents of *vinifera* grapes obtained by Marsh and Pitman (201). These workers reported an average pectin content of 0.12% for grapes at maturity with a range from 0.09 to 0.14%. They also found that the pectin content for each variety decreased during ripening. For the *labrusca* variety Concord, Carter (202) studied the changes in total pectic substances during maturation. Throughout the three-month period prior to harvesting, the total pectin content decreased from 0.76 to 0.28% at harvest. Similar findings were obtained for Concord grapes by Hopkins and Gourley (203), but the total pectic content at maturity was about 0.60%. This level of pectin was reported by Willaman and Kertesz (204) in *labrusca* grape juice. Recently, Smit and Couvillon (205) showed that the total pectic substances in Concord grapes from two vine training systems increased with an increase in soluble solids until a level of 8–10% soluble solids was reached in the grapes. Then, a decline in total pectin content occurred during further maturation. At harvest, the grapes from the modified Munson training system contained a greater total pectin content than the four-arm Kniffin system, approximately 0.33 and 0.40%, respectively.

An excellent summary of pectic substances in fruits has been given by Pilnik and Voragen (206). Further extensive studies of the chemical compositions of several fruits, including pectic materials, have been conducted by Money and Christian (33) and Strachan et al. (4). Results of both investigations found varietal differences in the total pectin content within the various fruits. The average total pectin content of several tree fruits in Canada, such as apples (0.55%), pears (0.59%), peaches (0.74%), and cherries (0.25%), was obtained by Strachan et al. (4). These values were quite similar to those reported in England by Money and Christian (33). The total pectin content of blackberries (cultivated varieties) averaged 0.63% and ranged from 0.40 to 1.19% (33). The same workers reported that the average pectin contents of strawberries and black raspberries were 0.54 and 0.40%, respectively.

Volatile Compounds

Volatile constituents of fruits contribute to the flavor and aroma of wines, and their detection and levels have received considerable attention from many researchers. This is particularly true in recent years with the development of gas chromatography and mass spectrometry which have permitted accurate volatile analysis of fruits.

Certain aroma components from fruits are responsible for the characteristic odor of many wines. An example is the typical aroma of many *V. labrusca* grapes which often gives their wines a distinctive aroma and flavor. Furthermore, some volatile constituents are characteristic of certain varieties and their wines are recognized by these distinctive aromas. This is significantly important for varietal wines (grape) where their distinct aromas are a prerequisite in rating the quality at the highest level. In addition to the importance of variety in the aroma and flavor of wines, other variables known to influence the volatile components within the fruit variety are maturity, location, climatic conditions, and cultural practices.

Methyl anthranilate was reported as the major contributor to the aroma of the Concord grape (*207*). Later, Robinson *et al.* (*208*) reported methyl anthranilate increased during ripening of Concord grapes and decreased when the fruit became over-mature. They found that the methyl anthranilate content in Concord juice was 5.28 mg/liter at normal harvest maturity. This ester does not provide the full characteristic odor of Concord grapes, and thus studies have been quite detailed in determining additional volatiles in this variety. Holley *et al.* (*209*) confirmed the presence of methyl anthranilate in Concord juice and identified other volatiles: ethanol, methanol, ethyl acetate, methyl acetate, acetone, acetaldehyde, and acetic acid. Several low-boiling constituents in Concord grape essence were examined by Stevens *et al.* (*210*). These workers identified 16 volatiles and found that ethanol, ethyl acetate, and 2-methyl-3-buten-2-ol were the major constituents. Neudoerffer *et al.* (*211*) performed an extensive analysis of Concord grape essence and tentatively identified 32 volatile compounds and reported that the predominant compounds were: acetaldehyde, propionaldehyde, methyl acetate, 2-propanol, ethyl propionate, and isoamyl alcohol. More recently, Stern *et al.* (*212*) examined Concord grape juice essence and identified 60 volatile constituents. They found that the esters ethyl acetate and methyl anthranilate predominated, and they identified methyl butyrate, ethyl isobutyrate, ethyl 2-hexenoate, ethyl benzoate, and methyl benzyl acetate which had not been previously reported in Concord grapes. They also were the first to report the presence of a series of crotonates in Concord juice with ethyl crotonate being predominant.

In contrast to most varieties of *V. labrusca,* the aroma characteristic to *V. vinifera* varieties is usually less pronounced, although many varieties are readily differentiated by their aromas. Haagen-Smit *et al.* (*213*) identified several components in the volatile oil from Zinfandel grapes. The major compounds were as follows: ethyl alcohol, *n*-butyl phthalate, acetaldehyde, leaf aldehyde, glyoxylic acid, acetic acid, acetylmethyl-carbinol, and *n*-caproic acid. Webb and Kepner (*214*) investigated the intense aromas in the volatile essence components of the Muscat of Alexandria. They found the following compounds: methanol, ethanol, 1-butanol, 3-methylbutanol, 1-hexanol, *cis*-3-hexenol, ethyl caproate, methyl acetate, several other esters, and acetals. Later, Stevens *et al.* (*215*) identified 60 volatile components from grapes of Muscat of Alexandria with the major constituents being 1-hexanol, geraniol, and linalool. The two latter compounds possess a floral aroma and are probably significant contributors to the aroma of Muscat grapes. The volatile components extracted from eight Muscat varieties were compared by Webb *et al.* (*216*).

The volatile components of other *V. vinifera* varieties have been extensively investigated: the variety Sauvignon blanc by Chaudhary *et al.* (*217*), White Riesling by Drawert and Rapp (*218*) and Van Wyk *et al.* (*219*), and Grenache by Stevens *et al.* (*220, 221*). The major components of White Riesling as reported by Van Wyk *et al.* (*219*) were ethanol, 2-methyl-1-propanol, 2-methylbutanol, 3-methylbutanol, 2-hexenal, 1-hexanol, *trans*-2-hexen-1-ol, and 2-phenethanol. In addition to these constituents, several alcohols, acids, and a few esters (ethyl acetate and isoamyl acetate) were identified as volatile components of White Riesling grapes. Amerine and Joslyn (*2*) provide an excellent review of volatile compounds in grapes, including *V. rotundifolia.*

An extensive list of volatile compounds in apples and other fruits was included in a review by Nursten (*222*). White (*223*) reported that the principal components of the aroma of apples were alcohols (92%): methanol, ethanol, 1-propanol, 2-propanol, 1-butanol, 2-methyl-1-propanol, 2-methyl-1-butanol, and 1-hexanol. The other constituents included 6% carbonyl compounds and 2% esters. Later, MacGregor *et al.* (*224*) tentatively identified 30 volatile components of McIntosh apple juice including four aldehydes, one ketone, 11 alcohols, 10 esters, and four fatty acids. The major organic volatiles in several different extractants of Delicious apple essence were identified and quantitatively estimated by Schultz *et al.* (*225*). They reported from sensory tests that low molecular weight alcohols contributed little to apple aroma. Flath *et al.* (*226*) identified

56 volatile compounds in an extract of Delicious apple essence. The principal components of the characteristic apple-like aroma included ethyl 2-methylbutyrate, hexanal, and 2-hexanal. Volatiles of several apple varieties were identified and compared by Flath *et al.* (*227*). Their results showed that apple varieties usually contained the same constituents but differed in the relative proportions of individual components.

Jennings *et al.* (*228*) fractionated an extract from Bartlett pears into 32 volatile components of which five were found to contribute significantly to the characteristic pear aroma. Later studies indicated that esters of *trans*-2:*cis*-4 decadienoic acid and hexyl acetate were significant components of the Bartlett pear aroma (*229, 230*). More recently, numerous volatiles of Bartlett pears were separated and identified including esters of methyl, ethyl, propyl, butyl, and hexyl alcohols, and C_{10} to C_{18} fatty acids (*231, 232*).

Nelson and Curl (*233*) reported that benzaldehyde was the major volatile constituent in Montmorency cherry juice. Also, an alcohol, probably geraniol, was found in trace amounts. Recently, Stinson *et al.* (*234*) found low-boiling neutral compounds in Montmorency cherry essence. The most abundant compounds were ethanol and methanol, comprising 9.0 and 0.5% of the essence. Other compounds in the essence were acetaldehyde, diethyl ether, propionaldehyde, acetone, isobutyraldehyde, methyl acetate, and ethyl acetate. For the high-boiling components of Montmorency cherry essence, Stinson *et al.* (*235*) found that the major compounds were *n*-propyl alcohol, isobutyl alcohol, isoamyl alcohol, and benzaldehyde. Several other minor components were also identified in the essence.

A comprehensive study of the volatiles from strawberries was conducted by McFadden *et al.* (*236*). Over 150 components were detected, and the major volatiles were identified from strawberries. The major components included alcohols, esters, acetals, aldehydes, furfural, methyl furfural, aromatic aldehydes, and ketones.

Volatile components from blackberries have received little attention. However, Scanlan *et al.* (*237*) recently identified 16 volatile components from a commercial essence of Evergreen blackberries. The identified compounds included acetals, esters, alcohols, ketones, terpenes, and an aromatic.

Enzymes

Certain enzymes present in grapes and other fruits may be responsible for wine disorders such as clouding, darkening, or an oxidized taste. These defects may result from enzymic action in musts or wine. Although enzymic browning may occur in wines, studies have shown that enzymes play a minor role in darkening wines (238, 239, 240). This is contrary to fresh juice where enzymic browning usually occurs rapidly, and its control is important in producing a high quality product, particularly white wines. To prevent this enzymic action, wineries routinely treat musts and wines with sulfur dioxide. Sulfur dioxide, in addition to its antimicrobial activity, has an antioxidative property which prevents browning and taste defects.

Enzymes found in grapes include peroxidase, polyphenoloxidases, catalases, invertase, tannases, and pectic enzymes (2). Within this list, polyphenoloxidase has received considerable attention in grapes because of its detrimental effect on wine quality.

Practices which reduce the undesirable effects of polyphenoloxidases include: selecting the proper variety, obtaining sound and mature fruit, pressing not too long or hard, centrifuging or settling musts, and bentonite fining (241, 242, 243, 244). Demeaux and Bidan (245) studied the effect of heat stability of polyphenoloxidase in grape musts and reported that 70°C for approximately 3 min nearly inactivated the enzyme.

In addition to the undesirable effects of some enzyme systems in musts and wine, enzymes may be responsible for the formation of certain desirable esters which would be essential to the aroma or bouquet of the wine (2). These enzymes may be present in the raw material, but their effects may become evident only during fermentation and aging.

Ivanov (246) reported that the main oxidizing system of grapes involved polyphenoloxidase. The polyphenoloxidase activity was greatest in grape skins and least in the juice of mature grapes. He also showed that varieties varied in the amount of polyphenoloxidase activity. The results indicated that the juices from the varieties Riesling and Aligoté had a greater activity than the varieties Muscat rouge and Dimiat. Studies by Cassignard (241) and Poux (242) found also that varieties differed in the amount of polyphenoloxidase. The variety Merlot blanc contained about twice the activity of other grape varieties studied by Cassignard (240). Poux's (242) results showed that the varieties Grenache blanc and Maccabeo had the highest polyphenoloxidase activity in his study. The occurrence and significance of other enzymes in grapes were reviewed by Amerine and Joslyn (2).

An excellent discussion of the many enzymes occurring in fruits is provided by Dilley (247). Of the numerous enzymes present in fruits, and as in the case of grapes, oxidizing enzymes have received considerable attention from researchers because their activity is usually accompanied by undesirable changes in raw product quality. A complete account of enzyme-catalyzed oxidative browning of fruits and their products is presented by Joslyn and Ponting (248).

Literature Cited

1. Joslyn, M. A., Amerine, M. A., "Dessert Appetizer and Related Flavored Wines; The Technology of Their Production," University of California, Berkeley, Calif., 1964.
2. Amerine, M. A., Joslyn, M. A., "Table Wines. The Technology of Their Production," University of California Press, Berkeley and Los Angeles, Calif., 1970.
3. Amerine, M. A., Berg, H. W., Cruess, W. V., "The Technology of Wine Making," AVI, Westport, Conn., 1972.
4. Strachan, C. C., Moyls, A. W., Atkinson, F. E., Britton, J. E., *Can. Dept. Agr. Publ.* (1951) 862.
5. Zubeckis, E., *Rept. Hort. Exp. Sta. Prod. Lab. Vineland Sta.* (Ontario, Canada) (1962) 90–116.
6. Amerine, M. A., Winkler, A. J., *Proc. Amer. Soc. Hort. Sci.* (1940) 38, 379.
7. Amerine, M. A., Winkler, A. J., *Calif. Univ. Agr. Exp. Sta. Bull.* (1963) 794.
8. Ough, C. S., Singleton, V. L., *Amer. J. Enol. Viticult.* (1968) 19, 129.
9. Ough, C. S., Alley, C. J., *Amer. J. Enol. Viticult.* (1970) 21, 78.
10. Amerine, M. A., Roessler, E. B., Filipello, F., *Hilgardia* (1959) 28, 477.
11. Ough, C. S., Amerine, M. A., *Hilgardia* (1963) 34, 585.
12. Ough, C. S., *Amer. J. Enol. Viticult.* (1966) 17, 20.
13. Gray, W. D., *J. Bacteriol.* (1945) 49, 445.
14. Amerine, M. A., Roessler, E. B., Wine Institute, San Francisco, Calif., 1964.
15. Amerine, M. A., Pangborn, R. M., Roessler, E. B., "Principles of Sensory Evaluations of Food," Academic Press, New York and London, 1965.
16. Amerine, M. A., Ough, C. S., *Amer. J. Enol. Viticult.* (1967) 18, 121.
17. Berg, H. W., Filipello, F., Hinreiner, E., Webb, A. D., *Food Technol.* (1955) 9, 23.
18. Berg, H. W., Filipello, F., Hinreiner, E., Webb, A. D., *Food Technol.* (1955) 9, 138.
19. Melamed, N., *Ann. Technol. Agr.* (1962) 11, 5.
20. Kliewer, W. M., *Plant Physiol.* (1966) 41, 923.
21. Amerine, M. A., *Advan. Food Res.* (1954) 5, 353.
22. Amerine, M. A., Thoukis, G., *Vitis* (1958) 1, 224.
23. Kliewer, W. M., *Amer. J. Enol. Viticult.* (1967) 18, 33.
24. *Ibid.*, (1965) 16, 101.
25. *Ibid.*, (1967) 18, 87.
26. Lott, R. V., Barrett, H. C., *Vitis* (1967) 6, 257.
27. Carroll, D. E., Hoover, M. W., Nesbitt, W. B., *J. Amer. Soc. Hort. Sci.* (1971) 96, 737.
28. Johnson, L. A., Carroll, D. E., *J. Food Sci.* (1973) 38, 21.
29. Kliewer, W. M., *Amer. J. Enol. Viticult.* (1965) 16, 168.
30. Amerine, M. A., Root, G. A., *Amer. J. Enol. Viticult.* (1960) 11, 137.
31. Caldwell, J. S., *J. Agr. Res.* (1925) 30, 1133.
32. Gallander, J. F., unpublished data, 1970.
33. Money, R. W., Christian, W. A., *J. Sci. Food Agr.* (1950) 1, 8.
34. Osborn, R. A., *J. Ass. Off. Agr. Chem.* (1964) 47, 1068.
35. Caldwell, J. S., *J. Agr. Res.* (1928) 36, 391.
36. Lee, C. Y., Shallenberger, R. S., Vittum, M. T., *N.Y. Food Life Sci. Bull.* (1970) 1.
37. Lott, R. V., *Proc. Amer. Soc. Hort. Sci.* (1943) 43, 56.
38. Widdowson, E. M., McCance, R. A., *Biochem. J.* (1935) 29, 151.
39. Hulme, A. C., *Advan. Food Res.* (1958) 8, 297.

40. Green, A., in "The Biochemistry of Fruits and Their Products," Vol. 2, Ch. 11, A. C. Hulme, ed., Academic, London and New York, 1971.
41. Whiting, G. C., in "The Biochemistry of Fruits and Their Products," Vol. 1, Ch. 1, A. C. Hulme, ed., Academic, London and New York, 1970.
42. Amerine, M. A., Roessler, E. B., Ough, C. S., *Amer. J. Enol. Viticult.* (1965) **16**, 29.
43. Winkler, A. J., Amerine, M. A., *Food Res.* (1938) **3**, 439.
44. Amerine, M. A., Winkler, A. J., *Food Res.* (1941) **6**, 1.
45. Kunkee, R. E., *Advan. Appl. Microbiol.* (1967) **9**, 235.
46. Ingram, M., *J. Soc. Chem. Ind.* (1948) **67**, 18.
47. Joslyn, M. A., Braverman, J. B. S., *Advan. Food Res.* (1954) **5**, 97.
48. Amerine, M. A., *Advan. Food Res.* (1958) **8**, 133.
49. Cruess, W. V., *J. Ind. Eng. Chem.* (1932) **24**, 648.
50. Cruess, W. V., Irish, J. H., *J. Bacteriol.* (1932) **23**, 163.
51. Vas, K., Ingram, M., *Food Mfr.* (1949) **24**, 414.
52. Berg, H. W., Keefer, R. M., *Amer. J. Enol. Viticult.* (1958) **9**, 180.
53. Berg, H. W., Akiyoshi, M., *Amer. J. Enol. Viticult.* (1971) **22**, 127.
54. Amerine, M. A., Winkler, A. J., *Proc. Amer. Soc. Hort. Sci.* (1942) **40**, 313.
55. Amerine, M. A., *Food Technol.* (1951) **5**, 13.
56. Amerine, M. A., *Wines Vines* (1956) **37**, 1.
57. Kliewer, W. M., Howarth, L., Omori, M., *Amer. J. Enol. Viticult.* (1967) **18**, 42.
58. Kliewer, W. M., Lider, L. A., Schultz, H. B., *Amer. J. Enol. Viticult.* (1967) **18**, 78.
59. Kliewer, W. M., *J. Amer. Soc. Hort. Sci.* (1971) **96**, 372.
60. Philip, T., Nelson, F. E., *J. Food Sci.* (1973) **38**, 18.
61. Gore, H. C., *J. Ind. Eng. Chem.* (1909) **1**, 436.
62. Robinson, W. B., Avens, A. W., Kertesz, Z. I., *N.Y. Agr. Exp. Sta. Tech. Bull.* (1949) **285**.
63. Crowther, R. F., Bradt, O. A., *Hort. Res. Inst. Ontario Rept 1970* (1970) 121-128.
64. Beelman, R. B., *Proc. 4th Penn. Wine Conf. Penn. State Univ.*, University Park (1971) 52-55.
65. Einset, J., Kimball, K. H., Robinson, W. B., Bertino, J. J., 1969-70 Vineyard and Cellar Notes, N.Y. State Exp. Sta. (Geneva), Cornell University Special Report No. 4, 1971.
66. Gallander, J. F., unpublished data, 1973.
67. Hartman, B. G., Hillig, F., *J. Ass. Off. Agr. Chem.* (1934) **17**, 522.
68. Robinson, W. B., Shaulis, N., Smith, G. C., *Food Res.* (1959) **24**, 176.
69. Mattick, L. R., Shaulis, N. J., Moyer, J. C., *Amer. J. Enol. Viticult.* (1972) **23**, 26.
70. Amon, A., Markakis, P., *Mich. Quart. Bull.* (1968) **50**, 485.
71. David, J. J., Luh, B. S., Marsh, G. J., *Food Res.* (1956) **21**, 184.
72. Jorysch, D., Sarris, P., Marcus, S., *Food Technol.* (1962) **16**, 90.
73. Burroughs, L. F., Carr, J. G., *Amer. Rep. Long Ashton Res. Sta.* (1956), **162**.
74. Du Plessis, L. De W., *S. Afr. J. Agr. Sci.* (1963) **6**, 485.
75. Radler, F., *Zentr. Bakteriol. Parasitenk.*, Abt. II (1966) **120**, 237.
76. Tressler, D. K., Celmer, R. F., Beavens, E. A., *J. Ind. Eng. Chem.* (1941) **33**, 1027.
77. Yang, H. Y., *J. Agr. Food Chem.* (1953) **1**, 331.
78. Fuleki, T., *Food Technol.* (1965) **19**, 1287.
79. Ough, C. S., *Amer. J. Enol. Viticult.* (1964) **15**, 167.
80. Ough, C. S., Kunkee, R. E., *Appl. Microbiol.* (1968) **16**, 572.
81. Ough, C. S., Lider, L. A., Cook, J. A., *Amer. J. Enol. Viticult.* (1968) **19**, 213.
82. Ough, C. S., *Amer. J. Enol. Viticult.* (1970) **21**, 213.

83. Lafon-Lafourcade, S., Peynaud, E., *Vitis* (1959) **2**, 45.
84. Lafon-Lafourcade, S., Guimberteau, G., *Vitis* (1962) **3**, 130.
85. Gallander, J. F., unpublished data, 1971.
86. Burroughs, L. F., *J. Sci. Food Agr.* (1957) **8**, 122.
87. *Ibid.*, (1960) **11**, 14.
88. Castor, J. G. B., Archer, T. E., *Food Res.* (1959) **24**, 167.
89. Castor, J. G. B., *Food Res.* (1953) **18**, 146.
90. Webb, A. D., Ingraham, J. L., *Advan. Appl. Microbiol.* (1963) **5**, 317.
91. Rankine, B. C., *J. Sci. Food Agr.* (1967) **18**, 583.
92. Ough, C. S., Bustos, O., *Wines Vines* (1969) **50**, 50.
93. Castor, J. G. B., *Food Res.* (1953) **18**, 139.
94. Luthi, H., Vetch, U., *Der Deutsch Weinbau* (1953) **7**, 33.
95. Silva, F. P., Petinga, O. S., *Agros.* (1962) **45**, 195.
96. Flanzy, C., Poux, C., *Ann. Technol. Agr.* (1965) **14**, 87.
97. Van Wyk, C. J., Venter, P. J., *S. Afr. Agr. Sci.* (1965) **8**, 57.
98. Gallander, J. F., Cahoon, G. A., Beelman, R. B., *Amer. J. Enol. Viticult.* (1969) **20**, 140.
99. Kliewer, W. M., *J. Food Sci.* (1969) **34**, 275.
100. *Ibid.*, (1970) **35**, 17.
101. Ough, C. S., *Vitis* (1968) **7**, 321.
102. Gallander, J. F., unpublished data, 1972.
103. Drawert, F., *Vitis* (1963) **4**, 49.
104. Nassar, A. R., Kliewer, W. M., *Proc. Amer. Hort. Soc.* (1966) **89**, 281.
105. Kliewer, W. M., *Amer. J. Enol. Viticult.* (1968) **19**, 166.
106. Burroughs, L. F., in "The Biochemistry of Fruits and Their Products," Vol. 1, Ch. 5, A. C. Hulme, ed., Academic, London and New York, 1970.
107. Burroughs, L. F., *Nature* (London) (1957) **179**, 360.
108. Silber, R. L., Beckler, M., Cooper, M., Evans, P., Fehder, P., Gray, R., Gresham, P., Rechsteiner, J., Searles, M., *J. Food Sci.* (1960) **25**, 675.
109. Kean, C. E., Marsh, G. L., *Food Technol.* (1956) **10**, 355.
110. Koch, J., Sajak, E., *Amer. J. Enol. Viticult.* (1959) **10**, 123.
111. Rankine, B. C., Emerson, W. W., *J. Sci. Food Agr.* (1963) **14**, 685.
112. Berg, H. W., *Proc. Amer. Soc. Enol.* (1951) **90**, 147.
113. Cooke, G. M., Berg, H. W., *Amer. J. Enol. Viticult.* (1971) **22**, 178.
114. Bayly, F. C., Berg, H. W., *Amer. J. Enol. Viticult.* (1967) **18**, 18.
115. Koch, J., Sajak, E., *Weinberg Keller* (1963) **10**, 35.
116. Beelman, R. B., Ph.D. Thesis, The Ohio State University, Columbus, Ohio, 1970.
117. Hansen, E., in "The Biochemistry of Fruits and Their Products," Vol. 1, Ch. 6, A. C. Hulme, ed., Academic, London and New York, 1970.
118. Hulme, A. C., *J. Sci. Food Agr.* (1951) **2**, 160.
119. Hansen, E., *Proc. Amer. Soc. Hort. Sci.* (1955) **66**, 118.
120. Boland, F. E., Blomquist, V. H., Estrin, B., *J. Ass. Off. Agr. Chem.* (1968) **51**, 1203.
121. Ribéreau-Gayon, P., *Ann. Physiol. Vegetale* (1964) **6**, 119, 211, 259.
122. Singleton, V. L., Esau, P., "Phenolic Substances in Grapes and Wine, and Their Significance," Suppl. 1, Advan. Food Res., Academic, London and New York, 1969.
123. Bate-Smith, E. C., *Advan. Food Res.* (1954) **5**, 261.
124. MacKinney, G., Chichester, C. O., *Advan. Food Res.* (1954) **5**, 301.
125. Joslyn, M. A., Goldstein, J. L., *Advan. Food Res.* (1964) **13**, 179.
126. Mattick, L. R., Weirs, L. D., *Dept. Food Sci. Technol., N.Y. Agr. Exp. Sta.* (Geneva), Cornell University, Res. Cir. **14**, 1968.
127. Webb, A. D., *Proc. Plant Phenolics Group Symp. No. 4* (1964), 21-37.
128. Van Buren, J., in "The Biochemistry of Fruits and Their Products," Vol. 1, Ch. 11, A. C. Hulme, ed., Academic, London and New York, 1970.

129. Clore, W. J., Neubert, A. M., Carter, G. H., Ingalsbe, D. W., Brummund, V. P., *Wash. Agr. Exp. Sta. Tech. Bull.* (1965) **48**.
130. Singleton, V. L., *Amer. J. Enol. Viticult.* (1966) **17**, 126.
131. Amerine, M. A., Winkler, A. J., *Proc. Amer. Soc. Hort. Sci.* (1947) **49**, 183.
132. Robinson, W. B., Weirs, L. D., Bertino, J. J., Mattick, L. R., *Amer. J. Enol. Viticult.* (1966) **17**, 178.
133. Ingalsbe, D. W., Neubert, A. M., Carter, G. H., *J. Agr. Food Chem.* (1963) **11**, 263.
134. Rice, A. C., *J. Ass. Off. Agr. Chem.* (1965) **48**, 525.
135. Hrazdina, G., *J. Agr. Food Chem.* (1970) **18**, 243.
136. Fitelson, J., *J. Ass. Off. Agr. Chem.* (1967) **50**, 293.
137. Hinreiner, E., Filipello, F., Berg, H. W., Webb, A. D., *Food Technol.* (1955) **9**, 489.
138. *Ibid.*, (1955) **9**, 351.
139. Rossi, J. A., Jr., Singleton, V. L., *Amer. J. Enol. Viticult.* (1966) **17**, 240.
140. Berg, H. W., *Food Res.* (1953) **18**, 399.
141. Singleton, V. L., Draper, D. E., Rossi, J. A., Jr., *Amer. J. Enol. Viticult.* (1966) **17**, 206.
142. Kramling, T. E., Singleton, V. L., *Amer. J. Enol. Viticult.* (1969) **20**, 86.
143. Berg, H. W., Akiyoshi, M., *Amer. J. Enol. Viticult.* (1956) **7**, 1.
144. Pilone, B. F., Berg, H. W., *Amer. J. Enol. Viticult.* (1965) **16**, 195.
145. Balakian, S., Berg, H. W., *Amer. J. Enol. Viticult.* (1968) **19**, 91.
146. Powers, J. J., Somaatmadja, D., Pratt, D. E., Hamdy, M. K., *Food Technol.* (1960) **14**, 626.
147. Hamdy, M. K., Pratt, D. E., Powers, J. J., Somaatmadja, D., *J. Food Sci.* (1961) **26**, 457.
148. Somaatmadja, D., Powers, J. J., Wheeler, R., *Amer. J. Enol. Viticult.* (1965) **16**, 54.
149. Albach, R. F., Kepner, R. E., Webb, A. D., *Amer. J. Enol. Viticult.* (1958) **9**, 164.
150. Rankine, B. C., Kepner, R. E., Webb, A. D., *Amer. J. Enol. Viticult.* (1959) **10**, 105.
151. Somers, T. C., *Vitis* (1968) **7**, 303.
152. Somaatmadja, D., Powers, J. J., *J. Food Sci.* (1963) **28**, 617.
153. Akiyoshi, M., Webb, A. D., Kepner, R. E., *J. Food Sci.* (1963) **28**, 177.
154. Albach, R. F., Webb, A. D., Kepner, R. E., *J. Food Sci.* (1965) **30**, 620.
155. Smith, R. M., Luh, B. S., *J. Food Sci.* (1965) **30**, 995.
156. Chen, L. F., Luh, B. S., *J. Food Sci.* (1967) **32**, 66.
157. Liao, F. W. H., Luh, B. S., *J. Food Sci.* (1970) **35**, 41.
158. Hennig, K., Burkhardt, R., *Amer. J. Enol. Viticult.* (1960) **11**, 64.
159. Cantarelli, C., Peri, C., *Amer. J. Enol. Viticult.* (1964) **15**, 146.
160. Ito, S., Joslyn, M. A., *Nature* (London) (1964) **204**, 475.
161. Ribereau-Gayon, P., *Wines Vines* (1965) **46**, 26.
162. Durkee, A. B., Poapst, P. A., *J. Agr. Food Chem.* (1965) **13**, 137.
163. Ito, S., Joslyn, M. A., *J. Food Sci.* (1965) **30**, 44.
164. Van Buren, J. P., Senn, G., Neukom, H., *J. Food Sci.* (1966) **31**, 964.
165. Nortje, B. K., *J. Food Sci.* (1966) **31**, 733.
166. Durkee, A. B., Johnston, F. B., Thivierge, P. A., Poapst, P. A., *J. Food Sci.* (1968) **33**, 461.
167. Duggan, M. B., *J. Agr. Food Chem.* (1969) **17**, 1098.
168. Ryan, J. J., *J. Food Sci.* (1971) **36**, 867.
169. Kean, C. E., Marsh, G. L., *Food Res.* (1956) **21**, 441.
170. Kenworthy, A. L., Martin, L., in "Nutrition of Fruit Crops," Ch. 24, N. Childers, ed., Horticultural Publications, Rutgers, New Brunswick, N. J., 1966.

171. Puissant, A., *Ann. Technol. Agr.* (1960) **4**, 321.
172. Garino-Canina, E., *Ann. Chem. Appl.* (1941) **31**, 342.
173. Zubeckis, E., Siron, G., *Rept. Hort. Exp. Sta. Prod. Lab. Vineland Sta.* (Ontario, Canada) (1963) 77-81.
174. Byrne, J., Saywell, L. G., Cruess, W. V., *J. Ind. Eng. Chem. Anal. Ed.* (1937) **9**, 83.
175. Mrak, E. M., Fessler, J. F., *Food Res.* (1938) **3**, 307.
176. Négre, E., Cordonnier, R., *Comp. Rend. Acad. Agr. France* (1953) **39**, 52.
177. Wickerham, L. J., *J. Bacteriol.* (1946) **52**, 293.
178. Williams, R. J., Saunders, D. H., *Biochem. J.* (1934) **28**, 1887.
179. Schultz, A. S., Atkin, L., Frey, C. N., *J. Amer. Chem. Soc.* (1937) **59**, 948.
180. Burkholder, P. R., McVeigh, I., Moyer, D., *J. Bacteriol.* (1944) **48**, 385.
181. Rainbow, C., *Nature London* (1948) **162**, 572.
182. Robinson, F. A., "The Vitamin B Complex," Wiley, New York, 1951. New York, 1951.
183. Morgan, A. F., Nobles, H. L., Wiens, A., Marsh, G. L., Winkler, A. J., *Food Res.* (1939) **4**, 217.
184. Perlman, L., Morgan, A. F., *Food Res.* (1945) **10**, 334.
185. Castor, J. G. B., *Appl. Microbiol.* (1953) **1**, 97.
186. Hall, A. P., Brinner, L., Amerine, M. A., Morgan, A. F., *Food Res.* (1956) **21**, 362.
187. Lafon-Lafourcade, S., Blouin, J., Sudraud, P., Peynaud, E., *Compt. Rend. Acad. Agr. France* (1967) **60**, 1046.
188. Ournac, A., *Ann. Technol. Agr.* (1965) **14**, 341.
189. Smith, M. B., Olmo, H. P., *Amer. J. Botany* (1944) **31**, 240.
190. Matthews, J., *Vitis* (1959) **2**, 57.
191. Ournac, A., Flanzy, M., *Ann. Technol. Agr.* (1957) **6**, 257.
192. Peynaud, E., Lafourcade, S., *Ann. Technol. Agr.* (1957) **6**, 301.
193. *Ibid.*, (1955) **4**, 381.
194. Lafon-Lafourcade, S., Peynaud, E., *Ann. Technol. Agr.* (1958) **7**, 303.
195. Mapson, L. W., in "The Biochemistry of Fruits and Their Products," Vol. 1, Ch. 13, A. C. Hulme, ed., Academic, London and New York, 1970.
196. Hansen, E., Waldo, G. F., *Food Res.* (1944) **9**, 453.
197. Kirk, M. M., Tressler, D. K., *Food Res.* (1941) **6**, 395.
198. Berg, H. W., *Amer. J. Enol. Viticult.* (1959) **10**, 130.
199. Flanzy, M., Loisel, M. Y., *Ann. Technol. Agr.* (1958) **6**, 311.
200. Pollard, A., Kieser, M. E., *J. Sci. Food Agr.* (1959) **10**, 253.
201. Marsh, G., Pitman, G. A., *Fruit Prod. J.* (1930) **9**, 187.
202. Carter, G. H., *Proc. Amer. Hort. Soc.* (1968) **92**, 319.
203. Hopkins, E. F., Gourley, J. H., *Proc. Amer. Hort. Soc.* (1930) **27**, 164.
204. Willaman, J. J., Kertesz, Z. I., *N.Y. Agr. Exp. Sta. Tech. Bull.* (Geneva) (1931) **178**.
205. Smit, C. J. S., Couvillon, G. A., *J. Amer. Soc. Hort. Sci.* (1971) **96**, 547.
206. Pilnik, W., Voragen, G. J., in "The Biochemistry of Fruits and Their Products," Vol. 1, Ch. 3, A. C. Hulme, ed., Academic, London and New York, 1970.
207. Power, F. B., Chesnut, V. K., *J. Amer. Chem. Soc.* (1921) **43**, 1741.
208. Robinson, W. B., Shaulis, J., Pederson, C. S., *Fruit Prod. J. Amer. Food Mfr.* (1949) **29**, 36.
209. Holley, R. W., Stoyla, B., Holley, A. D., *Food Res.* (1955) **20**, 326.
210. Stevens, K. L., Lee, A., McFadden, W. H., Teranishi, R., *J. Food Sci.* (1965) **30**, 1006.
211. Neudoerffer, T. S., Sandler, S., Zubeckis, E., Smith, M. D., *J. Agr. Food Chem.* (1965) **13**, 584.
212. Stern, D. J., Lee, A., McFadden, W. H., Stevens, K. L., *J. Agr. Food Chem.* (1967) **15**, 1100.

213. Haagen-Smit, A. J., Hirosawa, F. N., Wang, T. H., *Food Res.* (1949) **14**, 472.
214. Webb, A. D., Kepner, R. E., *Food Res.* (1957) **22**, 384.
215. Stevens, K. L., Bomben, J., Lee, A., McFadden, W. H., *J. Agr. Food Chem.* (1966) **14**, 249.
216. Webb, A. D., Kepner, R. E., Maggiora, L., *Amer. J. Enol. Viticult.* (1966) **17**, 247.
217. Chaudhary, S. S., Kepner, R. E., Webb, A. D., *Amer. J. Enol. Viticult.* (1964) **15**, 190.
218. Drawert, F., Rapp, A., *Vitis* (1966) **5**, 351.
219. Van Wyk, C. J., Webb, A. D., Kepner, R. E., *J. Food Sci.* (1967) **32**, 660.
220. Stevens, K. L., Bomben, J. L., McFadden, W. H., *J. Agr. Food Chem.* (1967) **15**, 378.
221. Stevens, K. L., Flath, R. A., Lee, A., Stern, D. J., *J. Agr. Food Chem.* (1969) **17**, 1102.
222. Nursten, H. E., in "The Biochemistry of Fruits and Their Products," Vol. 1, Ch. 10, A. C. Hulme, ed., Academic, London and New York, 1970.
223. White, J. W., *Food Res.* (1950) **15**, 68.
224. MacGregor, D. R., Sugisawa, H., Matthews, J. S., *J. Food Sci.* (1964) **29**, 488.
225. Schultz, T. H., Flath, R. A., Black, D. R., Guadagni, D. G., Schultz, W. G., Teranishi, R., *J. Food Sci.* (1967) **32**, 282.
226. Flath, R. A., Black, D. R., Guadagni, D. G., McFadden, W. H., Schultz, T. H., *J. Agr. Food Chem.* (1967) **15**, 29.
227. Flath, R. A., Forray, R. R., Teranishi, R., *J. Food Sci.* (1969) **34**, 382.
228. Jennings, W. G., Leonard, S., Pangborn, R. M., *Food Technol.* (1960) **14**, 587.
229. Jennings, W. G., Creveling, R. K., Heinz, D. E., *J. Food Sci.* (1964) **29**, 730.
230. Jennings, W. G., Sevenants, M. R., *J. Food Sci.* (1964) **29**, 158.
231. Heinz, D. E., Jennings, W. G., *J. Food Sci.* (1966) **31**, 69.
232. Creveling, R. K., Jennings, W. G., *J. Agr. Food Chem.* (1970) **18**, 19.
233. Nelson, E. K., Curl, A. L., *J. Amer. Chem. Soc.* (1939) **61**, 667.
234. Stinson, E. E., Dobley, C. J., Filipic, V. J., Hills, C. H., *J. Food Sci.* (1969) **34**, 246.
235. *Ibid.*, (1969) **34**, 246.
236. McFadden, W. H., Teranishi, R., Corse, J., Black, D. R., Mon, T. R., *J. Chromatogr.* (1965) **18**, 10.
237. Scanlan, R. A., Bills, D. D., Libbey, L. M., *J. Agr. Food Chem.* (1970) **18**, 744.
238. Hussein, A. A., Cruess, W. V., *Food Res.* (1940) **5**, 637.
239. Hussein, A. A., Cruess, W. V., *Fruit Prod. J. Amer. Vinegar Ind.* (1940) **20**, 271.
240. Caputi, A., Jr., Peterson, R. G., *Amer. J. Enol. Viticult.* (1965) **16**, 9.
241. Cassignard, R., *Vignes Vins* (1966) numéro spécial, 13.
242. Poux, C., *Ann. Technol. Agr.* (1966) **15**, 149.
243. Ivanov, T., *Ann. Technol. Agr.* (1967) **16**, 81.
244. Peri, C., Pompei, C., Montedoro, G., Cantarelli, C., *J. Sci. Food Agr.* (1971) **22**, 27.
245. Demeaux, M., Bidan, P., *Ann. Technol. Agr.* (1967) **16**, 75.
246. Ivanov, T., *Ann. Technol. Agr.* (1967) **16**, 35.
247. Dilley, D. R., in "The Biochemistry of Fruits and Their Products," Vol. 1, Ch. 8, A. C. Hulme, ed., Academic, London and New York, 1970.
248. Joslyn, M. A., Ponting, J. D., *Advan. Food Res.* (1951) **3**, 1.
249. Hulme, A. C., *Biochem. J.* (1953) **53**, 337.

250. Buch, M. L., Dryden, E. C., Hills, C. H., *J. Agr. Food Chem.* (1955) **3**, 960.
251. Whiting, G. C., Coggins, R. A., *J. Sci. Food Agr.* (1960) **11**, 337.
252. Phillips, J. D., Pollard, A., Whiting, G. C., *J. Sci. Food Agr.* (1956) **7**, 31.
253. Li, P. H., Hansen, E., *Proc. Amer. Soc. Hort. Sci.* (1964) **85**, 100.
254. Williams, M. W., Patterson, M. E., *J. Agr. Food Chem.* (1964) **12**, 80.
255. Nelson, E. K., *J. Amer. Chem. Soc.* (1925) **47**, 568.
256. Whiting, G. C., *J. Sci. Food Agr.* (1958) **9**, 244.
257. Anet, E. F. L. J., Reynolds, T. M., *Nature* (London) (1953) **172**, 1188.
258. Castor, J. G. B., Archer, T. E., *J. Amer. Enol. Viticult.* (1956) **7**, 19.
259. Hulme, A. C., Arthington, W., *Nature* (London) (1950) **165**, 716.
260. Hulme, A. C., Arthington, W., *Nature* (London) (1952) **170**, 659.
261. Kliewer, W. M., Ough, C. S., *Vitis* (1970) **9**, 196.
262. Rockland, L. B., *J. Food Sci.* (1959) **24**, 160.
263. Tinsley, I. J., Bockian, A. H., *Food Res.* (1959) **24**, 410.
264. Brown, W. L., *J. Amer. Chem. Soc.* (1940) **62**, 2808.
265. Bockian, A. H., Kepner, R. E., Webb, A. D., *J. Agr. Food Chem.* (1955) **3**, 695.
266. Sun, B. H., Francis, F. J., *J. Food Sci.* (1967) **32**, 647.
267. Timberlake, C. E., Bridle, P., *J. Sci. Food Agr.* (1971) **22**, 509.
268. Francis, F. J., *Hortscience* (1970) **5**, 42.
269. von Elbe, J. H., Bixby, D. G., Moore, J. D., *J. Food Sci.* (1969) **34**, 113.
270. Dekazos, E. D., *J. Food Sci.* (1970) **35**, 237.
271. Harborne, J. B., Hall, E., *Phytochemistry* (1964) **3**, 453.
272. Co, H., Markakis, P., *J. Food Sci.* (1968) **33**, 281.
273. Wrolstad, R. E., Putnam, T. B., *J. Food Sci.* (1969) **34**, 154.
274. Wilkinson, B. G., Perring, M. A., *J. Sci. Food Agr.* (1961) **12**, 174.
275. Wilkinson, B. G., *J. Hort. Sci.* (1957) **32**, 74.
276. Watt, B. K., Merrill, A. L., *U.S. Dept. Agr. Handbook No. 8* (rev. ed.) 1963.
277. McCance, R. A., Widdowson, E. M., "Chemical Compositions of Foods," Chemical Publishing, New York, 1940.

RECEIVED May 29, 1973. Ohio Agricultural Research and Development Center Journal Article No. 58-73.

3

The Chemistry of Red Wine Color

PASCAL RIBÉREAU-GAYON

Institut d'Oenologie, Université de Bordeaux II, Talence, France

Anthocyanins and tannins are the main pigments in red wine. These compounds give red wine its color and organoleptic character; all the differences between white and red wines depend on these compounds. The chemical changes in the coloring materials constitute the basic process of aging in red wines; these changes include the disappearance of red anthocyanins from grapes and the chemical transformation of tannins which gives them the yellow-red color typical of old wines. Wines from V. vinfera grapes and wines from hybrid grapes can be differentiated by anthocyanins content; these compounds exist as mono- and diglucosides, and the latter are found only in hybrid grapes.

Chemical study of the color of black grapes and red wines is doubly interesting. Theoretically, the substances involved are complex; their identification and quantitation present analytical problems not yet resolved. In addition, the particular chemical properties of these pigments allow for many structural changes with different colors as a function of the conditions of the medium. Technologically, the pigments in red wine not only give it its color but also its organoleptic character; the differences between white and red wines depend on these compounds. Chemical changes in the coloring materials constitute the basic process of aging in red wines. Significant progress has been made in this area during the past 20 years mainly because of chromatographic techniques, particularly paper chromatography.

A statement of the problems relative to red wine color should include a description of the compounds involved which belong to plant phenolics. These compounds include the anthocyanins of grapes and the tannins which are important in the color of wines, especially old wines.

The identification of grape anthocyanins leads us to describe a special application of our work which has practical importance. Actually, different species in the genus *Vitis* do not contain the same anthocyanins and corresponding characteristics which are transmitted by hybridization and reflected in the wines. One can thus differentiate between the wines of V. *vinifera*, the only species producing quality wines, and wines from interspecific hybrids. Anthocyanin analysis can detect the fraud of mixing wines of the second type with those of the first—a practice which is both frequent and injurious to wine quality.

With the first qualitative study on pigments accomplished by chromatography, it is apparent in retrospect that quantitative methods were needed. We describe the methods most applicable today although more work is needed, especially on tannins.

The theoretical knowledge of the chemical nature of the phenolic compounds responsible for the color of red wines has been applied to technological problems. It was possible to follow their change during maturation and to show the predominant role of climatic factors. For example, in the Bordeaux region, the levels of anthocyanins and tannins in the grapes could double from year to year in the same vineyard—*i.e.*, they vary much more than any other constituent in the fruit. Also, the dissolution of phenolics from the skins and seeds is complex and is accompanied by a large pigment loss.

Finally, we try to develop a chemical interpretation for the color of red wine—*i.e.*, the specific role of different substances and the changes they undergo which cause the evolution of wine color, especially during aging.

This report is based essentially on the author's work during 20 years at the Institute of Enology in the University of Bordeaux. The pigment work (*1, 2*), a general review on phenolic compounds (*3, 4*), and theses of the author's colleagues (*5, 6, 7*) have been published. A very well documented publication on this subject has been done in English by Singleton and Esau (*8*).

Anthocyanins

Nature of the Substances Present. The anthocyanins are red and blue pigments widely distributed in plants. Until 1952 the bulk of our knowledge on the chemical composition of the anthocyanins in grapes was based principally on the work of Willstatter and Zollinger (*9*), Karrer and Widmer (*10*), and Robinson and Robinson (*11*). These studies have made possible the elucidation of the chemical structure of anthocyanins, but they have not allowed a rigorous study of their distribution in plants because the methods are hard to apply and not sensitive enough. There are usually many related anthocyanins which are difficult to separate and identify in any given plant organ.

Techniques using paper chromatography, first applied to grape studies in 1953, resulted in my doctoral thesis in 1959, "Studies on Plant Anthocyanins. Application to the Genus *Vitis*" completely changed the knowledge on this subject. It produced, as an indirect result, an analytical method for detecting the differences between wines according to their genetic origins, *V. vinifera* or hybrids. This method, today universally accepted and made official, gave impetus to many publications in other countries.

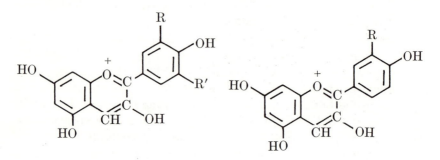

1 R = R' = OH; delphinidin
2 R = OCH₃; R' = OH; petunidin   4 R = OH; cyanidin
3 R = R' = OCH₃; malvidin   5 R = OCH₃; paeonidin

There are five anthocyanidins in the grape: delphinidin **1**, petunidin **2**, malvidin **3**, cyanidin **4**, and peonidin **5**. These aglycones exist in different heterosidic forms or as anthocyanins: 3-monoglucosides, 3,5-diglucosides, and acylated heterosides, whose structures, in the case of malvidin, are represented by formulas **6, 7,** and **8**. In the acylated anthocyanins, one molecule of cinnamic acid, more generally *p*-coumaric acid, is esterified with the —OH group in the sixth position of a glucose molecule (*12*).

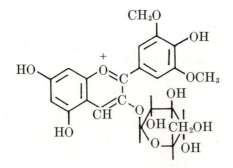

6 Monoglucoside-3-malvidin

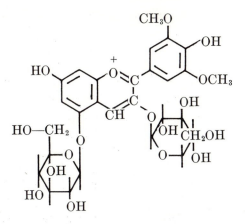

7 Diglucoside-3,5-malvidin

Acetic acid is one of the acylating agents of the 3-monoglucosides of *V. cinerea* (*13*).

The distribution of anthocyanins in grapes is very complex; it varies in the genus *Vitis* as a function of the species which may contain from six to 17 members from this family of pigments. The method of characterization for hybrid wines which contain anthocyanin diglucosides, always absent in wines from *V. vinifera*, is based on these pigment differences between species.

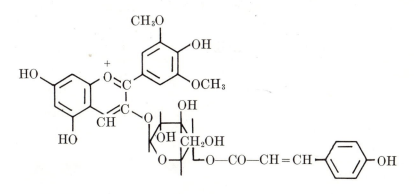

8 (*p*-coumaryl-4-glucoside)-3-malvidin

The only known chemical differences between grape varieties concerns the red pigments: grapes with or without anthocyanins or grapes with or without anthocyanins diglucosides. These differences form the basis for the methods of characterizing vine products: white wines and

red wines in one group, and red wines from *V. vinifera* and red wines
from hybrids in the other. Although in the first case the difference is
self-evident and in the second chemical analyses are necessary, the differences are essentially identical.

Properties of Anthocyanins. Several chemical or physicochemical
properties of the anthocyanins influence their structure and consequently
their coloration; they are thus important in interpreting the color of red
wine.

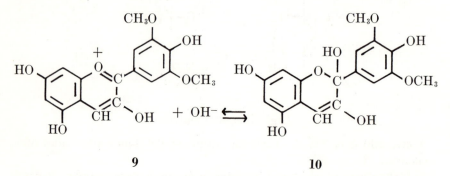

9   10

Under weakly acid conditions the red oxonium form **9** is in reversible
equilibrium with a colorless pseudo-base **10**; the position of the equilibrium depends on the pH. In a test trial in a synthetic medium a solution of anthocyanins is six times more colored at pH 2.9 than at pH 3.9.

**Table I.   Absorption of Cyanidin-3,5-diglucoside as a
Function of the Acidity of the Mixture**

pH	λ_1, nm	ε_1	λ_2, nm	ε_2
2.4	510	12000	278	17000
2.9	510	6000	278	17000
3.9	510	1000	278	16000
4.9		0	278	16000

The importance of this reaction on the color of anthocyanins is
shown in Table I. When the pH increases, the ultraviolet absorption
($\lambda = 278$ nm) from aromatic compounds is not affected; on the other
hand, the absorption caused by the central heterocycle ($\lambda = 510$ nm)
varies greatly.

With A^+ as the oxonium form and AOH as the pseudo-base, the equilibrium is written:

$$A^+ + H_2O \rightleftharpoons AOH + H^+$$

With $[H_2O] = 1$, at equilibrium one has: $K = \dfrac{[AOH]\,[H^+]}{[A^+]}$

which is log $[AOH]/[A^+] = pH - pK$

Experimentally, Sondheimer (*14*) for pelargonidin-3-monoglucoside and Berg (*15*) for malvidin-3-monoglucoside found the values of pK to be close to 3 which is close to the pH of wine. At this pH about 50% of the anthocyanin molecules are in the red form and 50% are in the colorless form.

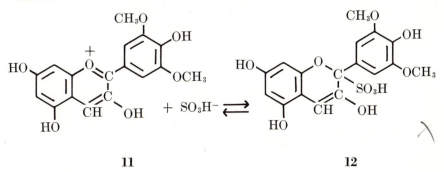

11   12

Bisulfite ions, HSO_3^-, condense with anthocyanins. This reversible reaction decreases the color by forming a colorless compound (12) (*16*). This effect is less evident in strongly acid media because the bisulfite ions are not as numerous since they are being converted to the undissociated acid. This property explains the decolorization of red wines following sulfite treatment; but, since it is reversible, the color gradually reappears as the free SO_2 (bisulfite ions) disappears. The major role of tannins in the color of old wines explains their insensitivity to color change with SO_2.

The anthocyanins are also decolorized by reduction, and again the reaction is reversible. The mechanism for this reaction has not been explained, but it may be supposed to occur through a flavene structure (14) analogous to the classic reaction for the reduction of biological molecules. This reaction must explain the light color of some wines fresh from fermentation tanks where this strongly reducing process occurs. The color may deepen with progressive oxidation of anthocyanins and

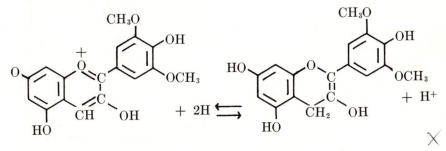

13   14

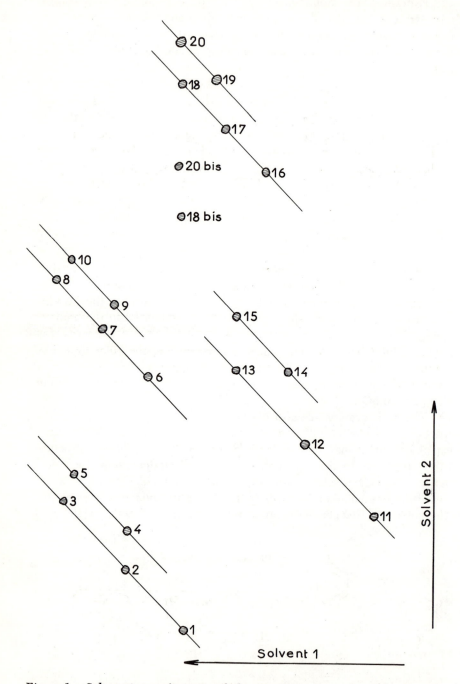

Figure 1.   Schematic two-dimensional chromatogram of grape anthocyanins (1)

probably of tannins, a reaction which is more rapid in 225-liter wooden casks than in large air tight tanks.

The anthocyanins with two OH in the ortho position of the lateral (side) aromatic ring (petunidin, delphinidin, cyanidin) complex with heavy metals (ferric iron and aluminum) forming intricately structured blue compounds. These reactions occur less readily in acid media. This property applies to the coloration of red and blue flowers. This complexing appears also as iron *casse* of red wines consisting of the formation of insoluble complexes of iron with the coloring compounds and tannins. However, there are few precise indications pinpointing the exact influence of these complexes on the true color of red wines; they probably don't interfere much. Nevertheless, the addition of iron to young wines rich in anthocyanins increases the color while producing a slight iron *casse*. This event must occur at the same time as the reoxidation of the reduced pigments during fermentation to explain the increased color in some wines in the weeks after vinification. The progressive oxidation of the ferrous to ferric ions makes the formation of these complexes possible.

Poorly understood mechanisms coincide with the destruction of anthocyanin molecules that occur during wine storage (15, 17). These phenomena are probably oxidative; they are catalyzed by Fe^{3+} ions and occur at increasing rates with increasing temperature. They may account for the enzymatic destruction of anthocyanins by anthocyanases. These enzymes do not seem to exist in grapes but are secreted in abundance by *Botrytis cinerea*, the grey rot fungus of grapes. This is responsible for the oxidative casse damage of red wines which destroys color, an accident common in wines made from grapes spoiled by mold.

Distribution of Anthocyanins in the Different Species of *Vitis*. Paper chromatography is well adapted for separting and identifying the anthocyanins in grapes (1, 18). The skins are obtained by crushing the berries individually between the thumb and index finger; extraction is done by maceration in 1% HCl. The resulting solution is rich enough in pigments that it may be directly chromatographed without preliminary concentration.

Table II. Pigments of Figure 1

	Aglycones				
	Delphin-idin	*Petun-idin*	*Malvi-din*	*Cyani-din*	*Peoni-din*
Diglucoside	1	2	3	4	5
Acylated diglucoside	6	7	8	9	10
Monoglucoside	11	12	13	14	15
Acylated monoglucoside (1)	16	17	18	19	20
Acylated monoglucoside (2)			18'		20'

Table III.   Distribution of Anthocyanins among

Anthocyan Pigments	Spot No., Figure 1	Species of Vitis					
		rot-undi-folia	ri-par-ia	ru-pes-tris	la-brus-ca	ariz-oni-ca	ber-land-ieri
Total		5	15	12	11	11	10
% Each Pigment:							
Cyanidin							
monoglucoside	14		2		5	8	8
acyl. monoglu.	19						
diglucoside	4	9	5	2		1	
acyl. diglu.	9						
Peonidin							
monoglucoside	15				10	14	16
acyl. monoglu.[b]	20				3	1	1
acyl. monoglu.[b]	20'						
diglucoside	5	6	2	8	1	10	2
acyl. diglu.	10						
Delphinidin							
monoglucoside	11		14	9	21	13	23
acyl. monoglu.	16		1	3			
diglucoside	1	38	12	34			
acyl. diglucoside	6		2	6			
Petunidin							
monoglucoside	12		10	3	15	10	20
acyl. monoglu.	17		1				
diglucoside	2	29	17	22	1	1	1
acyl. diglucoside	7		3	2			
Malvidin							
monoglucoside	13		6	2	34	29	25
acyl. monoglu.[b]	18		2		7	3	2
acyl. monoglu.[b]	18'				1		
diglucoside	3	18	21	8	2	10	2
acyl. diglu.	8		2	1			

[a] *V. vinifera* (a) = Muscat Hamburg, *V. vinifera* (b) = all other varieties studied.

Two-dimensional chromatography is performed (*1*) using the freshly prepared two-phase solvent system of butyl alcohol–acetic acid–water (4:1:5). The lower aqueous phase is solvent 1, and the upper phase is solvent 2. Using this method, all the pigments found in the different grapes studied separate on the chromatograms as shown in Figure 1; Table II gives the identity of each substance. The pigments represented by 18 and 18A, and 20 and 20A have not been identified chemically. The data in Table II were obtained by observing the color of the spots, their fluorescence under UV light, and the relationship between chemical structure and position of the pigment on the chromatogram.

the Different Species of the Genus *Vitis* (2)[a]

Species of Vitis

mon-ti-cola	cor-di-folia	ru-bra	lin-cecu-mii	aesti-valis	cor-ia-cea	amu-ren-sis	vinifera (a)	vinifera (b)
6	11	12	17	9	7	5	5	9
3	10	20	29	31	58		20	3
			7					
		1	2	3	4			
			5					
5	11	5	7	11	6	13	45	15
		1						2
								2
	2	1	3	4	4	15		
			2		4			
36	15	30	17	31	20		6	12
			6					
		1	1					
			3					
26	18	20	8	10	4	5	9	12
	2	2	1					
			3					
27	30	16	4	6		27	20	36
3	4			2				9
	1	1						9
	5	2	1	2		40		
	2		1					

The difference in structure of these two pigments has not been explained.

Each compound identified had to be isolated; this was done by a series of chromatographic separations using many sheets of heavy paper. Thus, for each anthocyanin, the chemical characterization of the aglycone, the sugar, and eventually the acylated residue could be achieved. All of these points, as well as the techniques for determining the proportions of the individual pigments, were developed in detail (*1*). The advantages of these methods, especially those of the chromatographic chart in Figure 1, have been fully discussed by Ingalsbe *et al.* (*19*) in their study of the anthocyanins in the Concord variety. The results of the identification of the anthocyanins in 14 species of *Vitis* are assembled in Table III; Table IV shows the place of each species within the genus.

Table IV.   Species

Vitis

	Euvitis planch
Muscadinia planch.	*American*

Muscadinia planch.	Euvitis planch — *American*
V. munsoniana SIMPSON	[a] *V. Labrusca* L.
[a] *V. rotundifolia* MICHX.	*V. californica* BENTHAM.
	V. caribaea CAND.
	V. coriacea SHTTL.
	[a] *V. Lincecumii* BUCKLEY
	V. bicolor LEC.
	[a] *V. aestivalis* MICHX.
	[a] *V. Berlandieri* PLANCH.
	[a] *V. cordifolia* MICHX.
	V. cinerea ENGELM.
	[a] *V. rupestris* SCH.
	[a] *V. monticola* BUCKLEY.
	[a] *V. Arizonica* ENGELM.
	[a] *V. riparia* MICHX.
	[a] *V. rubra* MICHX.
	V. candicans ENGELM.

[a] Denotes those studied.

The data in Table III show the following: First, malvidin-3-mono-glucoside (oenin) is the principal constituent of the grape coloring matter in V. *vinifera,* but it does not represent the majority of the pigment since it comprises only 36% (even less in Muscat Hambourg) of the total pigment. Second, there is no predominance of monomethylated derivatives as reported for American species. Third, we have never found anthocyanidins in the coloring matter of grapes. Fourth, anthocyanin diglucosides occur frequently in American species but not at all in V. *vinifera.* Before our work this fact had only been reported by Brown (*20*) for V. *rotundifolia.* This observation is very important because the determination of these diglucosides makes possible the differentiation between the grapes and vines of V. *vinifera* and hybrids. This has been confirmed by many researchers. Fifth, the presence of cyanidin and peonidin derivatives (two OH on the side ring) among the anthocyanins of grapes is common; they are systematically found in all the species although they are abundant in only a few. This explains why this fact went unnoticed for so long. Note that except for Muscat Hambourg the two species which best illustrate this are V. *lincecumii* and V. *aestivails* which, according to specialists in ampelography, are at least closely related if not identical. Sixth, the absence of acylated anthocyanins is demonstrated in the Pinots (*18, 21*) and this characteristic is specific to these varieties of V. *vinifera.*

of the Genus *Vitis*

Vitis

Euvitis planch

Asiatic	*European*
V. coignetiae PULL.	[a] *V. vinifera* L.
V. thumbergii SIEB.	
V. flexuosa THNBG.	
V. romaneti ROM.	
V. pagnucii ROM.	
[a] *V. amurensis* ROM.	
V. lanata ROXB.	
V. pedicellata LAWS.	
Spinovitis davidii ROM.	

Many works have been published on the comparison of species and on the presence of diglucosides in *V. vinifera;* they are reported in previous publications (*1, 2, 22*). Some of these publications report the presence of diglucosides in *V. vinifera,* but their conclusions have been proved false; there are no diglucosides in *V. vinifera.* The chromatographic methods have allowed the reclassification of some varieties reported to belong in the species *V. vinifera.* Notably, at the University of California, Davis, Bockian *et al.* (*23*) reported malvidin-3,5-diglucoside in Cabernet Sauvignon, but this has not been confirmed in further work by the same group (*18*). In the Soviet Union, Dourmichidzé and Noutsoubizé (*24*) and Dourmichidzé and Sopromadzé (*25*) have identified these diglucosides in some *V. vinifera;* we have not confirmed this result on the same vines grown in France nor on the grapes sent by S. V. Dourmichidzé except for Asuretuli Shavi. This variety was believed to be a *vinifera,* but now it appears to belong to another species. The same conclusion applies to the more recent results of Cappelleri (*26*) and Getov and Petkov (*27*) who also suggest the presence of diglucosides in *V. vinifera.* Since we have been unable to reproduce their results, we suggest that they arise from poor analyses.

People have vigorously debated the possible presence of traces of malvidin-3,5-diglucoside in *V. vinifera,* the identification of which is more often made in wines than in grapes. In fact, most of the work in this field has dealt more with the differentiation of wines from *vinifera* and hybrids, by analyzing their coloring materials, than with the physiology of the vine. The results are obtained by applying very sensitive methods which necessitate concentration of the coloring compounds but which can be shown to produce artifacts by carrying a weak spot along on the chromatograms which has never been identified but which is located at the same place as the anthocyanin diglucosides.

Finally, some work implies that the same vine may or may not, depending on conditions, produce anthocyanin diglucosides. This would be the same as saying that the same vine could produce white grapes or black grapes. Even if this were true, could a vine which produced black grapes continue to be called a white vine? The presence of diglucosides can be thought of as a physiological constant of the variety. Hence all the plants must be identical. Mutations do occur, but they are definitive, and they also lead to new varieties.

Tannins

Definition. Some confusion exists even in the definition of tannin; included under this name are many substances of varying structure, but having the common ability to transform fresh hides into rot-proof and barely permeable leather. The phenolic nature of these substances has often caused confusion between tannins and phenolic compounds of plants.

The tannins are special phenolic compounds characterized by their ability to combine with proteins and other polymers such as polysaccharides. This characteristic explains their tanning properties as arising from a tannin–collagen matrix and then astringency caused by precipitation of the proteins and glycoproteins from saliva. Also tannins are used in fining wines because they combine with proteins. Finally, they inhibit enzymes by combining with their protein fraction.

To maintain stable complexes with proteins, phenolic molecules must be relatively large so that they can form enough bonds with the protein molecules; however, if the tannin molecule is too large, it will be unable to bind with the active sites on the protein, and bonding will be less likely. Finally, protein–tannin complexes are more stable, and the tannin qualities are highest when the molecular weight of the tannins is between 500 and 3000.

Thus one arrives at a valid definition of tannins for plant chemistry, insofar as their characteristic properties are concerned, without reference to the leather industry. Such a definition has been given by Swain and Bate-Smith (28):

It would seem more realistic to define as tannins all naturally occurring substances which have chemical and physical properties akin to those which are capable of making leather. This means that they would be water-soluble phenolic compounds, have molecular weights lying between 500 and 3000, and, besides giving the usual phenolic reactions, have special properties such as the ability to precipitate alkaloids and gelatin and other proteins.

It is obvious that such a definition will include molecules which are not tannins in the commercial sense of being economically important in the tanning of hides, but will exclude a large number of substances whose only relationship to the tannins is their capacity to reduce alkaline permanganate or give colors with ferric salts.

This definition well demonstrates that to identify the tannins in a plant sample, it is not sufficient to know their amount. Their structures and particularly their molecular weights must also be known.

From a chemical standpoint tannins are formed by the polymerization of elementary phenolic molecules; according to the nature of these molecules, one can distinguish hydrolyzable tannins (gallics) and condensed tannins (catechins).

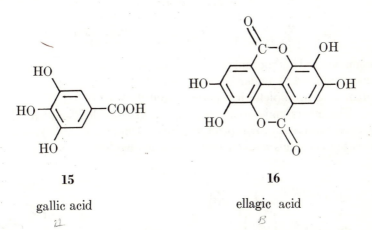

15

gallic acid

16

ellagic acid

Hydrolyzable Tannins. The hydrolyzable tannins are composed of one glucosidic molecule to which are bonded different phenolic moieties; the most important of these are gallic acid (15) and the lactone of its dimer, ellagic acid (16). These are not the natural tannins of grapes, but they are the principal commercial tannins (tannic acid) authorized by legislation to be added to wines. The oak tannins also belong to this family and may be added to wines stored in wooden casks. The presence of ellagic acid in wine, reported in the literature, needs to be confirmed.

Condensed Tannins. The tannins found in grapes and wines are condensed polymers from 3-flavanols (catechins) (17, 18, 19) and from 3,4-flavandiols (leucoanthocyanidins) (20, 21, 22). The monomeric leucoanthocyanidins, like their polymerized forms, display the characteristic, which differentiates them from the catechins, of trans-

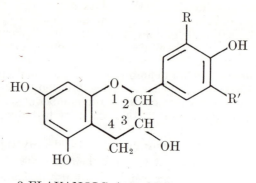

3-FLAVANOLS (catechin)
17 R = R' = H; afzelechin
18 R = OH,R' = H; catechin
19 R = R' = OH; gallocatechin

3,4-FLAVANDIOLS (leucoanthocyanidin)
20 R = R' = H; leucopelargonidin
21 R = OH, R' = H; leucocyanidin
22 R = R' = OH; leucodelphinidin

forming into red anthocyanidins when heated in acid media. For example, leucocyanidin (23) under these conditions yields cyanidin (24). This reaction is not complete; it converts about 20% of the elementary molecules. The others undergo a rapid condensation which leads to the phlobaphenes, insoluble brown-black products. Under the same heating conditions the catechins convert entirely to phlobaphenes.

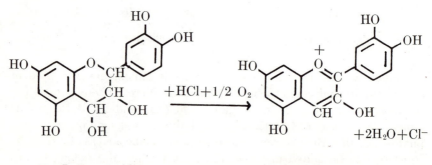

23   Leucocyanidin   24 Cyanidin

Plants contain leucoanthocyanins which convert to anthocyanins upon heating in acid. These anthocyanins were reported in grapes and wine as early as 1910 by Laborde (Agronomy and Enology Station at Bordeaux). However, only in the past 15 years has it been certain that this conversion to anthocyanins is not, from the biological point of view, the most important property of these substances, and they should not be considered as anthocyanin derivatives. Even though they are known as the main components of tannins today, leucoanthocyanin should not replace the word tannin which is accepted usage and which represents compounds of well defined properties.

The polymerization of flavan molecules is thus the essential feature in tannin structure which governs their properties, especially their different properties in wine. During preservation and aging, changes in the condensation state affect the color of tannins in solution and their organoleptic characteristics. The state of condensation can be appreciated on examining the average molecular weights of the tannins (500–700 for young wines and 2000–3000 for old wines). The chemical mechanisms responsible for this polymerization must then be considered in interpreting red wine color.

Among the hypotheses formulated, the most likely mechanism calls for the formation of a covalent bond between carbon 4 of the 3,4-flavandiol (leucoanthocyanidin) and carbons 6 or 8 of another flavan molecule. Benzylic alcohol is a reactive electrophile (loss of OH^-), and it donates readily in acid media; leucoanthocyanin (25) functions similarly at position 4. The phenolic group is a mesomeric structure which displays negatively charged nucleophilic centers in the ortho and para positions; analogous centers may be found at positions 6 and 8 of flavan molecules. This would allow the possibility of covalent bond formation between carbon 4 of 25 and carbons 6 or 8 of 26a or 26b. This bond is attributable to the elimination of a water molecule.

Based on these principles, Jurd (29) described the most likely reactions (Figure 2). One leucocyanidin molecule 25 reacts with a flavan molecule 26a or 26b yielding dimers 27a, 27b, 28a, or 28b. If the second flavan molecule is also a leucocyanidin (R = OH), the dimer again contains a benzylic alcohol (OH) and the condensation can continue. If this second molecule is a catechin (R = H), condensation cannot go beyond the dimer, which is a proanthocyanin. Proanthocyanins in plant tissues have long been known (30).

Structures 29 and 30 are such dimers (31). Furthermore, Bhatia et al. (32) have identified a proanthocyanidin 32, composed of three molecules of leucocyanidin 32, in the seeds of white grapes.

Finally, Weinges et al. (31, 33) postulated the formation of dimers such as 27 or 28, 29 or 30 directly from catechins without involving 3,4-flavandiols (leucoanthocyanidins). This process has never been demonstrated in fruits directly and specifically not in grapes. Only the proanthocyanin dimers have been positively identified through the formation of catechins and anthocyanidins during acid hydrolysis. Dimer formation proceeds by enzymatic oxidation of two molecules of catechin

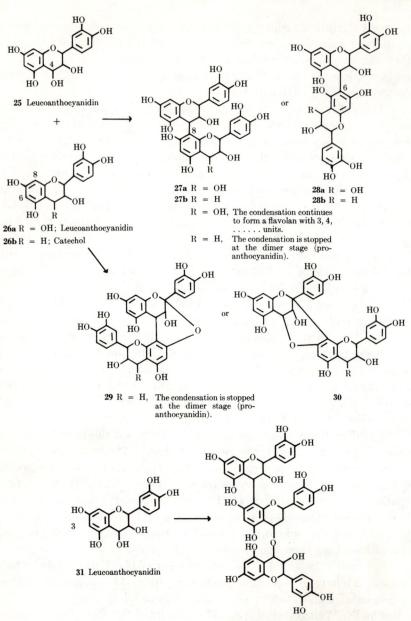

25 Leucoanthocyanidin

+

26a R = OH; Leucoanthocyanidin
26b R = H; Catechol

27a R = OH
27b R = H

R = OH, The condensation continues
to form a flavolan with 3, 4,
. . . . . . units.

R = H, The condensation is stopped
at the dimer stage (pro-
anthocyanidin).

28a R = OH
28b R = H

29 R = H, The condensation is stopped
at the dimer stage (pro-
anthocyanidin).

30

31 Leucoanthocyanidin

32 Condensation of **27a** with
a third molecule of leucoanthocyanidin. The con-
densation is stopped at the
trimer stage (proanthocy-
anidin).

*Figure 2.   Condensation of elementary flavin molecules: formation of proantho-
cyanidins (two or three elementary molecules) or of flavolans (three to 10
elementary molecules); flavolans are the constituents of condensed tannins*

(**26b**); the reaction mechanism described in Figure 2 does not inter-
vene. Upon heating, dimers and higher polymers form anthocyanins
in acid media; the 3,4-flavandiols formed under these conditions
should not interfere with the formation of the tannins. However,
Weinges' theory is based on the fact that the 3,4-flavandiols, capable of
occurring as dimers, have never been found in nature in monomeric
form. This argument is perhaps not enough; when these compounds
were synthesized in the laboratory (*34*), they were quite unstable and
very prone to condense, and this may be what happens in the plant.

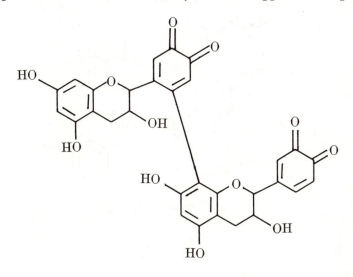

33

Another possibility for the condensation of flavan molecules, calling
for oxidative processes, is envisaged by Hathway and Seakins (*35*). They
have shown that the oxidation of catechin by oxygen in the air or by
polyphenoloxidase leads to a polymer of quinoidal structure, making a
bond between carbon 6 of one molecule and carbon 6 or 8 of the other
possible (*33*). These polymers are yellow-brown in color, which deepens
as the condensation increases. Even if this reaction did not occur naturally in grapes, its occurrence during wine storage seems quite possible;
it could also explain the browning of white wines. This oxidative reaction
should explain the catalytic role of iron (Fe^{3+}) in the changes of the
tannins in wine described elsewhere in this volume.

In conclusion, the characteristic properties of tannins (especially
their enological properties) are related to their ability to bind protein—a
quality directly related to the polymerization of tannin molecules which
arise from two to 10 simple flavan molecules with molecular weights

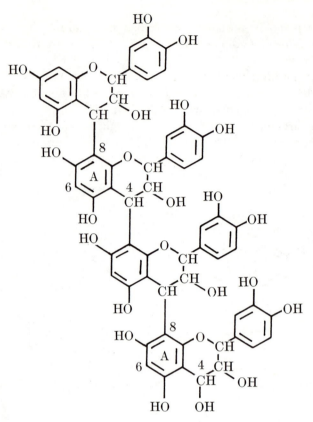

34   Flavolan   (tetramer)

around 20. These polymers are known as flavolans, as the tetramer in **34.**
The term proanthocyanidin is used for the dimers and possibly the
trimers.

The tannin in plant samples, grapes or wine for example, is a mixture of flavolans of varying structures which determine the properties of
the tannin. Consequently, to be positive about the tannin in a sample,
two tests should be made: (a) total tannin; and (b) amount of polymerization or the proportions of the different flavolans.

This second test is especially difficult to perform and has not been
perfected. This difficulty is inherent in the characteristics proper of the
tannins and in their ability to bind to proteins and other macromolecules
(cellulose, polyamide) used as chromatographic supports. This explains
why no truly effective fractionation of flavolans has been reported. It is
equally possible that the flavolans associate with each other as a function
of the media conditions and that all the necessary manipulations during
extraction allow some structural changes.

Classification of Wines by their Anthocyanins

Principle. A practical method for differentiating red wines of V. *vinifera* and those of hybrids has been developed. This method has had profound effects not only in wine analysis but also in the regulation of commercial transactions. Before this method was perfected, the planting of hybrid vines was increasingly common in some of the large Appellations controlées, making it more difficult to detect objectively the wines from the hybrids in the mixtures. The ensuing adulteration, threatening the reputations of the Appellations controlées while decreasing the wine quality and increasing the volume, has completely disappeared. The methods formerly used to detect this fraud were difficult if not impossible to apply. The new method depends on the following observations (*1*): (1) One never finds anthocyanins in the diglucoside form in the berries of V. *vinifera*. (2) Anthocyanin diglucosides in the berries is a property of V. *riparia* and V. *rupestris*. In addition, the diglucosides trait is genetically dominant.

V. *riparia* and V. *rupestris* are the two most used species in hybridization; the majority of the hybrids used commercially contain diglucosides in their coloring matter, but diglucosides trait in V. *vinifera* is recessive; it may appear after many crosses. Thus a cross between V. *riparia* and V. *vinifera* yields an F_1 population of hybrids, all having the diglucosides character, but if one of these F_1 hybrids is crossed with V. *vinifera*, half the F_2 hybrids will not have the diglucosides trait. The fruit of these last vines will have the same coloring matter as V. *vinifera* (*1*). There is, therefore, a limit inherent in the method.

Most of the common producing hybrids contain anthocyanin diglucosides in their coloring matter, but a few are the same as V. *vinifera* (Seibel 5455 for example). The differentiation method, then, consists of identifying the anthocyanin diglucosides: their presence proves the hybrid parentage of a grape of wine, but their absence is not proof that a grape or wine is V. *vinifera*. The absence of diglucosides in grapes from V. *vinifera* has been discussed in the Anthocyanins section. We now consider the mechanism for transmitting the diglucosides trait in successive hybridizations.

Transmission of the Diglucosides Character. We have shown that the diglucosides character of V. *riparia* and V. *rupestris* is dominant over V. *vinifera*. Any cross between V. *riparia* (or V. *rupestris*) and V. *vinifera* yields F_1 hybrids, all containing diglucosides in the coloring materials of their grapes. However, if an F_1 heterozygote is selfed, it should show the disjunction of characters and the reappearance of the recessive characters of V. *vinifera*. In other words, such a cross would give rise to a second generation, F_2, in which some individuals would produce no diglucosides in their grapes just like *V. vinifera*. The theoretical possibility for character disjunction in a heterozygote was shown for anthocyanin diglucosides of grapes (*2*). Two hybrid families were used, one from the selfing of Oberlin 595, and the other from the selfing of Seyve-Villard 18402. Oberlin 595 and Seyve-Villard 18402 are both heterozygous for the diglucosides character. In effect they contain anthocyanin diglucosides in their coloring matter from a cross in which one parent had no diglucosides (*V. vinifera* as one parent, Seibel 5455 as the other).

In the two cases described, individuals producing anthocyanin diglucosides in their grapes and individuals not doing so appeared, with a large majority of the first type. This totally agrees with genetic law, but we have been unable to determine the coefficient of disjunction (theoretically 75% should have diglucosides). In short, the number of individuals was too small in each case. This possibility of disjunction in the coloring material has also been verified by Boubals *et al.* (*36*) who confirmed the dominance of the diglucosides and their mode of transmission. Using chromatographic techniques, they studied the anthocyanins in 105 individuals from a cross between *V. vinifera* and a homozygous hybrid which was diglucosides dominant (Seyne-Viallard 2318). They all contained anthocyanin diglucosides in the coloring matter of their fruits. The same authors also studied 77 individuals arising from a cross between *V. vinifera* and a recessive homozygote (Seyve-Villard 23353); all lack diglucosides. Finally, a cross between *V. vinifera* and a heterozygote (Seyve-Villard 18402) produced 17 progeny with anthocyanin diglucosides in their fruit and 15 without them. This is close to 50% for each possibility which is the theoretical percentage predicted by genetic law.

Methods. The originally described method calls for a simplified form of paper chromatography to detect the anthocyanin diglucosides. This method separates the monoglucosides and diglucosides without separating the different substances in each family. Numerous variations have been suggested, all of which use some modification of paper chromatography; they are not original methods. Dorier and Verrelle's method (*34*) uses the same principles of differentiation, but evidence of diglucosides no longer depends on chromatography but on the specific formation, beginning with malvidin-3,5-diglucoside through the action with sodium nitrite, of a compound fluorescing green in ammoniacal medium.

Spot the wines to be studied on a piece of Arches 302 chromatography paper 6 cm from the lower edge, in bars 2-cm long and 2-cm apart. Ten applications need to be made, letting the spot dry between applications. The paper is kept vertical by appropriate means with its lower edge in the solvent, which is a weak solution of a non-volatile acid (citric acid at 6 grams/liter). The paper may also be positioned as an inverted V with its two ends dipping in the solvent (Figure 3), making it possible

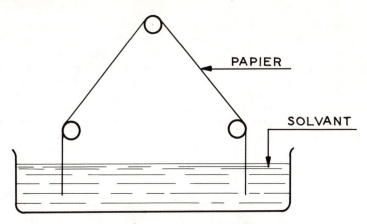

Figure 3. Device for separating red wine anthocyanins on paper to detect wines from hybrid grapes

to spot wines on both sides of the paper. The tank should have a cover to prevent excessive evaporation. The dimensions of the paper depend on the equipment used (18 × 44 cm for example).

The solvent rises the length of the paper by capillarity, carrying the diglucosides faster than the monoglucosides. Thus the desired separations are obtained. The method results in two types of separations (Figure 4). First, absence of diglucosides which corresponds to chromatogram 1 (*V. vinifera*). Second, presence of diglucosides which corresponds to chromatogram 2 (majority of the hybrids). The presence of diglucosides is confirmed by examining the chromatogram under ultraviolet light where the malvidin-3,5-diglucoside (malvoside or malvin) fluoresces a bright brick red. This property is very sensitive to this test.

The use of this method spread rapidly, mainly for the quality control of wines from the French Appellations controlées which must not contain any hybrid wine. Also, Germany legislated against the production and sale of hybrid wines; all the wines exported to France must first be analyzed by chromatography. In addition, the report on the work of the third reunion (April 10–12, 1961) and of the fourth reunion (May 7–9, 1962) of the Sub-Committee on Analytical Techniques of the International Office of Vine and Wine reported the exceptional interest in this method in enological circles. The method has been made official by French legislation. It is included in the official methods of wine analysis published in the *Journal Officiel* of September 20, 1963 (official order of the 24th of June, 1963) and is commonly used in all laboratories.

Anthocyanin and Tannin Analyses, Determination of Wine Color

Anthocyanin Measurements. Quantitative analysis of concentration by direct colorimetry in the visible range is not practical with red wines because of substances (tannins in particular) which absorb at the maximum for anthocyanins (from 500 to 550 nm). Such a simple method may be used if need be for plant parts especially rich in anthocyanins, such as grape skins.

Certain properties of these pigments (*e.g.*, color changes as a function of pH or transformation to colorless derivatives through action with, for example, sodium bisulfite) allow a solution to this problem.

The first method uses the fact that in acid media anthocyanins exist in a colored and a colorless form (**9** and **10**) in equilibrium, with the position of the equilibrium depending on pH. Consequently, the difference in color intensity between the two pH values (0.6 and 3.5 for example) is proportional to the pigment concentration. Since the phenol function is not affected by this variation, other phenolic compounds, especially tannins, do not interfere since their absorption at 550 nm is the same at both pH values.

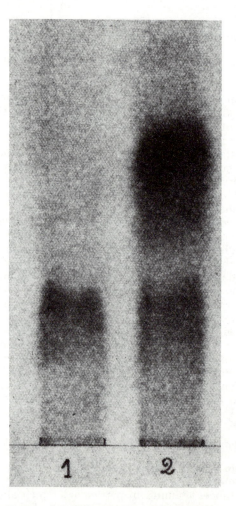

Figure 4.   Differentiation of wines according to their anthocyanins; 1 is V. vinifera, 2 is majority of the hybrids

The second method utilizes the fact that anthocyanins form colorless compounds with bisulfite ion (12). Again, the other wine constituents do not interfere, and the variation in color, after a large excess of bisulfite is added, is proportional to the concentration of anthocyanins.

Ribéreau-Gayon and Stonestreet (38) have described ways to apply these methods. The agreement between the results obtained by using thse methods, based on different principles, are close although the figures from the first are usually lower.

Thus the amounts of anthocyanin in young wines from the Bordeaux region may vary between 0.2 and 0.8 gram/liter, depending on yearly conditions. This percentage decreases progressively during aging, and at the end of *ca.* 10 years is around 20 mg/liter. It is not impossible that this value could fall to zero and that the minimum value (20 mg/liter) is an artifact of the assay. From another point of view it is not the anthocyanins but the tannins which play the leading role in the color of old wines.

Quantitative Analysis of Tannins. Beginning with determinations for total phenolic compounds, all attempts to isolate compounds related to tannins—*i.e.*, those forming stable compounds with proteins—have been unsatisfactory. Hence, one cannot use this as the basis for an analytical method.

The best method for measuring condensed tannins from leucoanthocyanins takes advantage of the fact that these compounds are transformed into anthocyanins when heated in acid media. The reaction occurs with monomeric forms (for example, the leucocyanidin 23) as well as with the flavolans (the tetramer in 34, for example). In this last case the reaction ruptures the bonds between the 3,4-flavandiols which then become anthocyanidins. The reaction does not go to completion, and the yield, at about 20%, depends on the procedure used, which must be followed exactly. Specifically, the efficiency of the conversion is better in alcoholic solution than in aqueous solution, but the reproducibility is not as good. The analysis amounts to a colorimetric reading of the anthocyanins formed. As far as red wines are concerned, the color of the anthocyanins itself changes little during heating. The results should be compared with a condensed tannin as a standard reference. At the Institute of Enology at Bordeaux, we generally use a reference compound prepared in the laboratory of J. Masquelier (Pharmacy Faculty at Bordeaux) from the bark of the maritime pine (*Pinus pinaster*). We have verified that the behavior of such a sample is very close to that of the condensed tannin extract from grape seeds. This method has been applied to red wines by Ribéreau-Gayon (14), by Masquelier *et al.* (39), and by Ribéreau-Gayon and Stonestreet (40).

Concerning the properites of tannins, we have stressed the importance of their structure, even more than their state of condensation or molecular weight. As it is impossible to separate and identify the flavolans in a given plant sample, one must be content with finding the aggregate indexes expressing the median state of polymerization.

The method described by Swain and Hillis (41) applied to wine (40) gives V/LA which is related to the state of polymerization in condensed tannins. The tannin sample is put through two tests whose reactions depend on the state of condensation. These two reactions are the conversion of the leucoanthocyanins to anthocyanins (*LA*) and the reaction with vanillin (*V*). At first this ratio decreases as condensation increases. This index has made some results possible—for example, the demonstration of the structural differences between the tannins in the different parts of the cluster (skins, seeds, stalks) (*2*). These results are insufficient to predict in any precise way the structures of the tannins in wine.

More recently, Ribéreau-Gayon and Glories (*42*) tried to work out a direct determination on the mean molecular weight (mass) of tannins separated by chromatography on Sephadex G 25 (*43*). This test is based on a cryometric method (Rast's method) which consists of measuring the degrees the melting point of camphor is lowered when it is mixed with the substance under study. This gives a more precise characterization of the tannins in wine and has allowed their evolution to be followed during aging of wines. However, truly satisfactory results will not be reported until the different polymers in tannins are determined.

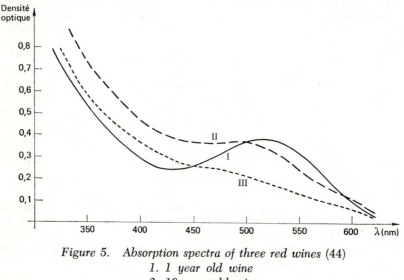

Figure 5. Absorption spectra of three red wines (44)
    1. 1 year old wine
    2. 10 year old wine
    3. 50 year old wine

Determination of Red Wine Color. Sudraud (*44*) has discussed the red wine absorption curves, notably as a function of aging which is the essential cause of color changes (Figure 5). Young red wines have a maximum absorption at 520 nm because of the grape anthocyanins which are responsible for the true red color. Between this absorption and the one in the ultraviolet, at about 280 nm, a minimum is found at about 420 nm. As the wine ages, the maximum at 520 nm tends to disappear, falling to a small shoulder in wines older than 10 years (Figure 5). This corresponds to an increase in yellow color (420 nm absorption) which explains why the true red turns to a tile-like red-orange.

Evaluation of these curves shows that it is not feasible to use the 520-nm absorption exclusively to characterize the intensity of wine color because young wines have exaggerated color in relation to old wines. However, the sum of the absorptions at 520 and 420 nm is a satisfactory expression of the color intensity.

To define the shade or tint of color, Sudraud (45) recommends using the ratio of the absorptions at 420 and 520 nm. These two indexes may be faulted for being conventional and for not defining the color in absolute terms, but they are a true advantage because they allow the wines to be compared among themselves in several processes (vinification, storage, aging). This is why tristimulus method seems to us to be an unnecessary complication, even in the simplified form recommended by the International Office of Wine and Vine (1962).

Wine cannot be diluted for measuring its color, probably because of the colloidal nature of the coloring matter. There is no proportionality between the dilution coefficient and optical density. Consequently, to get optical density readings on the order of 0.5 with precision, the measurements must be taken on very thin samples. For red wines, cuvettes 0.1-cm thick are used. With D_{420} and D_{520} representing the values for absorbancies at 420 and 520 nm in cells 1-mm thick, color intensity and tint are given by the two expressions: $I = D_{420} + D_{520}$ and $T = D_{420}/D_{520}$.

Table V. Anthocyanins and Tannins of Red Wines of Different Vintages from Two Vineyards of the Bordeaux Region[a]

| Vintage | Color Intensity $D_{420} + D_{520}$ | Anthocyanins, mg/liter | | Total Tannins, grams/ liter | V/LA |
		By pH difference	By reaction with $NaHSO_3$		
Vineyard 1					
1921	0.802	19	27	4.40	0.72
1926	0.690	16	20	3.30	0.65
1928	0.710	16	24	3.45	0.50
1929	0.846	20	26	3.05	0.42
1938	0.523	16	20	2.70	0.80
1952	0.607	18	20	2.50	0.98
1956	0.456	23	26	2.15	1.10
1959	0.545	97	93	2.05	1.43
1961	0.540	165	188	1.85	1.35
1962	0.487	305	330	1.50	1.40
Vineyard 2					
1953	0.720	30	42	3.70	0.84
1957	0.520	39	50	2.00	1.21
1960	0.590	60	69	2.25	1.18
1962	0.765	110	122	2.50	1.34
1963	0.359	133	125	2.25	1.70
1964	0.835	362	385	2.25	1.31

[a] Data from refs. *38, 40*.

Results. Table V gives the results from tests on red wines from different vintages and vineyards of the Bordeaux region. These data first showed the disappearance of anthocyanins during aging and the large role of tannins in the color of old wines. Old wine shows a much larger proportionality between tannins and color than between anthocyanins and color. Moreover, the older wines are also the richest in tannins. This fact is related to developments in vine culture and winemaking which tend to produce wines which are softer and hence less tannic. Finally, during aging the ratio V/LA shows an increase in the degree of polymerization.

Table VI.   Structures of Tannins of Different Wines (42)

	Year	Total Tannins grams/ liter	Mean Molecular Weight		Number of Flavan Units
			First Determination	Second Determination	
V. vinifera	1914	1.7	739 ± 49		2–3
wines	1952	4.5	3750 ± 600	4000 ± 811	10–14
	1957	2.9	2995 ± 400	3400 ± 600	8–11
	1962	2.5	2010 ± 211	2288 ± 268	6–7
	1966	3.5	2134 ± 303	1909 ± 181	6–8
	1967	2.9	2200 ± 230		6–8
	1968	1.9	1071 ± 58		3–4
	1969	2.0	895 ± 34		3
Hybrid	1967	1.7	2150 ± 360	2500 ± 350	6–8
wines	1968	1.1	1900 ± 235	1886 ± 189	5–7

The determination of tannin structure presented in Table VI was obtained by measuring molecular weight directly. Again, during aging, one finds an increase in average polymerization which includes from three to four elementary flavan molecules in young wines and from six to 10 in ten-year-old wines. Furthermore, at the end of a long aging period (wine from 1914), the coloring materials and the greatly polymerized tannins had precipitated; all that remain are slightly condensed forms.

Development of Pigments during Grape Maturation

Studies on grape maturation have included for simplicity, only determinations of the composition of grape juice (sugar and acidity). The necessity for completing these studies has long been recognized, especially the study of the solid constituents of the cluster (skins, seeds, stems) and particularly the phenolic compounds (anthocyanins and tannins) in red grapes. This investigation encounters double difficulties: measurements for complex substances, whose analyses have taken a long time to perfect, and a preliminary extraction which is indispensable.

Table VII. Development of Anthocyanins and Tannins as Fruit Matures

Cabernet-Sauvignon, vineyard P, results expressed in gram per 200 berries
following cold and hot extractions

Date	Anthocyanins	Tannins
25 August 1969	0.02	0.42
1 September 1969	.11	.70
8 September 1969	.27	.89
15 September 1969	.35	.93
22 September 1969	.37	.95
26 September 1969	.31	.75

Some flawless analytical methods do exist but more are needed. Quantitative extraction of all the phenolic compounds from skins and seeds is not possible; however we have developed a standard, reproducible procedure. Using a solvent with physicochemical properties similar to wine, one performs three cold extractions followed by two warm ones. This gives an estimate of the total phenols and an idea of their solubilities—an important technological factor. Our work, done in 1969–1972 involved the two principal red varieties of Bordeaux (Merlot and Cabernet-Sauvignon) from two vineyards. One (*P*) is a Grand Cru of the Médoc region characterized by fast maturation, and the other (*SC*) matures more slowly (*46, 47*).

Table VII shows the development of anthocyanins and tannins from the beginning of ripening to maturity. There are three stages: first, a rapid increase in all the substances; second, a slowing in the production

Table VIII. Phenolic Compounds in Grape Skins at Crop Maturity in Vineyard SC

Last sampling before harvest, 1969, 1970, 1971, and 1972

	Anthocyanins, gram per 200 berries		Tannins, grams per 200 berries		Phenol Index	
	Cold Extract	Total	Cold Extract	Total	Cold Extract	Total
Merlot						
1969, 22 Sept.	0.20	0.27	0.55	1.09	4.7	10.4
1970, 28 Sept.	0.32	0.41	0.66	1.03	6.0	10.0
1971, 27 Sept.	0.21	0.31	0.90	1.38	6.3	11.1
1972, 9 Oct.	0.21	0.24	0.44	0.77	4.0	7.9
Cabernet						
1969, 29 Sept.	0.22	0.31	0.36	0.86	4.2	9.2
1970, 28 Sept.	0.41	0.55	0.72	1.34	6.7	12.2
1971, 27 Sept.	0.24	0.35	0.57	1.12	5.5	13.2
1972, 16 Oct.	0.24	0.28	0.45	1.09	7.4	15.2

**Table IX. Phenolic Compounds in Grape Skins at
Crop Maturity in Vineyard P**

Last sampling before harvest, 1969, 1970, 1971, and 1972

	Anthocyanins, gram per 200 berries		Tannins, grams per 200 berries		Phenol Index	
	Cold Extract	Total	Cold Extract	Total	Cold Extract	Total
Merlot						
1969, 22 Sept.	0.23	0.30	0.50	0.96	4.6	9.6
1970, 28 Sept.	0.42	0.52	0.62	1.07	6.7	10.9
1971, 27 Sept.	0.32	0.41	0.59	0.96	6.3	11.6
1972, 2 Oct.	0.34	0.40	0.49	0.92	6.9	12.4
Cabernet						
1969, 26 Sept.	0.25	0.31	0.36	0.75	4.6	8.9
1970, 5 Oct.	0.34	0.42	0.53	0.87	5.6	9.2
1971, 27 Sept.	0.23	0.31	0.60	0.97	5.7	10.9
1972, 16 Oct.	0.19	0.20	0.41	0.82	6.4	11.7

of phenolic compounds; and last, a decrease at the end of maturation.
These phenomena were observed in all studies. The results also show
relatively high tannin levels at onset of ripening, the moment the anthocyanins appear. According to these results and contrary to general
thought, prolonging the maturation period does not increase anthocyanin
levels.

Another important fact is seen by comparing the data from four
successive years (Tables VIII, IX, X, and XI). The anthocyanin and

**Table X. Phenolic Compounds in Grape Seeds at
Crop Maturity in Vineyard SC**

Last sampling before harvest, 1969, 1970, 1971, and 1972

	Tannins, grams in 200 berries		Phenol Index	
	Cold Extract	Total	Cold Extract	Total
Merlot				
1969, 22 Sept.	0.04	0.81	0.4	3.1
1970, 28 Sept.	0.02	0.34	0.8	6.0
1971, 27 Sept.	0.08	1.03	1.1	10.1
1972, 9 Oct.	0.04	0.72	0.8	12.1
Cabernet				
1969, 26 Sept.	0.07	0.74	1.1	7.9
1970, 28 Sept.	0.06	0.35	1.2	6.0
1971, 27 Sept.	0.04	0.08	0.8	8.1
1972, 16 Oct.	0.05	0.68	1.1	8.7

tannin concentrations can almost double from year to year. If 1970 had been a good vintage year in Bordeaux, these three years are late years, so the variation must be a function of climatic conditions.

These tables show larger differences between years than between varieties. The comparison between the two vineyards, both cultivated by traditional methods, does not show significant differences. These experiments demonstrate the importance of the year and climatic conditions on the anthocyanin and tannin content. These variations are much more important than those of other chemical constituents (sugars and acids).

Table XI. Phenolic Compounds in Grape Seeds at Crop Maturity in Vineyard P

Last sampling before harvest, 1969, 1970, 1971, and 1972

	Tannins, grain in 200 berries		Phenol Index	
	Cold Extract	Total	Cold Extract	Total
Merlot				
1969, 22 Sept.	0.03	0.78	0.6	8.7
1970, 28 Sept.	0.04	0.32	1.1	6.3
1971, 27 Sept.	0.09	0.55	1.3	7.4
1972, 2 Oct.	0.06	0.76	1.1	14.7
Cabernet				
1969, 26 Sept.	0.07	0.75	0.6	6.9
1970, 5 Oct.	0.10	0.43	1.4	5.9
1971, 27 Sept.	0.07	0.54	1.2	7.9
1972, 16 Oct.	0.06	0.56	1.0	12.7

Evolution of Phenolic Compounds during Vinification

Calculation of the theoretical levels of anthocyanins and tannins in wines is possible, assuming complete extraction of these compounds from the parts of the cluster. Thus one sees only 20–30% of the possible grape pigments in wine.

Hence better extraction methods seem necessary, but this problem is not simple. The laws of extraction are complex; in the juice there is a simultaneous dissolution of phenolic compounds from the skins and seeds and their precipitation onto yeasts and other solid particles in suspension. Finally, the unstable anthocyanins are partially destroyed at this stage.

To verify the part played by yeasts in color fixation (48), we prepared a synthetic alcoholic mixture (10% alcohol, 5 grams/liter tartaric acid adjusted to a pH of 3.0 with concentrated sodium hydroxide) and added a solution of macerated Cabernet Sauvignon skins. We studied the

**Table XII.   Influence of Yeast Cells on the Phenolic Compound
Content of a Synthetic Medium**

	Tint	Color Intensity	Antho-cyanins gram/liter	Tannins, gram/liter
Control				
1st day	0.46	0.41		
17th day	0.50	0.43	0.19	0.7
Control + Yeast				
1st day	0.46	0.14		
17th day	0.60	0.13	0.05	0.2

behavior of a blank and a sample to which was added 2.0 grams of fresh
yeast (*Saccharomyces ellipsoideus*) per 20 ml, approximately 20 times
the normal amount in a fermenting must. The results (Table XII) show
a sharp increase in anthocyanin and tannin absorption which is practically
instantaneous.

The influence of the stems, also studied in model solutions, is summarized in Table XIII (49). Stem extract and stems themselves are
added to a solution of anthocyanins; the anthocyanin concentration
is the same in all cases. The stems decrease the anthocyanin levels
and color intensity which is not observed with stem extract alone. These
pieces exert their effects through absorption rather than through chemical
reactions of one of their constituents.

**Table XIII.   Influence of Stems on the Phenolic Compound
Content of a Synthetic Medium (49)**

	Tint	Color Intensity	Antho-cyanins, gram/liter	Tannins, grams/liter
Control				
1st day	0.46	0.41		
17th day	0.50	0.43	0.19	0.7
+ Stem Extract				
1st day	0.55	0.62		
17th day	0.64	0.59	0.17	6.8
+ Stems				
1st day	0.98	0.33		
17th day	1.66	0.20	0.03	4.0

Of course, adding stem extract (sample II) considerably increases
the tannin levels which in turn increases the color intensity—itself bound
to the increase in optical density at 420 nm. This increase in color is not
connected to the anthocyanin levels which vary little and even decrease
slightly in comparison with the blank. The increase in optical density

at 420 nm means the color shifts toward yellow; hence an increase in tint which grows as a function of time indicates change in tannins. In the presence of stems (sample II), an increase in tannins is also observed, but this is accompanied by a large decrease in the anthocyanins and in the tint as it moves towards red-orange. The large increase in the tint value (D_{420}/D_{520}) is, in this case, more related to the decrease in anthocyanins than to the increase in tannins.

Table XIV.   Changes in the Concentration of Phenolic Compounds as Function of Time of Skins–Juice Contact

Length of skin– juice contact, days	Tint	Color Intensity	Antho- cyanins, gram/ liter	Tannins grams/ liter	V/LA	Perman- ganate Index
1	0.78	0.46	0.19	0.75	1.2	18
2	0.56	0.89	0.46	1.77	1.3	30
3	0.56	1.24	0.50	1.96	1.4	37
4	0.52	1.52	0.63	2.42	1.6	45
6	0.53	1.43	0.67	2.63	1.9	48
8	0.56	1.62	0.61	3.18	1.8	50
10	0.52	1.41	0.61	3.39	1.9	60
14	0.51	1.36	0.59	3.55	1.9	62
20	0.59	1.21	0.48	3.65	1.9	62
30	0.67	1.20	0.39	3.74	2.0	67
40	0.67	1.22	0.38	4.26	2.1	70
50	0.71	1.23	0.37	4.30	2.2	72

The complete results are in Table XIV which shows the development of phenolic compound content during maceration. The anthocyanins increase until about the sixth day and then decrease. The tannins, on the other hand, increase continually, probably because they are more abundant, especially after the seeds are added. However, in some cases peculiar to vintages low in tannins, the production of these compounds is the same as that of anthocyanin.

The complexity of the problems relative to maceration explain how difficult it is to optimize the extraction of phenolic compounds during vinification.

Some say that adding sulfite will help this extraction. This is entirely true in synthetic media but is not valuable in fermenting musts, and rigorous experiments have never shown more intense color in normally sulfited vintages *vs.* nonsulfited vintages. In the crushed vintage, sulfurous acid combines with sugars quite fast, so its dissolving power dissipates rapidly. On the other hand, sulfurous acid exerts a large effect on the color of wines from moldy vintages in stabilizing the anthocyanins by inhibiting the oxidases capable of destroying them.

Interpretation of Wine Color and its Changes during Aging

Respective Roles of Anthocyanins and Tannins. The mediation of
tannins and anthocyanins in the color of red wine, covered in our work
(*38*) has been conducted in model solutions (*17*). The procedure is as
follows.

To an aqueous alcoholic solution (10% alcohol, 5 grams/liter tartaric
acid, adjusted to pH 3 with concentrated sodium hydroxide), add (a)
anthocyanins (abut 500 mg/liter), (b) tannins (about 5 grams/liter),
(c) a mixture of both at the same concentration. Other conditions call
for Fe^{3+} ions (5 mg/liter) and aeration (Table XV).

Table XV.   Anthocyanin and Tannin Coloration of
Model Solutions, Analytical Results

	Color		Antho-cyanins, mg/liter	Total Phenols Index	Tannins, grams/ liter	Tannin Conden-sation Index, V/LA
Conditions[a]	Tint	Intensity				
1. A f⁺o⁺	0.44	0.15	350	5	0	
2. T f⁺o⁺	1.92	0.50	25	55	4.4	2.5
3. M f⁺o⁺	1.96	0.74	60	62	4.8	2.4
4. A f⁺o⁻	0.32	0.15	375	6	0.2	
5. T f⁺o⁻	1.94	0.21	20	65	5.1	3.2
6. M f⁺o⁻	0.96	0.36	135	70	5.5	3.0
7. A f⁻o⁺	0.31	0.15	355	6	0.2	
8. T f⁻o⁺	2.03	0.29	20	60	4.5	3.4
9. M f⁻o⁺	1.85	0.51	70	70	5.4	2.6
10. A f⁻o⁻	0.27	0.15	375	6	0.2	
11. T f⁻o⁻	1.95	0.19	10	62	5.2	3.1
12. M f⁻o⁻	0.93	0.34	135	68	5.2	2.9

[a] A = anthocyanins; T = tannins; M = anthocyanins + tannins: f⁺ and f⁻ =
presence and absence of Fe^{3+}; o⁺ and o⁻ = presence or absence of oxygen of air.

The anthocyanin is malvidin-3,5-diglucoside (Fluka), and the tannins
are a sample of leucoanthocyanins extracted from the bark of the maritime pine (*Pinus pinaster*). The products are not identical to the natural
grape pigments; nevertheless, it may be assumed that the observed facts
correlate, as a first approximation, with red wine color and its changes.

The observation of results and analytical tests are done at the end of
two months using the previously described methods (Table XV). Color
development is shown in Figure 6. Without completely intepreting all
of the experimental data described elsewhere (*17*), the essential facts
are:

1. The anthocyanins by themselves are not the same color as wine (Figure 6, tube 4). This color is affected little by oxidation even when in contact with Fe^{3+} (compare tubes 8, 5, and 2).

2. Tannins by themselves, protected from air, are yellow (tube 5) and by oxidation become brown (tube 2). This oxidation is catalyzed by Fe^{3+} ions (compare tubes 8, 5, and 2).

3. The mixture of anthocyanins and tannins kept from air are a color reminiscent of young wines (tube 6). Upon oxidation they take on an orange-brown hue reminiscent of old wine. Again in this case Fe^{3+} catalyzes the oxidation (compare tubes 6, 3, and 9).

4. After oxidation, the solutions of anthocyanins plus tannins, and that of tannins alone are a similar color (tubes 3 and 2).

5. The data from Table XV show a decrease in anthocyanins in the presence of tannins. This level is 350 mg/liter in the tubes containing anthocyanin alone, 135 mg/liter in the tubes containing anthocyanins plus tannins protected from oxygen, and 60 to 70 mg/liter when the same mixture is oxidized.

As a first approximation, this experiment represents the evolution of the coloring matter in red wines during aging. It confirms the large role of tannins in the color of old wines and thus shows that tannins play an important part in the color of young wines. Taken as another point, the possibility of a reaction between the anthocyanins and tannins already postulated (50) is accurately described.

This reaction was confirmed in another experiment (17) using standard tests which followed the evolution of pigments in identical mixtures in tubes 1, 2, and 3 for two months. After allowing the tannins to react it appears that the anthocyanins contribute to the color of the solution to a limited extent.

The mechanism for his reaction may be interpreted as a copolymerization between the anthocyanins and the flavan centers in the tannins. No sediment can be seen in the solution, so the anthocyanins are not eliminated by precipitation. The mechanism proposed in Figure 7 was inspired by the work of Jurd (16) and Somers (50). The central anthocyanin ring is electrophilic, and its positive charge may be localized, either on the oxygen (35) yielding an oxonium ion or on carbon 4 (36) yielding a carbonium ion. A bond may thus be formed between the fourth carbon and the sixth or eighth positions of a flavan molecule (37). Conforming to the mechanism in Figure 2, one thus obtains the dimer 38 in which the anthocyanin is in the flavene form, oxidizable to flavylium, and giving a final product (39) which is in equilibrium, as a function of pH, with the anhydrous base (40). If R = OH, the polymerization may lead to a flavolene composed of condensed tannins.

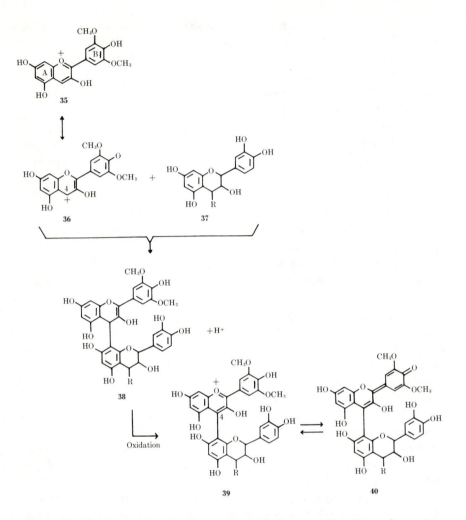

Figure 7. Condensation of anthocyanins with flavans. The dimer obtained is red, but as the 4 position of the anthocyanin is substituted, it does not react with bisulfite and its color does not vary with changes in pH.

According to Somers (*50*) this occurrence will lead to a positive wine pigment with the structure **39** in which the anthocyanin fraction gives the color. This pigment may be separated from proper anthocyanins by extraction with amyl alcohol. Otherwise, considering anthocyanins, the color of this pigment will be stabilized in relation to variations in pH and sulfites by substitution on carbon 4 (*51*). However, the model solutions studied in the first part of this article indicate that in the presence of anthocyanins, a limited color change occurs during oxidative condensation of tannins (tubes 2, and 3, Figure 6).

Influence of the Physicochemical State of Pigments on Wine Color. Red wine color does not depend exclusively on the anthocyanin and tannin levels; the physicochemical state of these pigments also exerts an influence. Related wine types, especially those of the same age, can have colors which may not be interpreted solely on the basis of anthocyanin and tannin levels.

Table XVI.   Influence of Type and Size of Container on Wine Color (46, 47)

	Color		Anthocy-anins, gram/ liter	Tannins, grams/ liter
	Tint	Intensity		
First Experiment[a]				
Wine stored in 225–1 new wood	0.76	0.67	0.14	2.7
Wine stored in 225–1 used wood	0.72	0.64	0.12	2.8
Wine stored in 100–1 stainless steel tank	0.87	0.52	0.17	2.7
Second Experiment[b]				
Wine stored in 300-hl tank	0.72	0.41	0.18	3.0
Wine stored in 300-hl tank	0.73	0.40	0.18	3.2
Wine stored in 300-hl tank	0.70	0.39	0.18	3.2
Wine stored in 225–1 wood	0.70	0.50	0.16	3.1

[a] 1966 vintage wine, analyzed January 1970.
[b] 1969 vintage wine, analyzed May 1970.

Table XVI shows two experiments which compared the same wine stored under different conditions (46). In both cases, the wine richer in anthocyanins is also that which is less colored. The mediation of tannins is not enough to explain the differences in color; these can only be explained by a different structuring of the anthocyanin molecules. More specifically, the anthocyanin molecules will be reduced to colorless flavenes (14) during fermentation, which is a reductive process. The reoxidation occurs more rapidly in wooden casks which allow better oxygen penetration than metal storage tanks or large capacity casks. However, the flavenes themselves are relatively instable and can be irreversibly hydrolyzed into dihydrochalcones (16). This explains the lack of relationship between anthocyanin concentration and color, inde-pendent, of course, of the eventual appearance of free sulfur dioxide.

Practical Applications

1. In young wines the anthocyanins play the most important role, but the tannins also add to the color. The anthocyanins are partially in the colorless form after their reduction during fermentation.

2. During the weeks that follow vinification several things happen simultaneously: (a) The reduced anthocyanins are reoxidized, producing a color increase; (b) The anthocyanins, in reduced or oxidized form, are partially destroyed by various chemical reactions or by condensation with tannins. One may thus explain why, during that period the color of some wines increases while that of others decreases, depending on the relative rates of the two reactions occurring at this time.

3. During aging the anthocyanins continue to disappear by condensing with tannins which themselves undergo an oxidative condensation resulting in a color shift from yellow to orange-brown. Finally it is these tannins which play the most important role in the characteristic color of old wines.

Literature Cited

1. Ribéreau-Gayon, P., "Recherches sur les anthocyanes des végétaux. Application an genre *Vitis*," *Libr. Gen. Enseignement*, Paris, 1959.
2. Ribéreau-Gayon, P., "Les composés phénoliques du raisin et du vin," Institut national de la Recherche agronomique, Paris (1964).
3. Ribéreau-Gayon, P., "Les composés phénoliques des végétaux," Dunod, Paris, (1968).
4. Ribéreau-Gayon, P., "Plant Phenolics," Oliver and Boyd, Edinburgh, 1972.
5. Stonestreet, E., "Contribution à l'étude des tanins et da la matière colorante des vins rouges," *Thèse doctorat 3ème cycle*, Bordeaux, 1965.
6. Milhé, J. C., "Recherches technologiques sur les composés phénoliques des vins rouges," Thèse doctorat, Bordeaux, 1969.
7. Glories, Y., "Essais de détermination de l'état de condensation des tanins des vins rouges," *Thèse doctorat* 3ème cycle, Bordeaux, 1971.
8. Singleton, V. L., Esau, P., "Phenolic Substances in Grapes and Wine and their Significance," Academic, New York, 1969.
9. Willstater, R., Zollinger, E. H., *Ann. Chim.* (1915) **83**, 408.
10. Karrer, P., Widmer, F., *Helv. Chim. Acta* (1927) **10**, 5.
11. Robinson, G. M., Robinson, R., *Biochem. J.* (1932) **26**, 1647.
12. Gueffroy, D. E., Kepner, R. E., Webb, A. D., *Phytochem.* (1971) **10**, 813.
13. Anderson, D. W., Gueffroy, D. E., Webb, A. D., Kepner, R. E., *Phytochem.* (1970) **9**, 1579.
14. Ribéreau-Gayon, P., *C.R. Acad. Agric.* (1957) **43**, 197, 596, 821.
15. Berg, H. W., *Ann. Technol. Agric.* (1963) **12**, n°hors série, 247.
16. Jurd, L., *J. Food Sci.* (1964) **29**, 16.
17. Ribéreau-Gayon, P., *Vitis* (1974) in press.
18. Webb, A. D., in "Phenolics in Normal and Diseased Fruits and Vegetables," *Plant Phenolic Group of North America*, United Fruit Co., Norwood, Mass., 1964.
19. Ingalsbe, D. W., Neubert, A. M., Carter, G. M., *J. Agr. Food Chem.* (1963) **11**, 263.

20. Brown, W. L., *J. Amer. Chem. Soc.* (1940) **62**, 2808.
21. Albach, F., Kepner, R. E., Webb, A. D., *Amer. J. Enol. Viticult.* (1959) **10**, 164.
22. Ribéreau-Gayon, P., *Ind. Agr. Aliment.* (1963) **80**, 1079.
23. Bockian, A. H., Kepner, R. E., Webb, A. D., *J. Agr. Food Chem.* (1955) .3, 695.
24. Dourmichidzé, S. V., Noutsoubidzé, N. O., *Dokl. Akad. Nauk SSSR* (1958) **46**, 1197.
25. Dourmichidzé, S. V., Sopromadzé, A. N., *Dokl. Akad. Nauk SSSR Géorgie* (1963) **30**, 163.
26. Cappelleri, G., *Riv. Viticult. Enol.* (1965) **8**, 350.
27. Getov, G., Petkov, G., *Mitt. Rebe Wein* (1966) **16**, 207.
28. Swain, T., Bate-Smith, E. C., in "Comparative Biochemistry," Vol. III, A. M. Florkin, H. S. Mason, Eds., Academic, New York, 1962.
29. Jurd, L., *Amer. J. Enol. Viticult.* (1969) **20**, 191.
30. Thompson, R. S., Jacques, D., Haslam, E., Tanner, R. J. N., *J. Chem. Soc., Perkin I* (1972) **11**, 1387.
31. Weinges, K., Wild, R., Kaltenhauser, W., *Z. Lebensm. Unters. Forsch.* (1969) **140**, 129.
32. Bhatia, V. K., Madhav, R., Seshadri, T. R., *Curr. Sci.* (1968) **37**, 582.
33. Weinges, K., Gorissen, H., Lontie, R., *Ann. Physiol. Veg.* (1969) **11**, 67.
34. Michaud, J., Masquelier, J., *Bull. Soc. Chim. Biol.* (1968) **50**, 1346.
35. Hathaway, D. E., Seakins, J. W. T., *Biochem. J.* (1957) **67**, 239.
36. Boubals, D., Cordonnier, R., Pistre, R., *C.R. Acad. Agr.* (1962) **48**, 201.
37. Dorier, P., Verelle, L. P., *Ann. Fals. Expert. Chim.* (1966) **59**, 1.
38. Ribéreau-Gayon, P., Stonestreet, E., *Bull. Soc. Chim.* (1965) **9**, 2649.
39. Masquelier, J., Vitte, G., Ortéga, M., *Bull. Soc. Pharm. Bordeaux* (1959) **98**, 145.
40. Ribéreau-Gayon, P., Stonestreet, E., *Chimie Anal.* (1966) **48**, 188.
41. Swain, T., Hillis, W. E., *J. Sci. Food Agr.* (1959) **1**, 63.
42. Ribéreau-Gayon, P., Glories, Y., *C.R. Acad. Sci.* (1971) **273 D**, 2369.
43. Somers, T. C., *Nature* (1966) **209**, 368.
44. Sudraud, P., "Etude expérimentale de la vinification en rouge," Doctoral Thesis, University of Bordeaux, 1963.
45. Sudraud, P., *Ann. Technol. Agr.* (1958) **7**, 203.
46. Ribéreau-Gayon, P., *Conn. Vigne Vin* (1971) **5**, 87.
47. *Ibid.* (1971) **5**, 247.
48. Ribéreau-Gayon, P., Sudraud, P., Milhé, J. C., Canbas, A., *Conn. Vigne Vin* (1970) **2**, 133.
49. Ribéreau-Gayon, P., Milhé, J. C., *Conn. Vigne Vin* (1970) **1**, 63.
50. Somers, T. C., *Phytochem.* (1971) **10**, 2175.
51. Timberlake, C. F., Bridle, P., *Chem. Ind.* (1968) 1489.

RECEIVED July 24, 1973.

4

Chemistry of Winemaking from Native American Grape Varieties

A. C. RICE

Taylor Wine Co., Hammondsport, N. Y. 14840

Native American grape varieties, primarily Vitis labrusca, *and direct producer hybrids differ from* Vitis vinifera *wine grapes in sugar, acid, and pectin content. Concord types contain methyl anthranilate and pronounced fruity flavors not present in* vinifera *grapes. Native grapes contain both mono- and diglucoside anthocyanins,* vinifera *grapes contain only monoglucosides. Special winemaking practices permitted or required with the high-acid, low-sugar native grapes are amelioration to reduce acidity and increase alcohol content, heat treatment of pulp to extract anthocyanin pigment, and use of pectolytic enzymes combined with cellulose fibers to facilitate pressing. Malo-lactic fermentation by* Leuconostoc *sp., following alcoholic fermentation, reduces acidity and is considered beneficial in native American and hybrid wines.*

The chemistry of winemaking can be divided into at least two distinct segments, the chemistry of the grape and the chemistry of fermentation. The chemistry of fermentation for native American and hybrid grapes is essentially the same as for *Vitis vinifera* grapes since similar yeasts and fermentation conditions are used. However, the chemistry of the native American and hybrid grapes differs in a number of respects from that of the *vinifera* grapes. These differences in grape chemistry result in differences in winemaking practices and in the wines.

Historical

The history of commercial winemaking in the eastern United States is directly related to the history of the development of acceptable grape varieties from the grapes native to America. Hedrick (*1*) relates that from the first colony founded in Virginia by Lord Delaware in 1619 until the early 1800's, numerous attempts were made in the several colonies to develop vineyards for wine production from grape cuttings imported from Europe. Vignerons were brought from Europe to oversee the new plantings. However, all attempts to propagate *V. vinifera* varieties failed.

Ca. 1800, several native varieties came into prominence as a result of the persistence of a number of vineyardists and horticulturists. The first native grape to achieve prominence was the Cape, claimed by James Dufour to be a European grape from the Cape of Good Hope. Later, it was determined that the Cape, also called Alexander, was a native grape of the species *Vitis labrusca*. At about the same time the Catawba grape was introduced by John Adlum. Although its origin is not positively known, it is reported to have come from a farm in North Carolina in the early 1800's. The Isabella, also of unknown origin, was being grown at this time. Both Catawba and Isabella are considered to be *Vitis labrusca* species with some *vinifera* in their ancestry (2).

By 1830, several other native grapes had achieved acceptance. The Scuppernong grape, *Vitis rotundifolia,* and the Norton, a *Vitis aestivalis* variety, were being grown in the southern states. In the northern states, the Clinton, a *Vitis riparia* variety, was gaining acceptance.

These native varieties grew well because of their resistance to disease and insects and their adaptation to climate. The early successful vineyards were set to Catawba and Isabella grapes in 1820–1830. A vineyard of 75 acres of these varieties was set at Croton Point on the Hudson in 1827. The first grapes planted in the Finger Lakes by the Rev. William Bostwick at Hammondsport in 1830 were the same two varieties. Many of these early grapes were marketed as fresh fruit. It was not until 1850–1860 that winemaking developed to the point that wineries became a significant factor in the use of native grapes.

There was a reluctance on the part of the colonists to use the wild native grapes for winemaking because of the flavor characteristics, particularly of *Vitis labrusca*. The term fox grapes was applied early to various native grapes, the exact reason being somewhat in doubt. It has been related that foxes ate the grapes, that the leaf resembles a fox's track, and that the musty odor of the grape suggested that of the fox. With the development of the Concord grape between 1840 and 1850 by E. Bull at Concord, Mass. from a native *V. labrusca* grape and its subsequent widespread planting, the term foxy has become a common term for describing the flavor characteristic of the *V. labrusca* type grape.

During this same period several other grapes destined to become important in winemaking were discovered. The Delaware, thought to be a *labrusca–vinifera–bourquiniana* hybrid, was first brought to notice about 1849 and was thought to have originated in a garden of foreign grapes in New Jersey. The Niagara grape, produced from a cross between Concord and Cassidy in Lockport, N. Y. in 1868, was highly promoted and gained rapid acceptance. The Ives, also a *labrusca*, came from seed planted in 1840 by Henry Ives who claimed that the seed was from Madeira grapes sent to him from abroad.

These native grapes together with several others became the foundation on which the Eastern wine industry developed. As the industry grew and the demands for all wines increased, it became apparent to many eastern producers that new varieties of grapes were needed to supplement the native varieties being used. In the 1940's and early 1950's, a number of wine producers (3) and nursery operators began to look seriously at some of the French–American hybrids produced by hybridizers in Europe. An excellent review of the development of French–American hybrids is given by Wagner (4). A number of these hybrids, such as Seibel 5279 (Aurore), Seibel 1000 (Rosette), Seibel 10878 (Chelois), Baco ‡1 (Baco noir), and Kuhlmann 188–2 (Marechal Foch) possessed many of the desirable characteristics of both *vinifera* and native American species. The wines had the vinous characteristics of *vinifera*, and the vines inherited the disease resistance and winter-hardiness of the American species. Since new red varieties were especially sought, the more adaptable hybrids were propagated in quantity and by the 1960's had become a significant factor in the production of Eastern wines.

Although the hybrids lost the pronounced *labrusca* flavor, they generally had low sugar and high acid content. These French–American hybrids generally could be vinified by the same processes used for the native varieties.

The fruit of the native species differs markedly from the fruit of *Vitis vinifera*. Vinification processes in the eastern United States have developed to accommodate these characteristics. Differences are noted in sugar, acid, pectin, color, and flavor.

Sugars

Native American grapes (5) contain less sugar than *Vitis vinifera* varieties. A publication on California wine grapes (6) revealed that the average Balling of must from the recommended wine varieties for California ranged from 20° to 25° with a few exceptions above and below these limits.

Early studies of Concord and other native grape juices (7, 8) found soluble solids content of these juices in the 12°–20° Brix range and found them to be variable with location and season. The chemical composition of New York State grapes and juices was studied in 1938–1939 (9). Brix of Concord juice from ripe fruit from different areas within the state ranged from 14.6° to 20.7°. This range encompassed nearly all Brix values observed for other varieties including Niagara, Delaware, and Catawba. The location effect on quality of Concord must and wine under Canadian conditions was studied by Crowther *et al.* (*10*). Over a four-year period on seven locations, Brix of Concord must ranged from 10.2° to 18.5°.

Brix data for several native and hybrid grape musts produced for experimental wines at the New York State Agricultural Experiment Station (*11, 12, 13, 14, 15*) are reproduced in Table I. Soluble solids content varied from 12.8° Brix for Elvira, to 22° Brix for Delaware. The degrees Brix of Concord, Catawba, and Aurore exhibited the range 2.2°–2.4°. On a Canadian site, Aurore varied from 15.0°–19.5° over a five year period (*16*). An extreme range of soluble solids observed under Eastern growing conditions was illustrated by Ives for which the range was 6.5°. A recent study of the Scuppernong grape (*17*) indicated the soluble solids content to be 11.9% which would place this native variety among the lowest in sugar content.

Results in commercial operations were similar but did not duplicate those reported for experimental lots. Average Brix values of must for commercially processed fruit in New York State (*18*) for the years

Table I.   Brix Values of Experimental Eastern Grape Musts[a]

Variety	Year	Harvest Date	°Brix
Delaware	1965	10/20	22.0
	1967	9/25	18.8
	1968	9/19	18.5
Elvira	1965	10/10	12.8
Niagara	1966	10/15	17.6
Aurore (S-5279)	1966	9/14	14.9
	1967	9/25	15.6
	1968	9/06	17.0
	1969	9/16	17.2
Catawba	1967	10/11	17.0
	1968	10/08	18.2
	1969	10/23	19.2
Ives[b]	1965	10/05	14.0
	1968	10/08	20.5
Concord	1966	10/06	15.6
	1967	10/05	15.8
	1968	10/02	17.8
	1969	10/09	15.4

[a] Data taken at New York State Agricultural Experiment Station, Geneva, N. Y.
[b] Hot press.

Table II. Average Brix and Sugar Content of Commercial Eastern Grape Musts

Variety	Year	°Brix	Glucose, grams/ 100 ml	Fructose, grams/ 100 ml	Ratio	Average Glucose/ Fructose Ratio
Aurore	1972	16.30	6.80	8.24	0.825	
	1971	14.71	7.20	6.89	1.045	
	1970	15.26	6.60	7.05	0.936	
	1969	15.89	7.12	7.48	0.952	
	1968	15.74	7.03	7.52	0.935	
	1967	13.00	6.19	6.71	0.923	0.936
Delaware	1972	18.50	7.72	9.86	0.783	
	1971	16.10	7.75	8.39	0.924	
	1970	18.90	8.56	10.18	0.841	
	1969	19.90	9.12	10.52	0.867	
	1968	19.60	8.87	10.10	0.878	
	1967	17.00	7.80	9.03	0.864	0.860
Niagara	1972	13.90	5.70	6.75	0.844	
	1971	12.40	5.81	6.02	0.965	
	1970	13.90	6.32	6.96	0.908	
	1969	14.00	6.23	6.90	0.903	
	1968	15.70	6.90	7.71	0.895	
	1967	13.00	5.97	6.69	0.892	0.901
Catawba	1972	16.70	6.94	8.32	0.834	
	1971	17.20	8.10	8.61	0.941	
	1970	17.90	8.16	8.99	0.908	
	1969	18.00	—	—	—	
	1968	19.21	8.69	9.48	0.917	
	1967	15.10	7.08	6.84	1.035	0.927
Concord[a]	1972	12.57	4.80	5.80	0.828	
	1971	12.00	5.60	5.43	1.031	
	1970	12.27	5.32	5.53	0.962	
	1969	12.98	5.54	5.91	0.937	
	1968	13.91	5.82	6.26	0.930	
	1967	11.50	5.11	5.58	0.916	0.934
Concord[b]	1972	16.70	6.36	7.96	0.799	
	1971	17.03	7.14	8.18	0.873	
	1970	16.96	6.85	8.17	0.838	
	1969	17.43	—	—	—	
	1968	18.26	7.73	8.73	0.886	
	1967	15.56	6.46	7.07	0.914	0.862
Ives[b]	1972	14.60	4.86	7.31	0.665	
	1971	13.00	4.71	6.28	0.750	
	1970	14.60	5.16	7.37	0.700	
	1969	15.71	—	—	—	
	1968	17.81	6.45	8.74	0.738	
	1967	14.40	4.95	7.21	0.687	0.708

Table II. Continued

Variety	Year	°Brix	Glucose, grams/ 100 ml	Fructose, grams/ 100 ml	Ratio	Average Glucose/ Fructose Ratio
Baco noir[b]	1972	19.80	8.19	9.21	0.889	
	1971	17.15	7.93	7.64	1.038	
	1970	19.40	8.94	8.80	1.016	
	1969	20.27	—	—	—	
	1968	20.50	9.14	9.36	0.977	
	1967	17.00	8.26	7.60	1.088	1.002

[a] Cold press.
[b] Hot press.

1967–1972 are shown in Table II. These data were obtained by Brix hydrometer from filtered must samples representative of each day's total production.

None of these varieties, which together account for approximately 90% of the grape tonnage processed by Eastern wineries, had an average Brix of 20° or greater. Varieties such as Niagara, Ives, Concord (cold pressed), and Aurore were consistently below 16° Brix and were generally less than 15°. There were wide variations in sugar content from one season to another, *e.g.*, Delaware; 19.90° in 1969 and 16.1° in 1971. An approximate average Brix for all varieties processed would be 16°.

Glucose and fructose are the predominant natural sugars in *V. vinifera* grapes. A number of investigators have reported the proportions of these sugars, and generally at maturity the ratio is close to 1:1 although wide variation has been noted (*19, 20*).

Little information is available on the proportions of natural sugars in the native American varieties. Lee *et al.* (*21*) reported concentrations of glucose and fructose (as well as other sugars) for four grape varieties designated *V. labrusca* which permit calculation of glucose/fructose ratios of 1.02 for Concord and of 0.96 for Delaware. Lott (*22*) and Lott and Barrett (*23*) analyzed glucose and fructose in Concord grape juice during fruit maturation and in 39 clones at maturity. Calculation of the glucose/fructose ratio in Concord juice at weekly intervals over a 28-day period showed a gradual decrease from 0.80 to 0.70. At maturity experimental musts of Concord, Delaware, and Catawba had ratios of 0.74, 0.74, and 0.80, respectively. Calculated glucose/fructose ratio for mature Scuppernong fruit (*17*) was 1.03.

Analyses of glucose and fructose in musts of commercially processed fruit have been made for several years (*18*). Glucose oxidase (Worthington Glucostat reagent) has been the method of choice for glucose analysis. Acid diphenylamine reagent (*24*), which is specific for the keto group, has been used for fructose analysis. Because of the sensitivity of these reagents, must is diluted 1:2000 to bring the glucose and fructose concentrations into the proper range for analysis.

Average concentrations of the two sugars and the calculated glucose/fructose ratios are given in Table II. The whites generally had a ratio slightly less than 1.0. Delaware exhibited the greatest divergence, having an overall average for the six seasons of 0.86. The two reds had lower ratios, with Ives having an average for four seasons of 0.71. Between the Concord cold-press and hot-press harvests, fructose concentration increased more than the glucose concentration, resulting in a lower ratio in the mature Concord must.

Kiewer (20) indicated that for the *vinifera* grapes studied, glucose breakdown apparently occurred when grapes were in an overripe condition which resulted in a sharp decrease in the glucose/fructose ratio. Normal ratio during the ripe period was considered to be *ca.* 1.00 ± 0.10. This was in accord with the observation with Concord (25) that sucrose was synthesized in the leaves and translocated to other vine parts where hydrolysis resulted in 1:1 ratio of glucose to fructose.

Most of these varieties conform to this observation. The somewhat lower ratio for hot-press Concord must could be attributed to an increased rate of glucose breakdown, as reported for *vinifera* fruit since some overripe fruit may be processed near the end of the harvest. However, the Ives appears to be an exception because of its consistently low ratio. The vineyard characteristics of this variety are such that it is normally harvested at the early-ripe to ripe stage. Thus the low glucose concentration found in the must may be the result of factors other than the glucose breakdown observed in overripe fruit. The low glucose content in the juice may be related to the extremely high anthocyanidin diglucoside content of the berry.

The native grape varieties, although developing the same sugars and in generally the same proportions, have considerably less total sugar than *vinifera* varieties.

Acids

The native American grape varieties, in addition to being low in sugar, have high total acidity. Early research (7, 8) on the juice characteristics of Concord and other native varieties indicated total acidity to be high and variable and the pH to be low. Studies in New York (9) and Illinois (22) confirmed the great variability of total acid which occurs in these native grapes in different locations and seasons. A recent report (26) notes that total acidity of Concord juice apparently has been increasing in recent years as a result of changes in cultural practices.

Data on total acid and pH of must of the recommended California wine varieties (6) indicate that average total acid content generally ranges from 0.40 to 1.0 gram/100 ml with most values falling within the range 0.5–0.9 gram/100 ml. The range of average pH values is from 3.10 to 3.90 with most values in the range 3.25–3.80.

The range of pH and total acid values found in commercial musts of five native and two hybrid varieties for a six-year period (18) are given in Table III. Total acidity and pH were determined in filtered must samples representative of each day's total production, first noting pH then titrating to pH 8.2 using a Beckman Zeromatic pH meter (27). Only the Niagara grape consistently had a total acid content below 1.000

Table III.   Average pH and Total Acid Content of Commercial Eastern Grape Musts

Variety	Year	pH	Total Acid, grams/100 ml as tartaric		
			Average	Maximum	Minimum
Aurore	1972	3.08	1.159	1.233	1.055
	1971	3.26	1.039	1.222	0.900
	1970	3.37	1.114	1.170	1.063
	1969	3.04	1.025	1.213	0.930
	1968	3.10	1.008	1.207	0.880
	1967	3.11	1.285	1.299	1.268
Delaware	1972	3.11	1.148	1.219	1.090
	1971	3.20	0.901	0.936	0.885
	1970	3.30	1.148	1.275	0.957
	1969	3.14	0.918	0.960	0.870
	1968	3.15	0.883	0.960	0.840
	1967	3.21	0.982	1.050	0.937
Niagara	1972	3.03	0.826	0.952	0.720
	1971	3.15	0.740	0.893	0.645
	1970	3.28	0.822	0.892	0.735
	1969	3.07	0.771	0.788	0.735
	1968	3.15	0.630	0.675	0.585
	1967	3.16	0.830	0.855	0.795
Catawba	1972	2.77	1.718	1.822	1.602
	1971	3.04	1.170	1.273	1.035
	1970	3.00	1.113	1.170	1.072
	1969	3.01	1.190	1.260	1.087
	1968	2.95	1.118	1.200	1.080
	1967	2.95	1.612	1.655	1.525
Concord[a]	1972	2.81	1.642	1.778	1.500
	1971	3.03	1.152	1.328	1.077
	1970	3.13	1.403	1.650	1.233
	1969	2.80	1.348	1.425	1.252
	1968	2.89	1.160	1.328	0.975
	1967	2.92	1.605	1.702	1.447

Table III.   Continued

| Variety | Year | pH | Total Acid, grams/100 ml as tartaric | | |
			Average	Maximum	Minimum
Concord[b]	1972	3.12	1.366	1.545	1.197
	1971	3.27	1.089	1.200	0.945
	1970	3.35	1.124	1.485	0.990
	1969	3.23	1.244	1.320	1.108
	1968	3.22	1.002	1.118	0.907
	1967	3.22	1.326	1.402	1.213
Ives[b]	1972	3.22	1.175	1.207	1.147
	1971	3.14	1.151	1.242	1.093
	1970	3.43	1.157	1.263	1.087
	1969	3.22	1.114	1.140	1.080
	1968	3.18	1.068	1.259	0.975
	1967	3.18	1.210	1.260	1.148
Baco noir[b]	1972	3.28	1.530	1.538	1.513
	1971	3.31	1.428	1.462	1.395
	1970	3.45	1.675	—	—
	1969	3.27	1.335	1.342	1.327
	1968	3.19	1.328	—	—
	1967	3.23	1.598	—	—

[a] Cold press.
[b] Hot press.

gram/100 ml. Delaware must generally averaged between 0.90–1.15 gram/100 ml. The other native and hybrid varieties ranged from an average of 1.00–1.718 gram/100 ml. Maximum and minimum values indicate the seasonal range of total acidity for the daily must samples which was from less than 0.05 to 0.50 gram/100 ml. Low pH values accompany the high acidities. The range of average pH values, from 2.77 to 3.45, is nearly 0.4 pH unit below the range indicated for *vinifera* varieties (6).

Yearly fluctuations are evident, indicating an effect of climatic conditions on acid production. In 1967 acid values were very high and sugar values were low. Conversely, in 1968, acid values were quite low and sugar values were generally high. However, in 1970 acid values were quite high, and for most of the varieties sugar values were also high. Shaulis (28) reported that soluble solids and total acid development during maturation are affected by mean ambient air temperature and augmented by solar radiation to raise fruit and leaf temperatures. Soluble solids and yields are increased and acidity is decreased by opening the canopy to sunlight. Crowded vines with reduced leaf exposure resulted in decreased soluble solids and yield and increased total acidity. Climatic conditions affecting ambient air temperatures and the total amount of sunlight presumably would affect entire crops in much the same manner that clusters were affected on individual vines.

Malic and tartaric acids in the must samples were determined by gas chromatography (29). Data for four seasons, 1969–1972, are given in Table IV. There is great variation in both tartaric and malic acids.

Table IV.   Average Tartaric and Malic Acid Content of Commercial Eastern Grape Musts

Variety	Year	Tartaric, grams/ 100 ml	Malic grams/ 100 ml	Tar/ Malic Ratio	Average Tar/ Malic Ratio
Aurore	1972	0.36	0.49	0.73	
	1971	0.34	0.26	1.31	
	1970	0.79	0.69	1.14	
	1969	0.49	0.61	0.80	1.00
Delaware	1972	0.40	0.68	0.59	
	1971	0.77	0.50	1.54	
	1970	0.91	0.50	1.82	
	1969	0.49	0.56	0.88	1.21
Niagara	1972	0.50	0.39	1.28	
	1971	0.81	0.29	2.79	
	1970	0.83	0.37	2.24	
	1969	0.44	0.23	1.91	2.06
Catawba	1972	0.64	0.87	0.74	
	1971	0.80	0.61	1.31	
	1970	1.17	0.56	2.09	
	1969	0.77	0.58	1.33	1.37
Concord[a]	1972	0.60	0.67	0.90	
	1971	0.60	0.39	1.54	
	1970	1.22	0.60	2.03	
	1969	0.78	0.70	1.11	1.40
Concord[b]	1972	0.76	0.49	1.55	
	1971	1.11	0.37	3.00	
	1970	1.30	0.38	3.42	
	1969	0.89	0.41	2.17	2.54
Ives[b]	1972	0.85	0.51	1.67	
	1971	1.28	0.35	3.66	
	1970	1.24	0.35	3.54	
	1969	1.12	0.35	3.20	3.02
Baco noir[b]	1972	0.48	0.46	1.04	
	1971	1.04	0.85	1.22	
	1970	1.31	0.91	1.44	
	1969	0.52	0.54	0.96	1.17

[a] Cold press.
[b] Hot press.

This is evident within a variety as well as among the several varieties tabulated. Generally the years 1969 and 1972 produced results that were similar within each variety. The results for the two intervening years were more variable and, in particular, the tartaric acid content was greater.

Since the grapes came from diverse growing areas throughout central and western New York, it was impossible to relate tartaric and malic acid content in particular years to specific climatic conditions. However, an interesting relationship with crop size was observed. The five-year average crop size (1968–1972), as determined by the average tons per acre for all grapes delivered to the winery, was 4.06 tons/acre. The crop size for 1970 and 1971 was above average, 4.68 and 5.71, respectively. For 1969 and 1972 the crop size was average or below, 4.11 and 2.58, respectively. High tartaric acid in must was observed in years of large crop size. It is noteworthy that the lowest concentrations of tartaric acid occurred in 1972, a year in which the crop size was perhaps the smallest on record and in which the total acidity for most varieties was extremely high.

Peynaud and Maurie (30) attempted to correlate tartaric and malic acid formation in *vinifera* grapes to general meteorological conditions over a five-year period with little success. They concluded that there were years of tartaric acid formation and years of malic acid formation. The same observation would probably be appropriate for the native American and hybrid grapes.

Research on acid in maturing *vinifera* varieties has shown that the concentration of both tartaric and malic acid decreases during the early ripening stages of fruit maturity (31). Mattick (32) demonstrated a similar pattern in native American varieties as exemplified by Concord grapes, Figure 1. In commercial musts the tartaric acid content of immature (cold press) Concord was lower than the tartaric acid content of mature (hot press) Concord. The tartaric acid was not completely extracted from the immature fruit by the pressing operation and a portion remained in the pomace. The tartrates were more completely extracted in the hot press operation which resulted in a high concentration in the must.

The malic acid relationship between cold press and hot press musts conformed to the experimental observations on maturing fruit. As the fruit matured, the malic acid content decreased. The malic acid content of hot press Concord must was nearly one-third less than that of cold press Concord must.

The ratio of tartaric acid to malic acid was generally greater than 1.0. Exceptions were noted in 1969 and 1972, primarily among the whites. The ratio was highest for the two native red varieties which were hot

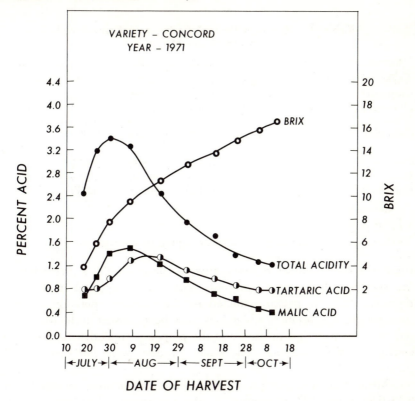

VARIETY - CONCORD
YEAR - 1971

BRIX

PERCENT ACID

BRIX

TOTAL ACIDITY
TARTARIC ACID
MALIC ACID

10 20 30 9 19 29 8 18 28 8 18
|◄JULY►|◄——AUG——►|◄——SEPT——►|◄OCT►|

DATE OF HARVEST

New York Food and Life Science Quarterly

Figure 1. Changes in Brix and acid contents of Concord grapes during maturation (32)

pressed, Concord and Ives. The Baco noir ratio was comparable with that of the whites. The four-year average ratio for the whites ranged from 1.00 to 2.0 while for the two native reds, it was 2.5 and 3.0. These ratios, for both whites and reds, were within the range previously reported for *V. vinifera* musts in California (*33, 34*).

Pectin

Winkler (*31*) reported that *vinifera* grapes contain only small amounts of pectin in relation to American grapes such as Concord which are rich in pectin. Willaman and Kertesz (*35*) reported pectin content of Concord to be 0.60%. The range of average pectin values for *vinifera* grapes summarized from the literature by Amerine and Joslyn (*36*) was 0.05–0.26%.

Pectin content of Eastern grape musts, determined by acetone precipitation (*37*), was observed over a five-year period, Table V. Daily

Table V. Average Pectin[a] Content of Commercial Eastern Grape Musts[b]

Variety	1968	1969	1970	1971	1972	Average
Aurore	0.121	0.213	0.243	0.294	0.226	0.219
Concord[c]	0.283	0.294	0.425	0.406	0.463	0.374
Delaware	0.363	0.309	0.433	0.384	0.513	0.400
Niagara	0.398	0.322	0.535	0.339	0.499	0.419
Catawba	0.234	0.791	0.299	0.575	0.692	0.518
Concord[d]	0.449	0.831	0.599	0.802	0.975	0.731
Ives[d]	0.396	0.632	0.621	0.786	0.785	0.644
Baco noir[d]	—	0.267	0.480	0.203	0.335	0.321

[a] Acetone precipitable solids.
[b] In grams/100 ml.
[c] Cold press.
[d] Heated pulp treated with pectolytic enzyme prior to pressing.

composite must samples representative of the entire day's production of the specific variety were analyzed. The values reported in Table V represent only the pectin extracted by the pressing operation and not the total pectic substances which may have been present in the fruit. The red varieties were hot pressed and the pulp was treated with pectolytic enzyme prior to pressing. The white varieties were not treated with pectolytic enzyme.

Musts from the cold-press native varieties contained pectin in the 0.25–0.80 grams/100 ml range. The greatest seasonal variation was observed with Catawba which is late-ripening and generally harvested during the last two weeks in October. The hybrid varieties contained less pectin than the native varieties. However, the values observed for the hybrids are comparable with the highest values reported for *vinifera* varieties.

Studies of total pectic substances have indicated that in maturing fruit, total pectin gradually decreased (38, 39). For a particular variety, the harvest is generally completed within a week or less. For Concords,

Table VI. Pectin Content

Time, min	Early Season		Mid-Season	
	10/11/68	9/29/70	10/18/68	10/9/70
0 (no enzyme)[b]	.251	.330	.275	.735
15[b]	.398	.476	.344	.835
30[b]	.426	.433	.458	.790
40–50 (press juice)	.570	.572	.572	.976
Composite Juice[c]	.619	.672	.456	.649

[a] In grams/100 ml.
[b] Pulp pressed with small laboratory hydraulic press.
[c] Total day's production.

however, the harvest of mature fruit for red wine generally requires a two to three week period. Since the fruit continues to ripen during this period, changes in pectin content potentially could be significant.

The pectin content of mature Concords at early-, mid-, and late-season harvest was determined on specific lots of pulp (approximately 4 tons) by pressing small samples with a laboratory hydraulic press before enzyme addition and at several intervals after enzyme addition. The pectin content of the composite must sample from the day's production was determined for comparison. Results of these studies for 1968 and 1970 are given in Table VI.

The pectin content of Concord must from pulp pressed immediately after heating was comparable with that observed in cold press musts. As noted above, this represents only the pectin extracted by the pressing operation and not the total pectic constituents of the fruit. The 30 minute enzyme treatment used in this process increased the pectin content of must. Laboratory studies of pectin content of must pressed from pulp held up to one hour at processing temperature ($55°C$) with no enzyme added indicated a slight decrease in pectin content (from an average of 0.646–0.564 gram/100 ml). The increase noted in the commercial process has been attributed to a partial breakdown of pectic components in the grape tissue and to an increased solubilization of pectic material through partial hydrolysis.

Although there was a decreasing trend noted throughout the harvest period in the pectin content of the composite must samples, there was no definite trend in the pectin values for the must at zero time (no enzyme treatment). Variation was apparent from season to season as well as throughout a given season.

Flavor

Probably the most readily recognized differences between the native grape varieties and *vinifera* varieties are those related to the odorous

of Concord Must[a]

Late Season		Average	
10/25/68	*10/16/70*	*1968*	*1970*
.272	.589	.266	.551
.408	.674	.383	.662
.463	.720	.449	.648
.771	.718	.637	.755
.431	.601	.502	.641

constituents of the fruit. These components carry through to the wine and impart a strong fruity aroma and flavor characteristic of the grape. The foxy aroma related at least in part to the presence of methyl anthranilate is typical of Concord and related *V. labrusca* species. This ester, identified by Power (*40*) and Power and Chesnut (*41, 42*), was found in Concord, Ives, Delaware, Niagara, and other *labrusca* varieties but not in Catawba, *vinifera* varieties, or Scuppernong grapes.

The development of methyl anthranilate in Concord grapes was reported by Robinson *et al.* (*43*) to occur during the last stages of maturity and within the four week period from mid-September to mid-

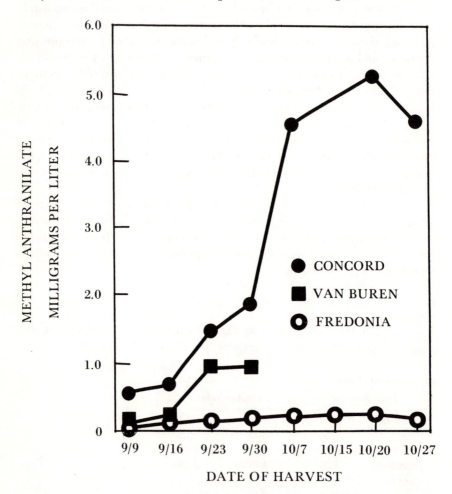

Fruit Products Journal and American Food Manufacturer

Figure 2. Methyl anthranilate development during ripening of labrusca *grapes* (43)

October, as shown in Figure 2. To update the information available on methyl anthranilate content of the commercially important native and hybrid varieties, composite must samples from the 1972 vintage were analyzed by both the AOAC procedure (44) and a fluorometric procedure developed by Casimir and Moyer (45).

Table VII.   Methyl Anthranilate Content of Commercial Eastern Grape Musts, 1972

	Methyl Anthranilate, mg/liter	
Variety	*AOAC Method*	*Fluorometric[a] Procedure*
Delaware	0.0	0.0
Niagara	0.3	0.3
Elvira	0.0	0.0
Catawba	0.0	0.0
Concord[b]	0.0	0.3
Concord[c]	1.4	1.75
Ives[c]	1.4	1.4
Aurore	0.0	0.0
Cascade (S-13053)[c]	0.0	0.0
Baco noir[c]	0.0	0.0

[a] Procedure from Ref. 45.
[b] Cold press.
[c] Hot press.

The results (Table VII) indicate detectable concentrations of methyl anthranilate only in Concord, Ives, and Niagara. The contrast between the methyl anthranilate content of cold press and hot press Concord is quite striking. As mentioned earlier, Concords for cold press are normally harvested about mid-September while fruit for hot press is harvested throughout the first three weeks of October. These results for 1972 musts would compare very closely in relative magnitude with the 1948 results of Robinson *et al.* (43).

Flavor characteristics of white Concord wine are completely different from those of red Concord wine. Although fruity, the foxy character is almost entirely lacking in the white wine. The white Concord wine is valuable in blends and in the production of baked dessert wines.

The Ives grape is of special interest both from the aspect of pigmentation (classed as *teinturier*) and its unusual flavor. Although it has a significant concentration of methyl anthranilate, other components distinguish its flavor from that of Concord.

Stern *et al.* (46) identified 59 compounds from Concord grape essence; esters represented the largest per cent of these. The esters had a pleasant odor which contributed to the overall pleasant aroma of Concord grapes. The fruity character of other native varieties probably is also attributable to the complement of esters. Thus far, little research has been directed toward identification of volatile flavor constituents of native varieties other than Concord.

The Scuppernong grape has a distinctive flavor characteristic. Kepner and Webb (47) analyzed the volatile essence of this grape and identified six alcohols, including β-phenylethyl alcohol, five aldehydes, and five esters. No nitrogen or sulfur containing compounds were found which indicated the absence of methyl anthranilate.

The fruity characteristics of the native and hybrid wines are accepted by an extensive segment of the American public. One explanation may be that Americans from early childhood have developed a love of fruit juices, including grape, apple, citrus, cranberry, and many others. The grape juice industry was based on the Concord grape. Grape juice, frozen grape concentrate, grape jelly, and jam were made entirely from Concord until recently. As a result Americans are well acquainted with the flavor characteristics of the native grapes and accept this fruity character in native American wines.

Pigments

The anthocyanin pigments of native and hybrid varieties have been studied extensively. Much of the impetus for this work came from efforts to detect the addition of hybrid wine to *vinifera* wine in Europe and, conversely, the addition of *vinifera* grape concentrate to Concord concentrate, juice, and other grape products (48, 49) in the United States.

The principal pigment of V. *vinfera* grapes is malvidin with lesser concentrations of peonidin, delphinidin, and petunidin. Ribéreau-Gayon (50, 51, 52) found no diglucosides in V. *vinifera* although V. *labrusca* contains 4%, V. *riparia* 27%, and V. *rupestris* 74% diglucoside. All of the pigments in V. *vinifera* occurred as the monoglucosides or acylated monoglucosides. Much research has been done on the V. *vinifera* pigments, and there are a few reports of the presence of diglucosides in some varieties. The pigments in *vinifera* grapes are discussed in detail by Amerine et al. (5) and Amerine and Joslyn (53).

The pigment content of Concord, Ives, and Baco noir grapes, among others, were reported by Robinson et al. (54). Principal components of Concord were delphinidin (43.4%), cyanidin (22.1%), and petunidin (8.5%) monoglucosides. Acylated pigments comprised about 18% and mixed diglucosides about 8% of the remaining components. Ives and the hybrid Baco noir differed greatly in that each of these grapes contained predominantly malvidin (and peonidin) diglucoside (45.8 and 73%, respectively). Ives contained a very high content of acylated pigment (42.5%).

Following the 1972 vintage, a comparison was made of the pigment content of commercial Concord wines prepared by hot pressing and fermenting on the skins. The grapes were harvested, pressed, and vinified at approximately the same time and thus represented essentially the same degree of maturity. Pigments were chromatographed and analyzed by Van Buren's methods (55) immediately after fermentation. The optical density at 525 nm was 5.00 for the fermented-on-the-skins lot and 6.46 for the hot-pressed lot. There was greater pigment extraction by the hot-press method, but the proportion of the several anthocyanins present was essentially the same in both wines. Each contained approximately 70% monoglucosides, 5% acylated monoglucosides, 12.5% diglucosides, and 12.5% acylated diglucosides. The dominant pigment was identified as delphinidin. The proportions of these components in the newly fermented wine very closely approximated the proportions reported in the Concord fruit (54).

Van Buren *et al.* (56) analyzed the anthocyanin pigment composition in experimental wines made at the New York State Agricultural Experiment Station from native and hybrid grapes approximately six months after vinification. Concord wine (fermented on the skins) had 40% monoglucosides, 40% diglucosides, and 20% acylated pigments. These observations support the data presented by Robinson *et al.* (54) which indicated that delphinidin monoglucoside was least stable of the pigments studied and that peonidin and malvidin proved most stable. The same general relationship was true for the diglucosides (57). Although monoglucosides showed a greater tendency to decolorize in storage than the diglucosides, the diglucoside molecule increased the tendency to brown (54). Further research with the acylated pigments (55) showed these to be most stable, particularly to the decolorizing effect of light.

From a practical standpoint, the winemaker must accept the pigment complex present in the fruit. However, final color of a blended wine may be a combination of the pigment complement of several varieties. The value of the Ives grape for color is apparent from the chromatographic analysis that revealed malvidin (and peonidin) diglucoside to comprise nearly half of the total pigment present. The identification of the components of the complex pigment mixtures present in the native grapes and the many direct producer hybrids is of great value both to the winemaker and to the viticulturist.

Modification of Winemaking Practices

The natural levels of sugars and acids in native grape varieties are unbalanced when compared with *vinifera* varieties. With an initial Brix of 20°, 22°, or 24°, *vinifera* varieties have been classified (53), in relation to Brix/acid ratios of 28.6, 31.4, and 34.3, respectively, as table wine grapes if below these ratios or dessert wine grapes if above. Brix/acid ratios of native varieties with Brix values (1972) ranging from 12.6° to 19.8° were in the range 7.7–16.8. However, these exceptionally low Brix/ acid ratios do not preclude the use of native and hybrid varieties for the production of stable, palatable, quality wines.

Federal regulations (58) controlling wine production in the United States take this imbalance into consideration. With high acid, low sugar fruit, amelioration (addition of water, sugar, or a combination of the two) is permitted to reduce the total acidity to 5 parts per thousand. However, amelioration may not exceed 35% of the total volume of ameliorated material regardless of total acidity of the initial juice or wine. Maximum amelioration may be used with acidities in excess of 7.69 grams/thousand milliliters of juice. Niagara is the only native variety used in quantity for winemaking that occasionally falls below this level (Table III); the other varieties greatly exceed this value every year.

As discussed by Carl (5), the degree of amelioration for specific musts is a decision to be made by the winemaker and is dependent on both the must characteristics and the type of wine to be made. It is necessary to re-evaluate each must each year to consistently produce high quality wines from the various native and hybrid varieties. Total acidity is not the only criterion on which this decision is based. Sugar content, color, flavor intensity, and ultimate wine type are important factors in determining the amount of amelioration to be used.

In practice, must is thoroughly mixed, sampled, and analyzed for total acid content immediately after pressing. The amelioration credits allowable under Federal wine regulations are calculated in terms of gallons of ameliorating material. These amelioration credits may be used before, during, or after fermentation (but must be used before finishing of the new wine, e.g., fortifying, blending with other wine, etc.). The low sugar content of the native and hybrid grapes makes some sugar addition necessary at the time of fermentation to produce a stable wine of 12–13% alcohol. A portion of the amelioration credits are therefore used at that time.

Commercial practice varies as to the timing of this addition. Some winemakers add sugar to the must just prior to fermentation. Others add it after fermentation is underway. The choice seems to depend largely on the fermentation equipment and individual preference.

Amelioration credits remaining after fermentation is complete are used or waived after organoleptic evaluation of the new wines. Each time credits are used, a report must be submitted to the Bureau of Alcohol, Tobacco, and Firearms, Department of the Treasury showing the amount used and the amount remaining. All credits remaining for a particular wine after it has been ameliorated by the winemaker must be waived and so reported to the BATF. Generally, only a portion of the total amelioration credits available are used. The per cent used depends on the must characteristics and the winemaker's judgment and varies considerably from year to year.

Table VIII. Average pH and Total Acid Content of Commercial Eastern Must and Wine, 1972

Variety	*Must*		*New Wine*		*After M–L Fermentation*	
	pH	TA[a], grams/ 100 ml	pH	TA[a], grams/ 100 ml	pH	TA[a], grams/ 100 ml
Aurore	3.08	1.159	3.39	0.889[b]	3.46	0.654
Delaware	3.11	1.148	3.30	0.946	3.30	0.705
Niagara	3.03	0.826	3.10	0.709	3.13	0.690
Catawba	2.77	1.718	2.97	1.309	—	—
Concord[c]	2.81	1.642	2.97	1.258	—	—
Concord[d]	3.12	1.366	3.20	0.910	3.35	0.724
Ives[d]	3.22	1.175	3.22	0.893	3.37	0.693
Baco noir[d]	3.28	1.530	3.38	1.129	3.50	0.818

[a] Total acid, as tartaric.
[b] Decrease includes the effects of partial amelioration and deposition of tartrates during fermentation.
[c] Cold press.
[d] Hot press.

The effects of initial amelioration are most readily discernible in the alcohol and acid content of the new wine. Instead of alcohol concentrations of 7–10%, the new wines contain 12–13%. The effect of partial amelioration on acid content is illustrated in Table VIII. The total acid values of the new wines are the result both of partial amelioration and deposition of tartrates during the alcoholic fermentation. Even so, most of the acid values remain high, and additional techniques must be employed to reduce the acid content to an acceptable level. Processes currently used include further amelioration, blending several native varieties and hybrids of the same or different vintages, blending a small proportion of low-acid *vinifera* wine, and encouraging the malo–lactic fermentation. Treatment with calcium carbonate is permitted by federal regulations and occasionally is used to reduce natural acidity in the absence of a malo–lactic fermentation.

Pectin in the musts of native and hybrid varieties has caused problems, primarily in the hot pressing of red varieties. In years past when the Eastern grape juice and wine industries were using the hydraulic rack and frame press for pressing Concords and other native varieties, problems in pressing the heated Concord pulp were solved by slowing down the pressing operation or sacrificing yield. However, in the late 1940's and early 1950's, a number of processors and individuals turned to a continuous press such as the Garola press for hot pressing Concords. The first attempts to press Concords in these presses were near failures because the pectic substances gave the heated pulp a slippery characteristic that made it practically incompressible in the continuous presses. As reviewed in Tressler and Joslyn (59) a process for treating the Concord pulp gradually evolved which combined the addition of pectolytic enzymes and a bulking agent, such as cellulose fibers or rice hulls, to the heated pulp and holding for a specified period. With the refinement of process time, temperature, enzyme concentration, and fiber content, the hot pressing of Concord pulp in continuous presses became practical.

Today several types of presses are used in the Eastern wine industry, among them continuous presses of the Coq type and batch presses of the Willmes and Vaselin types. For the hot pressing of Concord, Ives, and red hybrid varieties, the pulp is warmed to 49°–60°C (depending on the variety), treated with a pectolytic enzyme, and held for 30–60 min. Pectin content of the must actually increases during this period, but the nature of the pectic substances changes so that the slippery characteristic of the pulp is essentially lost, and the pulp becomes readily compressible. Cellulose fibers in sheet form are dispersed in the pulp as it is accumulated in a treatment tank. With continuous presses, treated pulp is frequently passed over a screening device to remove as much free-run juice as possible to increase the capacity of the press. With Willmes-type presses a de-juicing operation is not considered essential since the free-run juice is quickly expressed by revolving the cylinder after the press has been filled.

Treatment of pulp with pectolytic enzymes has two additional benefits. Free-run juice is released more readily from the pulp and overall juice yield is increased. Studies with heated Concord pulp containing cellulose fibers have shown that from 1000-gram portions of pulp, free-run juice obtained 5 min after placing the pulp on an 8-mesh screen was increased from 337 ml for controls with no enzyme to an average of 387 ml for lots containing pectolytic enzymes (at a concentration equivalent to that used in commercial operations). This represented an increase of 15% in free-run juice. Total yield, combining both free-run and hand-pressed juice, increased by 3%. (Equivalent yield on a commercial basis would have been 193.6 gal/ton.)

It is difficult to obtain this type of data from commercial operations during the activity of a vintage season. However, although actual figures are not available, it is suggested by a number of Eastern vintners that these effects of enzyme addition evaluated in the laboratory are realized in the commercial process.

In cold pressing pulp for white wines, present practices vary considerably within the industry. Some processors using continuous presses have found it an advantage to treat pulp with pectolytic enzymes prior to pressing. However, with the Willmes press, cold pulp to which cellulose fibers have been added press with no difficulty, and juice yields have been considered satisfactory. A general practice has been to add pectolytic enzymes to white must before yeast inoculation as an aid to rapid clarification of the new wine.

Table IX.   Comparison of Pectin Content in Commercial
Eastern Must and Wine

	1970			1971		
Variety	Must[a]	Wine[a]	% Reduction	Must[a]	Wine[a]	% Reduction
Aurore	0.243	0.011	95	0.294	0.041	86
Concord	0.425	0.066	84	0.406	0.040	90
Delaware	0.433	0.049	89	0.384	0.034	91
Niagara	0.535	0.084	84	0.339	0.035	90
Catawba	0.299	0.074	75	0.575	0.053	91
Concord[b]	0.599	0.045	93	0.802	0.101	87
Ives[b]	0.621	0.079	87	0.786	0.050	94
Baco noir[b]	0.480	0.062	87	0.203	0.051	75

[a] In gram/100 ml.
[b] Hot press.

During fermentation, 30–90% of the pectins are precipitated (5). An analysis of pectin content of wines from native and hybrid varieties (Table IX) for the years 1970 and 1971 indicate a reduction of 75–95% during fermentation. It is probable that the acetone-precipitable material remaining in the wine is gum or partially hydrolyzed pectin since the precipitate, a grey amorphous material, is entirely different from the translucent gelatinous mass typical of precipitated pectin. Although pectin is a problem in hot pressing native and hybrid varieties, the pectin content of the wine is comparable with that of *vinifera* wine.

Malo–Lactic Fermentation

As a result of the high natural acidity in the native and hybrid grapes, the malo–lactic (M–L) fermentation is of particular significance in the production of quality wine from this fruit. A survey of New York State wines in 1964 (60) revealed that 45% had undergone the M–L fermentation, including 68% of the reds and 27% of the whites. This was comparable with the incidence reported for *vinifera* wines in California (*61*) and Australia (*62*).

Cellar practices were shown to influence the extent of M–L fermentation in Eastern wines (*60*). The highest incidence occurred where must was fermented immediately after pressing. The lowest incidence was found where only a portion of the must was fermented immediately and the remainder was pasteurized, cooled, and held in cold storage until fermenter space became available. The effect of sulfur dioxide concentration on incidence of M–L fermentation was not evaluated in that survey.

The M–L fermentation causes several beneficial changes in these high acid, low pH wines, among them a decrease in acidity and an increase in the pH. The effect of the conversion of malic acid to lactic acid on the total acidity of native and hybrid wines is shown in Table VIII. The total acidity decreased to the range 0.6–0.8 gram/100 ml which is considered desirable in these native wines. This conversion is of particular significance in regard to flavor since lactic is less sour than malic at the same titratable acidity and the same pH (*63*).

The rise in pH which normally accompanies the M–L fermentation was observed. Although the pH change in Delaware and Niagara, was minimal in 1972, a study of specific lots of Delaware, Niagara, and Elvira in 1969 (*64*) indicated that the initial pH values of 2.90–3.10 increased to 3.25–3.50 on completion of the M–L fermentation. This shift is important since the very tart sensation imparted by wines of low pH changes significantly as the pH is raised (*63*).

A further consideration of the change in pH is the affect on tartrates. As the pH reaches 3.56–3.60 (the midpoint between the two pK_a's for tartaric acid) the precipitation of potassium bitartrate is increased (*65, 66*). This decrease in tartrate concentration in the wine is beneficial in the overall process of achieving tartrate stability. Although the pH of the 1972 wines did not reach this level, the pH does reach and exceed this range in some years (*64*).

During an extended period of study of the malo–lactic fermentation in one New York State winery, a pattern has emerged. In this operation all grapes received during a vintage were pressed immediately (either cold or hot press), the must was cooled to 20°–21°C, and sulfur dioxide was added only to white must to achieve a free SO_2 concentration of 20 ppm or less. The must was inoculated with yeast at the completion of the day's pressing.

Lactic acid organisms developed in the new wine following the alcoholic fermentation. Cell numbers were determined periodically by plate count, in a TGYE–tomato juice–wine agar at pH 4.5 (*60*). Population reached a level of 10^6 to 10^7 cells per ml after three to seven weeks at cellar temperatures in the range 12°–22°C. A periodic check of the organic acids by paper chromatography indicated that conversion of malic to lactic acid occurred when the cell population approximated 10^6 cells per ml and was usually complete within a period of 7 to 14 days. The well-documented effects of the M–L fermentation (66) were observed in these wines, *e.g.*, increase in pH and volatile acidity, decrease in titratable acidity, and a mellowing of the tart flavor of the wine.

The causative agent has been studied in detail, and the same organism was responsible for the M–L fermentation in the new wines each year. Lactic acid organisms isolated from new wines during the M–L fermentation were gram-positive, heterofermentative, cocci/short rods and were identified as *Leuconostoc* sp. (67, 68). These organisms are exceptionally acid tolerant and have a pH optimum of 4.5–4.7 in a laboratory broth medium (tryptone–glucose–fructose–yeast extract–salts). Perhaps as a result of the consistently low pH of these native wines and the cool cellar temperatures, *Lactobacilli* and *Pediococci* were inhibited. [Pederson (69) has shown that *Leuconostoc* predominate in cucumber fermentations at low temperatures.]

Temperature and pH influence the rate at which the M–L fermentation proceeds. By determining cell numbers in the new wine at weekly intervals (60, 64), the generation time (g) of the lactic culture during the M–L fermentation was calculated. Extreme growth rates in terms of generation time (g) together with the related pH and temperature data are shown in Table X. Generation time varied widely with different combinations of temperature and pH.

Table X.   Extreme Growth Rates of Lactic Acid Organisms in Native American Wines

Wine	*T, °C*	*pH*	*Generation Time* (g), *hrs*
1970 Niagara	21.7	3.30	26.7
1968 Concord[a]	16.1	2.97	65.4

[a] Cold press.

The data for 28 lots of wine were subjected to linear regression and correlation analysis to determine the correlation of generation time to pH, to temperature, and to total acid. Regression lines and equations are given in Figure 3. The coefficients of correlation (r) for temperature −0.647 and for pH −0.546 are statistically significant although low. A correlation coefficient of 0.188 was obtained for total acid, indicating a lack of correlation between these two parameters. Five red wines underwent M–L fermentation at a mean temperature of 18.9°C. Generation times relative to pH are given in Table XI. Linear regression analysis yielded a correlation coefficient of −0.971 which is statistically significant.

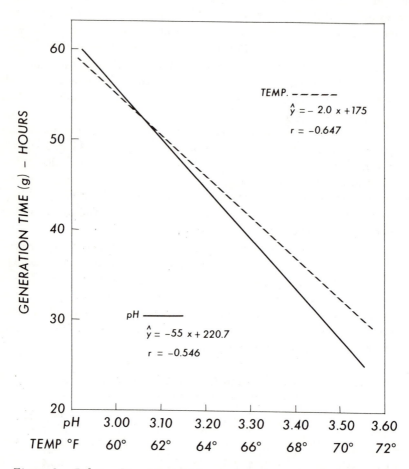

Figure 3. Relationship of bacterial growth, expressed as generation time (g) to pH and temperature of wine during M–L fermentation

The rate at which the M–L fermentation proceeds is a function of the generation time. As pH and temperature decrease, the generation time increases, and the M–L fermentation proceeds more slowly. These observations agree with the conclusions of Bousbouras and Kunkee (65) in relation to the effect of pH on the rate of M–L fermentation. Even at the low pH values observed in some of the native white wines, *Leuconostoc* organisms were able to develop and complete the M–L fermentation. The cool cellar temperatures did not inhibit the M–L fermentation

Table XI. **Relationship of Generation Time to pH in Red Wines**[a]

Wine	pH	General Time (g), hrs
Baco 1970	3.42	29.5
Baco 1968	3.42	32.2
Concord[b] 1970	3.30	35.0
Baco 1967	3.26	39.9
Ives 1966	3.11	53.1

Regression Equation $\hat{y} = -70.1X + 269.4 : r = -0.971$

[a] Mean temperature = 18.9°C.
[b] Hot press.

although the length of time required for completion increased significantly as the temperature decreased.

Rankine *et al.* (*70*) observed in Australian wines that the critical factor governing growth of lactics in wine was sulfur dioxide content. These studies of the M–L fermentation in native American and hybrid wines seem to confirm this observation. With the range of pH and temperature noted for these wines, the M–L fermentation occurred quite regularly as long as the free sulfur dioxide content was low (20 ppm or less). However, no growth of lactics was found if levels of free sulfur dioxide approached or exceeded 50 ppm.

Acknowledgment

We thank L. R. Mattick for organic acid, methyl anthranilate, and anthocyanin analyses; W. B. Robinson for advice and technical assistance; and R. F. Belscher for must and wine analyses.

Literature Cited

1. Hedrick, U. P., "The Grapes of New York," J. B. Lyon, Albany, 1908.
2. Munson, T. V., "Foundations of American Grape Culture," Orange Judd, New York, 1909.
3. Mendall, S. C., N.Y.S. Hort. Soc. (1964).
4. Wagner, P. M., *Amer. J. Enol. Viticult.* (1955) **6**, 10–17.
5. Amerine, M. A., Berg, H. W., Cruess, W. V., "The Technology of Wine-making," 2nd ed., Avi, Westport, 1967.
6. Amerine, M. A., Winkler, A. J., *Calif. Agr. Exp. Sta. Bull.* (1963) 794.
7. Caldwell, J. S., *J. Agr. Res.* (1925) **30**, 1133–1176.
8. Hartmann, B. G., Tolman, L. M., *U.S. Dep. Agr. Bull.* (1918) 656.
9. Kertesz, Z. I., *N.Y. Agr. Exp. Sta. Geneva Tech. Bull.* (1944) **274**.
10. Crowther, R. F., Neudoerffer, N. C., Bradt, O. A., *Ann. Rep. Hort. Exp. Sta. Prod. Lab. Vineland Sta.*, Ontario, 1964.
11. Robinson, W. B., Einset, J., Kimball, K. H., Bertino, J. J., *N.Y. Agr. Exp. Sta. Res. Cir.* (1966) **7**.
12. *Ibid.*, (1967) **9**.

13. Einset, J., Kimball, K. H., Bertino, J. J., Robinson, W. B., N.Y. Agr. Exp. Sta. Res. Cir. (1969) 18.
14. Robinson, W. B., Bertino, J. J., Einset, J., Kimball, K. H., N.Y. Agr. Exp. Sta. Spec. Rep. (1970) 1.
15. Einset, J., Kimball, K. H., Robinson, W. B., Bertino, J. J., N.Y. Agr. Exp. Sta. Spec. Rep. (1971) 4.
16. Crowther, R. F., Truscott, J. H. L., Bradt, A. O., Ann. Rep. Hort. Exp. Sta. Prod. Lab. Vineland Sta., Ontario, 1964.
17. Johnson, L. A., Carroll, D. E., J. Food Sci. (1973) 38, 21–24.
18. Taylor Wine Co., laboratory records.
19. Amerine, M. A., Thoukis, G., Vitis (1958) 1, 224–229.
20. Kliewer, W. M., Amer. J. Enol. Viticult. (1965) 16, 101–110.
21. Lee, C. Y., Shallenberger, R. S., Vittum, M. J., N.Y. Food Life Sci. Bull. (1970) 1.
22. Lott, R. V., Hort. Res. (1967) 7, 126–134.
23. Lott, R. V., Barrett, H. C., Vitis (1967) 6, 257–268.
24. Melamed, H., Ann. Technol. Agr. (1962) 11, 5–31.
25. Swanson, C. A., El-Shishiny, E. D. H., Plant Physiol. (1958) 33, 33–37.
26. Mattick, L. R., Shaulis, N. J., Moyer, J. C., Amer. J. Enol. Viticult. (1972) 23, 26–30.
27. Amerine, M. A., "Laboratory Procedures for Enologists," Associated Students Store, University of California, Davis, 1965.
28. Shaulis, N., Mattick, L. R., Moyer, J. C., Proc. N.Y. Wine Ind. Tech. Adv. Panel (1972), N.Y. Agr. Exp. Sta., Geneva.
29. Mattick, L. R., Rice, A. C., Moyer, J. C., Amer. J. Enol. Viticult. (1970) 21, 179–183.
30. Peynaud, E., Maurie, A., Amer. J. Enol. Viticult. (1958) 9, 32–36.
31. Winkler, A. J., "General Viticulture," University of California, Berkeley and Los Angeles, 1962.
32. Mattick, L. R., Moyer, J. C., Shaulis, N., N.Y. Food Life Sci. Quart. (1973) 6, 8–9.
33. Amerine, M. A., Wines Vines (1956) 37 (10), 27–30, 32, 34–36.
34. Ibid., 37 (11), 53–55.
35. Willaman, J., Kertesz, Z. I., N.Y. Agr. Exp. Sta. Geneva Tech. Bull. (1931) 178.
36. Amerine, M. A., Joslyn, M. A., "Table Wines. The Technology of their Production in California," University of California Berkeley and Los Angeles, 1951.
37. Hinton, C. L., "Fruit Pectins, their Chemical Behaviour and Jelling Properties," Chemical Publications, New York, 1940.
38. Hopkins, E. F., Gourley, J. H., Proc. Amer. Soc. Hort. Sci. (1930) 27, 164–169.
39. Kertesz, Z. I., "The Pectic Substances," Interscience, New York, 1951.
40. Power, F. B., J. Amer. Chem. Soc. (1921) 43, 377–381.
41. Power, F. B., Chesnut, V. K., J. Amer. Chem. Soc. (1921) 43, 1741–1742.
42. Power, F. B., Chesnut, V. K., J. Agr. Res. (1923) 23, 47–53.
43. Robinson, W. B., Shaulis, N., Pederson, C. S., Fruit Prod. J. (1949) 29, 36–37, 54, 62.
44. Association of Official Agricultural Chemists, "Official Methods of Analysis," 9th ed., Washington, 1960.
45. Casimir, D. J., Moyer, J. C., Proc. N.Y. Wine Ind. Tech. Adv. Comm. (7/15/65), N.Y. Agr. Exp. Sta., Geneva.
46. Stern, D. J., Lee, A., McFadden, W. H., Stevens, K. L., J. Agr. Food Chem. (1967) 15, 1100–1103.
47. Kepner, R. E., Webb, A. D., Amer. J. Enol. Viticult. (1956) 7, 8–18.
48. Mattick, L. R., Weirs, L. D., Robinson, W. B., J. Ass. Off. Agr. Chem. (1967) 50, 299–303.

49. Rice, A. C., *J. Ass. Off. Agr. Chem.* (1968) **51**, 931–933.
50. Ribéreau-Gayon, J., Ribéreau-Gayon, P., *Amer. J. Enol. Viticult.* (1958) **9**, 1–9.
51. Ribéreau-Gayon, P., *Ann. Physiol. Vegetale* (1964) **6**, 119–147, 211–242, 259–282.
52. Ribéreau-Gayon, P., "Les composés phénoliques du raisin et du vin," Inst. Nat. de la Rech. Agran, Paris, 1964.
53. Amerine, M. A., Joslyn, M. A., "Table Wines. The Technology of their Production," 2nd ed., University of California, Berkeley and Los Angeles, 1970.
54. Robinson, W. B., Weirs, L. D., Bertino, J. J., Mattick, L. R., *Amer. J. Enol. Viticult.* (1966) **17**, 178–184.
55. Van Buren, J. P., Bertino, J. J., Robinson, W. B., *Amer. J. Enol. Viticult.* (1968) **19**, 147–154.
56. Van Buren, J. P., Bertino, J. J., Einset, J., Remailey, G. W., Robinson, W. B., *Amer. J. Enol. Viticult.* (1970) **21**, 117–130.
57. Hrazdina, G., Borzell, A. J., Robinson, W. B., *Amer. J. Enol. Viticult.* (1970) **21**, 201–204.
58. Code of Federal Regulations, Wine. Title 26, Part 240, Revised 7/1/70.
59. Tressler, D. K., Joslyn, M. A., "Fruit and Vegetable Juice Processing Technology," Avi, Westport, 1961.
60. Rice, A. C., *Amer. J. Enol. Viticult.* (1965) **16**, 62–68.
61. Ingraham, J. L., Cooke, G. M., *Amer. J. Enol. Viticult.* (1960) **11**, 160–163.
62. Fornachon, J. C. M., *Aust. J. Appl. Sci.* (1957) **8**, 120–129.
63. Amerine, M. A., Roessler, E. B., Ough, C. S., *Amer. J. Enol. Viticult.* (1965) **16**, 29–37.
64. Rice, A. C., Mattick, L. R., *Amer. J. Enol. Viticult.* (1970) **21**, 145–152.
65. Bousbouras, G. E., Kunkee, R. E., *Amer. J. Enol. Viticult.* (1971) **22**, 121–126.
66. Kunkee, R. E., *Advan. Appl. Microbiol.* (1967) **9**, 235–279.
67. Garvey, E. I., *J. Dairy Res.* (1960) **27**, 283–292.
68. Garvey, E. I., *J. Gen. Microbiol.* (1967) **48**, 431–438.
69. Pederson, C. S., Albury, M. N., *N.Y. Agr. Exp. Sta. Bull.* (1950) **744**.
70. Rankine, B. C., Fornachon, J. C. M., Bridson, D. A., Cellier, K. M., *J. Sci. Food Agr.* (1970) **21**, 471–476.

RECEIVED May 29, 1973.

5

Chemistry of Wine Stabilization: A Review

GEORGE THOUKIS

E. & J. Gallo Winery, Modesto, Calif. 95353

Wine derives its general quality and chemical composition from grapes and from the yeast fermentation which converts their juice to wine. The processing, including aging, which new wine undergoes before it becomes esthetically acceptable depends on the wine type. No two wines are exactly the same. The winemaker uses old and new technology coupled with artistic creativity to produce consistently palatable wines. Aside from the expected characteristics such as color, clarity, bouquet, flavor, and taste, modern wines must be physically, chemically, and biologically stable. Worldwide consumption of wine is increasing rapidly. Better wine is produced today because we have a better understanding of its chemical and biochemical nature.

Wine has been part of the diet of civilized man since the early settlements in the Tigris-Euphrates basin and in Egypt. From these areas the grape vine was introduced to the Mediterranean basin countries and Europe by pre-Christian traders where wine became a safe and healthful beverage and an important food adjunct. Since it induced euphoria and pleasing relaxation from the strains of life, wine eventually took on social importance where it was used for religious feasting and celebration as well as for entertaining.

Grape juice was fermented inadvertently into wine by natural yeast. To nomadic tribes, especially, it was welcome since it had flavor reminiscent of the fresh fruit or juice. In addition it could be stored and transported easily and remained drinkable from season to season. Eventually wine found its place as an article of commerce with necessary quality requirements. Not until the time of Louis Pasteur did the scientific foundation of winemaking become established and did enology become the science of wine. Since the early Pasteur experiments and discoveries winemaking has developed from a haphazard, ill understood, and risky process into a well defined scientific discipline. Likewise, the consumer has become sophisticated, and he will not accept inferior wines. The American wine producer has used his chemical and biological knowledge of grapes and wine to offer a stable beverage produced under exemplary hygienic conditions from grape varieties which have been carefully selected. The ultimate character and quality of wine results from the nature of the raw materials from which it is made, the fermentation process, and the changes which are induced or occur naturally during the post-fermentation period (*1, 2, 3*).

General

The term wine refers to the natural beverage produced from the juice of sound and ripe grapes, in strict accordance with federal and state regulations. The stabilization principles discussed will have equal application to fruit wines in general except for tartrate stabilization since tartaric acid, the primary organic acid of grapes, is not found in any other fruits commonly used in winemaking.

Amerine said that the enologist who is unfamiliar with his raw material, grapes, is like a blind painter (*4*). However, it is beyond the scope of this review to discuss the accepted viticultural principles for producing grapes. It will be assumed that grapes delivered to the winery for the purpose of wine production are sound and at their optimum maturity. No amount of technical skill or knowledge can be substituted for the lack of good grapes in the production of wine.

Wine stability is a very relative term and may denote various things to different people. Today's demand is for brilliantly clear wines. The presence of sediment or the development of haze during marketing is generally considered by American consumers as evidence of incipient spoilage or instability (*2*). Berg and Akiyoshi (*5*) defined wine stability as a state or condition such that wine will not, for some definite period of time, exhibit undesirable physical, chemical, or organoleptic changes. These undesirable changes that denote wine instability were listed by them as: (a) browning or other color deterioration, (b) haziness or very slight cloudiness, (c) cloudiness, (d) deposits, and (e) undesirable taste or odor. Modern winemaking depends on controlled pure yeast fermentations (*6*). Grapes, as they arrive at the winery, harbor many types of natural, but undesirable yeasts, lactic bacteria, and occasionally acetic acid bacteria and molds. Soon after crushing, conditions become ideal for the growth of such microorganisms, and these must be inactivated to prevent their proliferation. This can be accomplished by heating the must to a high enough temperature to inactivate all the native microflora which arrive with the grapes. Excessive heating, however, may tend to contribute undesirable cooked flavors to the wine. The method of choice for the inactivation of the native organisms is the judicious use of 100–125 mg/liter of sulfur dioxide at the time of crushing. In addition to its bacteriostatic properties, sulfur dioxide inhibits enzymatic browning and assists in preventing other oxidative changes because of its oxygen scavenging capacity. Sulfur dioxide is not foreign to wines since many yeasts known to ferment well will produce (during normal fermentation) up to 150 mg/liter of sulfur dioxide from other sulfur compounds, primarily sulfate, in grapes. Grapes contain natural pectolytic enzymes which hydrolyze the small amounts of natural pectins in grapes. However, certain grape varieties, particularly the native American species of *Vitis labrusca* (Concord types) contain such high pectin concentrations that they must be enriched with commercial preparations of pectolytic enzymes when crushed. Hydrolysis of grape pectins is essential to free as much of the juice as possible from the pulpy cells of the grape and to facilitate clarification and efficient filtration of juice or wine at some later processing stage. Care must be taken to select pectolytic enzymes on the basis of their efficiency in depectinizing grape juice with minimal side effects, such as oxidation or browning, which may be brought about by impurities in commercial preparations.

Grape Harvesting, Crushing, Stemming, Pressing, Settling, and Fermentation. The winemaking process must be strictly controlled from grape to glass. Such control involves expert knowledge of the necessary facts and procedures and the ability to modify each in the desired direction when necessary. Knowledge of the microbiological status, chemical composition, and sensory quality at every winemaking stage is essential. Modern wineries apply such knowledge through well-staffed technical departments.

Winemaking begins with the grape harvest. Grapes must be picked at their optimum stage of maturity for the particular type of wine for which they will be used. They should reach the winery in good condition, and they should be processed as soon as possible after picking. Several potential stability problems such as enzymatic and oxidative browning, wild yeast fermentation, acetification, mold proliferation, and associated off flavors, are prevented by careful and speedy handling of the grapes.

Processing grapes into wine begins immediately after delivery when the berries are separated from their stems and possible accompanying leaves. They are crushed to give the must which contains grape juice, skins, and seeds. It is imperative to eliminate all stems and leaves from the must and to avoid disintegration or breaking of the seeds; otherwise the secretion of tannins and other polyphenolics along with leaf and stem flavors, will produce wines which will be astringent, bitter, and grassy.

White wines are more prone to darkening from excessive enzymatic or oxidative browning, and all wines are difficult to clarify following fermentation if proper care is not given in the crushing and destemming operation.

Following destemming, crushing, addition of sulfur dioxide, and possibly pectolytic enzymes, the juice from white grapes must be mechanically separated from the grape skins and pulp solids in the shortest period of time to avoid the excessive extraction of tannins and other polyphenolics. The juice which is freely drained is commonly referred to as free-run and is fermented separately to produce the best quality white wine. The lesser quality press juice, which results from pressing the grape skins, is used to produce white wine of lower overall quality. Such wines may be darker, more astringent, and bitter because of higher concentration of tannins and polyphenolic compounds as a result of longer contact of the must with grape skins and solids. Wines from press juice are frequently distilled for use in brandy. Modern enological procedures call for the separation prior to fermentation of even the small amounts of pulp solids which remain in free-run grape juice after draining. Alcohol acts as a solvent to extract tannins from the remaining grape pulp solids which is not desirable in white wines. Separation of these finely dispersed pulp solids is not easy but may be accomplished to a degree by natural settling, filtration, or centrifugation. All these operations must be accomplished before the onset of fermentation for white wines.

For red (and rosé) wines, fermentation must be induced as soon as possible after the destemming and crushing. Such wines must be fermented in contact with the grape skins since the anthocyanin pigments which give red grapes their color are in the pigmented layers of cells on the inside periphery of the grape skins (except for a few minor varieties where the pigment is in the juice). Grape crushnng or pressing without fermentation will not release enough pigments into the juice for good color. Hot pressing of the must from red grapes has been practiced (especially in the eastern United States), but the practice of choice for producing red and pink wines from the *Vitis vinifera* (European) species has been to ferment the must in contact with the skins until the desired amount of color has been extracted. Extraction of the anthocyanin pigments is enhanced by the solvent action of the alcohol produced during fermentation and by the action of carbon dioxide on the anthocyanin-bearing cells of the grape skins. When sufficient color has been extracted, the free-run red wine is drained off to finish its fermentation separately from the press wine recovered from pressing the grape skins (pomace). Extreme care and skill are required during the color extraction phase of winemaking since the winemaker must extract just enough anthocyanin pigments without extracting too much tannin which would make such wines difficult to age or difficult to ready for consumption within a reasonable time. During the color extraction phase of fermentation the grape skins, buoyed by carbon dioxide, surface and form a cap. This cap must be kept submerged for efficient color extraction and to prevent it from becoming dry, hot, and saturated with air. Exposure to air invites infection with vinegar bacteria which results in high volatile acidity and quality deterioration.

In the United States and in most wine producing countries, the fermentation of both white and red wines is induced by inoculating the must with 2–3% selected pure yeast culture at its exponential stage of growth to provide a minimum of one million yeast cells per ml in the juice to be fermented. Wineries use commercially available pure yeast strains or propagate their own yeast starters under strict control to prevent contamination with wild yeasts or bacteria. Grape juice is an ideal medium for yeast growth, and very rarely if ever is it necessary to rely on yeast foods to promote the primary fermentation of fresh grape must or juice.

Most authorities advocate fermentation of white wines at lower temperatures, 60°–70°F, to preserve their delicate varietal character. Red wines are fermented between 65° and 75°F. All wine fermentations, and certainly those of table wines, must be temperature controlled. When the fermentation temperature is not controlled by refrigeration, the large amounts of heat energy generated during this process will overheat the wine to the point where volatile varietal flavors will be lost by evaporation, cooked off flavors will develop, and the yeast will die, causing a stuck fermentation and leaving behind a wine with residual sugar and low alcohol content. No amount of winemaking skill can salvage any wine so mistreated.

The great strides which have been made in the United States in recent years in improving wine quality are attributed to viticultural efforts to provide the best suited grape varieties for a given area or product, careful grape handling at the winery, pure yeasts, and temperature control during fermentation and aging.

Post-Fermentation Care of Wine. After the desired amount of fermentable sugar has been used during fermentation, the raw wine is ready for clarification, settling, racking (decanting), filtration, aging, cold stabilization, blending, bottling, and distribution.

It is imperative that wine be separated from its fermentation solids (lees) as soon as possible. The lees include yeast, seeds, finely divided grape pulp, and grape skin particles and impair the quality of the wine by prolonged contact. Dead yeast cells, especially, will begin their autolysis under the highly reducing or anaerobic conditions in the new wine, and detrimental off flavors from hydrogen sulfide, mercaptans, and other reduced or yeasty odors will develop.

Immediate filtration is the procedure of choice in separating the new wine from its fermentation lees. Unfortunately, the vast volumes of new wine produced during the relatively short grape crushing period, which is rarely more than 10 weeks during the fall, may not permit the timely filtration of all new wine as soon as fermentation is completed. Most new wines are pumped out of their fermentation vessels into holding tanks and allowed to settle their gross lees and yeast for a few days before they are racked (or decanted).

As soon as a wine leaves its original fermentation vessel, it can no longer be stored for any length of time in anything but full containers. This is absolutely essential in avoiding contact with oxygen which is the greatest enemy of most wines because its high reactivity with wine constituents leads to oxidation flavors, browning, and general quality deterioration when oxidized. Progressive winemakers purge their table wines and vessel headspaces with inert gases such as nitrogen or carbon dioxide to exclude oxygen during all subsequent movements or handling of wine.

Fining and Clarification

Depending on the type of grapes used, length of fermentation, and type of wine produced, the new wine after its first racking and rough filtration may still be cloudy or hazy because of suspended colloidal particles of grape or yeast components, and it may remain so for a long time. It is rare when a sound wine will become brilliantly clear by natural settling. Amerine (7) states that until 50 to 70 years ago most German white wines had to be aged several (five to six) years before they became brilliant on their own; by this time, of course, most of them became brown and deteriorated in quality because of overaging. With present technology, such clouds or hazes caused by grape or yeast proteins, peptides, pectins, gums, dextrans, unstable grape pigments, tannins, and other compounds, may be assisted in their separation from wine by the judicious use of small amounts of fining agents which adsorb or combine chemically and physically with the haze particles or colloids or neutralize their electric charges causing them to agglomerate and gravitate to the bottom in a reasonable time. Such treatment results in a relatively clear wine which tends to become brilliant by subsequent filtration. Bentonite, one of the fining agents most commonly used in winemaking, effectively removes protein or peptide materials. Activated carbon, gelatin, casein, and poly(vinylpyrrolidone) may also be used, and they assist in removing unstable tannins and other pigments. If excess protein is not removed from the new wine at this point, elevated temperatures either at the winery or after distribution may cause a visible haze or actual sedimentation of coagulated protein or its reaction products with other wine constituents such as tannins and inorganic cations.

U.S. regulations (8) limit the use of fining agents so that removal of the cloudiness, haze, or undesirable odor or flavor will not change the basic character of the wine or remove wine components that will change its basic character. The criterion for government approval of clarifying agents is that none of the agent remain in the wine. Generally government regulations do not permit the addition of any compounds to grapes or wine that will leave a residue in the wine if such are not already naturally present in grapes or wines.

The amount and type of fining agents used depend on the nature of the haze or cloud forming colloidal matter and the type and composition of the wine. Winemakers carry out small-scale fining tests in the laboratory before attempting to clarify commercial quantities of wine to determine the efficiency of the fining agent, its effect on the quality and flavor of the wine, and particularly, to establish accurately the minimum quantity of fining agent required to provide acceptable clarification (see Ref. 2).

Non-Microbial Disorders of Wine

Tartrate Instability. Tartaric acid is not in any of the major carboxylic acid cycles of plant metabolism. Aside from its presence in tamarinds, only grapes, among the economically significant fruit crops, contain relatively high concentrations of tartaric acid. In grapes it normally constitutes over 50% of the total organic acids. Grapes are also rich in potassium, and significant concentrations of potassium bitartrate, cream of tartar, are present in grape juice. After fermentation, wine, because of its alcohol content, becomes supersaturated with natural potassium bitartrate, and removel of the excess is necessary to avoid sedimentation after the wine is bottled. Potassium bitartrate is not harmful, but the American consumer has been conditioned to suspect any product with sediment, especially wines, and is inclined to discriminate against such products. The procedure of choice for stabilizing wine against the sedimentation of potassium bitartrate after the wine has been bottled and sealed is the cold stabilization technique where wine is chilled just above its freezing point (this will vary from 26°F to 15°F depending on alcohol and soluble solids content) and held there to precipitate the excess potassium bitartrate. The time required to hold the wine at these near-freezing temperatures depends on the type of wine and its alcohol content, total acidity, pH, potassium, total tartrates, etc. This time will vary between a few weeks and several months. When it has been established by chemical analysis that no more potassium bitartrate is dropping out of solution, the wine is filtered at low temperature to separate it from the crystalline potassium bitartrate. Filtering the wine before it has a chance to rise in temperature appreciably is essential if redissolving any of the potassium bitartrate into the wine is to be avoided. Wineries use various cold temperature tests to determine empirically whether wines have achieved their desired potassium bitartrate stability. One such test combines either freezing of the wine at about 0°F for 4–6 hrs after which the wine is thawed slowly at room temperature and observed for potassium bitartrate crystals at the point when the last floating trace of ice melts away or a less severe cooling of the wine to 32°F where it is kept for a period of several weeks and checked while cold for the presence or absence of potassium bitartrate crystals at weekly intervals. If a wine does not show any signs of potassium bitartrate precipitation by these tests the winemaker can be reasonably certain that severe climatic conditions during shipment or abuse of the wine by improper storage by the retailer or the consumer will not precipitate potassium bitartrate during the sales portion of the life of the wine.

Protective colloids, which prevent the crystallization of excess potassium bitartrate in a reasonable time, make a wine resistant to cold stabilization, even by prolonged refrigeration. In such cases ion exchange treatment of a small portion of the wine will remove enough potassium to render the entire lot of wine potassium bitartrate stable when blended back. The work of Berg and Keefer (9) provides the modern enologist with useful data from which to calculate the relative stability of potassium bitartrate in wines of various alcohol contents under different conditions of pH, temperature, and time. The tables they presented for using approximate concentration products to calculate potassium bitartrate stability should be used by winemakers to determine how much of the potassium and tartrate of the wine must be removed to cold stabilize wine while avoiding excessive ion exchange which may cause undesirable taste and quality defects. The possibility of using electrodialytic techniques to remove excess potassium and tartrate from wines has been explored. Initial experiments have proved successful in achieving cold stabilization of wines, but problems still remain with the irreversible loss in capacity of the electrodialysis membranes which are commercially available. Electrodialysis more closely resembles refrigeration for tartrate stabilization of wines since it removes the components present in supersaturation whereas no new components are introduced to the wine. With further research in the technology of the electro-dialysis membrane field, it is expected that this process will become an important tool for wine stabilization.

In addition to deposits of crystalline potassium bitartrate, infrequent calcium tartrate deposits occur in wines. The calcium level of carefully produced wines is seldom high enough to cause stability problems. Occasionally, however, wines may extract calcium from improperly prepared filter materials. Prolonged storage in uncoated concrete tanks also will release calcium into wine.

The cold stability tests used by wineries are of short duration and do not predict potential calcium tartrate instability which may occur several months after a wine has been packaged. Berg and Keefer (10) have published data on the solubility product of calcium tartrate in alcohol–water solutions at various pH values, and this information can be used to predict the probability of calcium tartrate instability in wines. They warn that such data must be interpreted with care since other components of wine can form complexes with calcium and tartrate ions. However, such complexing with other ionic moieties in wine will tend to eliminate a certain amount of calcium and tartrate ions from active participation in the formation of calcium tartrate crystals. Therefore, the prediction of stability/instability on the basis of solubility product calculations from Berg and Keefer data, collected from alcohol–water model systems will be conservative, and a wine predicted to be of borderline stability will prove to be stable and not subject to calcium tartrate crystallization in actual practice. Yamada and DeSoto (11), using Berg and Keefer data (9, 10) studied a wide range of wine types and determined the maximum concentration products that would permit a wine to remain tartrate stable for prolonged periods under commercial conditions. Wines which may defy prediction of their potential calcium tartrate instability on the basis of the above considerations are those where the natural dextrorotary tartrate has undergone autoracemization to yield optically inactive or racemic tartrate, of which the calcium salt is much less soluble according to the following data (50):

D-Calcium tartrate = 230 mg/liter
L-Calcium tartrate = 250 mg/liter
Racemic calcium tartrate = 30 mg/liter

The autoracemization of dextrorotary tartrate in wines is quite slow and requires elevated temperatures for prolonged periods. It can be assumed that under U.S. wine production practices, sherry, because of its prolonged heating to develop the desired baked or nutty character and flavor, would be one of the few wine types which could undergo autoracemization of tartrate. Martini (*12*) has proposed adding racemic tartaric acid prior to refrigeration to remove excess calcium from wine.

Protein Cloudiness. Of the various colloids known to be present in wines, peptides and proteins are of particular importance because of their influence on wine stability and clarification (*13*). According to Amerine and Joslyn (*2*) proteins serve as nuclei about which copper, iron, and salts deposit and form hazes, clouds, suspensions, or sediments on denaturation by heat, cold, or prolonged aging of wines. Proteins were shown by Ribéreau-Gayon (*14*) to be associated with ferric phosphate and cupric sulfide sediments, and this was confirmed by Kean and Marsh (*15, 16, 17*) and by Lukton and Joslyn (*18*). Protein clouds or hazes can also be the result of protein–tannin complex formation in the presence of traces of heavy metals, especially tin according to Krug (*19*). Kielhöfer and Aumann (*20*) reported tin–protein precipitates in tin-containing wines. The levels of protein present in wine as reported by various investigators and summarized by Amerine (*27*), varied from 0.015 to 0.143% of nitrogen as protein. Several studies on the nature of the proteins in German table wines have been cited by Amerine and Joslyn (*2*). These show higher protein concentrations in the wines of warmer seasons, and large variations are manifested in wines made from different grape varieties. Fermentation, heating, bentonite fining, and filtration removes the bulk of proteins from wine according to Moretti and Berg (*22*). Holden (*23*) proposed a combined heat and cold treatment for the protein stabilization of wine.

Unfortunately, in spite of the published literature on wine proteins, we do not know the actual protein levels at which table or dessert wines are stable. The changes in protein content during production and processing of wines are still not known with sufficient accuracy to predict their behavior. The winemaker has to depend on empirical tests if he is to produce protein stable wines. Early separation of new wines from their fermentation yeast greatly improves their chances for protein stability by decreasing the release of yeast autolysis products into the wine.

Iron and Copper Cloudiness. Wine cloudiness caused by colloidal complex formation involving cations, particularly copper and iron, are referred to in the enological literature as *casse*. Under normal conditions of vinification, the copper and iron content of wines as it originates from grapes is not high enough to cause stability problems, according to Thoukis and Amerine (*24*). They reported that up to 90% of the copper and up to 70% of the iron initially present in grape juice is eliminated during fermentation mainly by yeast assimilation. It is generally believed that wines containing an excess of 0.5 ppm copper or 10 ppm iron may be susceptible to clouding or sedimentation as well as flavor deterioration by the strong catalytic and oxidative properties of these cations. Modern wineries which use stainless steel or coated mild steel in all equipment contacting grapes, grape juice, or wine have eliminated the incidence of copper and iron cloudiness which was common in the pre-World War II wine industry.

Iron clouds forming in wine depends on a number of factors including the concentration of iron, the nature and concentration of the predominating organic acids, the pH, the oxidation–reduction potential, and the concentration of phosphates and tannins. Ribéreau-Gayon (14) made an extensive study of the conditions under which iron clouds will occur in wines and concluded that when these conditions are not at their optimum, iron clouds will not occur even with fairly high iron contents. According to Marsh (25, 26) the iron in wine exists in several ionic forms depending on the physical and chemical constitution of the wine. Usually, under reducing conditions, the ferrous (Fe^{2+}) state predominates, but under oxidative conditions the ferrous form is converted to the ferric (Fe^{3+}) state. Both of these ionic forms of iron may exist in their free state or as soluble complexes with wine constituents such as citrate. Under certain conditions, i.e., low acidity or considerable amounts of tannin or phosphate, the ferric state of iron formed by exposure of the wine to air will combine with the tannins and/or the phosphates to yield insoluble or colloidal ferric complexes. Under California conditions ferric tannate casse is quite rare whereas the ferric phosphate casse occurs primarily in white wines in the pH range 3.0–3.6.

The mechanism of copper casse and the factors influencing its development in wines have been investigated by Joslyn and Lukton (18, 27), Kean and Marsh (15, 16, 17), and Peterson et al. (28).

In heavily sulfited white wines containing over 0.5 ppm copper and stored in sealed containers, a reddish-brown deposit may form. This occurs in the absence of oxygen and ferric ions but redissolves readily upon exposure to oxygen. Its formation may be accelerated by exposure to sunlight or heat, and it is believed to consist of colloidal cupric sulfide (14, 29). More commonly, copper casse may arise from reactions between copper and sulfur-containing amino acids, peptides, and proteins (15, 16, 17).

Methods for removing excess copper and iron cations to prevent copper and iron turbidities in wine have been proposed by Joslyn and Lukton (30) and Joslyn et al. (31). One of the most efficient techniques for removing excess copper and iron from wines is to add potassium ferrocyanide, a technique developed in Germany by Moslinger around 1905 and described by Ribéreau-Gayon (32). This treatment is called bluefining, and its use is permissible in most wine producing countries including Germany, France, Italy, Austria, U.S.S.R., Yugoslavia, Hungary, and Luxembourg. When properly used, it is successful in achieving wine stability. Using potassium ferrocyanide as such to treat wines is not permitted in the U.S. although none remains in the wine when properly used. It is possible to decrease the copper and iron content of wines by ion exchange using selective resins. However, excessive copper and iron in wines is the result of careless contact of wine, after fermentation and during aging or processing, with mild steel or copper alloy surfaces. Eliminating such equipment from winery operatons will solve any problems of copper or iron instability which still persists in the wine industry.

Oxidation Defects. During clarification, filtration, aging, blending, and bottling, wine must be protected against aeration, except for types such as sherry, Marsala, and Madeira where controlled oxygenation is indicated. Table wines, and all white wines in particular, must not be unduly exposed to air to avoid discoloration, haze formation, or loss of flavor because of the oxidation of alcohol, tannins, pigments, and other wine constituents. Partially filled tanks, leaky pumps, use of compressed air in mixing, or improper transfer of wines contribute to oxygen pick up.

Oxidasic *casse* has been defined by Amerine and Joslyn (2) as the clouding of wines and their change of color on exposure to air. Red wines become brown, and white wines turn various shades of yellow, orange, or brown. Such wines may also develop a cooked or rancid flavor which may result from increased acetaldehyde content. These changes are aggravated by phenolase or polyphenoloxidase which, in small concentrations, is normally present in grapes. Its activity may be inhibited by moderate amounts of sulfur dioxide or by flash pasteurization. Grapes with high mold count will produce juice and wine with abnormally high concentrations of polyphenoloxidase, but this condition is more common in Europe than in the U.S. In California, some wine producers use third party grape inspections conducted by the State Department of Agriculture to determine and eliminate excessive mold counts from grapes. Berg (33) demonstrated a clear correlation of grape variety to the browning susceptibility of white wines, and this was confirmed by Berg and Akiyoshi (5).

Stripping with nitrogen or other inert gases (34, 35, 36) may prevent the wine from picking up oxygen during the processing stages. Bonetti (37) pointed out that air-tight tanks are needed to store table wines under a blanket of nitrogen, and he stressed the utility of glass-lined or stainless steel tanks for this purpose.

Microbial Disorders of Wine

Microorganisms can cause various degrees of haze, cloudiness, or sediment formation as well as changes in the composition of wine constituents by metabolizing components of the wine.

Only a few types of microorganisms are able to survive and proliferate under the relatively high alcohol and low pH conditions which characterize most wines. Of these, yeast, acetic acid bacteria, and lactic acid bacteria are the most common. No pathogenic organisms are able to survive in wine.

Yeast Disorders. Many genera and species of yeasts are found on grapes. Undesirable yeasts must be inhibited as soon as the grapes are crushed, and the primary fermentation must be controlled by inoculating the must with pure strains of yeast. This ensures a clean fermentation in a reasonable time with efficient conversion of the grape sugars to alcohol. Selecting pure yeasts is a matter of preference for individual wineries, but the preference in the United States has been for yeasts which do not produce undesirable by-products.

Rankine (6, 38) believes that the differences in the secondary products formed in wines during fermentation by various yeasts are quantitative rather than qualitative, and careful selection of pure yeast strains can eliminate wine disorders caused by large amounts of undesirable by-products such as hydrogen sulfide, mercaptans, acetaldehyde, acetic acid, ethyl acetate, higher alcohols, etc.

Hydrogen sulfide in wines is aggravated by elemental sulfur (from the grapes) in fermenting media as reported by Thoukis and Stern (39), Rankine (40), and Acree et al. (41). When elemental sulfur is not available, yeasts differ widely in their ability to produce hydrogen sulfide from sulfate, sulfur dioxide, sulfur containing amino acids, and other sulfur compounds. Selecting yeasts on this basis has largely eliminated this disorder in Australian wines (36).

Ethyl acetate, above threshold levels, is considered a spoilage defect by most enologists. This condition is often attributed to acetification of wine by vinegar bacteria, but quite often it is the result of contamination with spoilage yeasts belonging to the genera *Kloeckera, Hansenula, Pichia,* and *Saccharomycodes* in the early stages of alcoholic fermentation.

During storage and aging of wine, various spoilage yeasts from genera such as *Pichia, Hansenula, Torulopsis,* and *Candida* may grow as surface films and cause quality deterioration by by-product formation. Because of their affinity for oxygen and their limited tolerance for alcohol and sulfur dioxide, film forming spoilage yeasts seldom grow in the main body of wine. Their growth as surface films can be discouraged by keeping wine containers full or by preventing air from headspaces above the wine by sweeping with an inert gas.

The Spanish sherry type of film yeasts tolerate high alcohol concentrations (up to 16 percent) and cause compositional and taste changes which are considered to be serious defects in nearly all wines except Spanish-type flor sherries. The most obvious compositional changes in wines contaminated with flor sherry yeasts of the genus *Saccharomyces* are the large increases in acetaldehyde, acetal, diethyl succinate, and 2-phenethyl alcohol (42) and the insipid or oxidized flavor and taste characteristics which accompany these changes in table wines.

Wine spoilage by various genera of yeasts after bottling is one of the most serious problems that challenges winemakers in every wine producing country. Such yeasts may be those which brought about the initial fermentation of the wine and were not completely removed, or they may be spoilage yeasts which infected the wine during fermentation, aging, or bottling operations. Marketing such wines becomes impossible since the consumer will reject them on the basis of their cloudiness, sediment, and in the case of wines with residual reducing sugar, fermentation, and bottle breakage. The general quality of such wines is usually impaired to the point where they must be withdrawn from commercial distribution. Important research in the area of sterile filtration and aseptic bottling technology is continuing.

Acetic Acid Bacterial Disorders. A small amount of acetic acid is produced by yeasts during alcoholic fermentation and new wines normally contain about 0.02 to 0.03 gram of acetic acid per 100 ml. Higher concentrations of acetic acid in wines are usually the result of contamination with various species of *Acetobacter* which may invade the wine during fermentation or aging. Acetic acid bacteria require a great deal of oxygen to proliferate and to oxidize ethanol to acetic acid. With pure yeast fermentations and storing wine in full containers to ensure minimal contact with air, acetification of wines is no longer the disorder it once was. A sound wine will not undergo acetification after bottling because the high oxygen demand of *Acetobacter* cannot be satisfied.

Lactic Acid Bacterial Disorders. The widely distributed acid-tolerant organisms which account for widespread spoilage of wines are gram-positive bacteria which produce lactic acid. The lactic acid bacteria which are involved in wine spoilage belong to the general group comprising *Pediococcus, Leuconostoc,* and *Lactobacillus* (6) and may be either hetero- or homofermentative, depending their ability to form carbon dioxide from glucose.

Lactic acid bacteria isolated from wine may use residual sugars or alcohol, or decompose organic acids as a source of carbon for growth and energy. Malic, citric, and tartaric acids may be metabolized, depending on conditions.

The bacterial conversion of malic acid to lactic acid and carbon dioxide has been recognized since 1890 and is referred to as the malolactic fermentation. This conversion has been promoted under controlled conditions in the cooler viticultural regions of the world where grapes mature with excessive amounts of malic acid which causes taste imbalance because of high total acidity. Lüthi (43) emphasized that malo-lactic acid fermentation is necessary for the production of acceptable Swiss wines.

The extensive literature relating to the historical development of our knowledge of the bacteria involved in and the chemistry of the malo-lactic fermentation has been cited by Amerine and Joslyn (8). Vaughn and Tchelistcheff (44) reported methods to control the malo-lactic fermentation in California. Under California conditions, the desirability of promoting the malo-lactic fermentation has not been established clearly, and the nature of any improvement in sensory quality of such wines is not fully evident. Pilone and Kunkee (45) demonstrated differences in the sensory characteristics of wines fermented with several strains of malo-lactic bacteria, but their taste panel was not able consistently to rank wines with or without malo-lactic fermentation in order of their quality. Amerine and Joslyn (2) point out that if the activity of such bacteria were limited to a net decrease in acidity, this would rarely be desirable under normal California conditions. Amerine (46) stressed that the malo-lactic fermentation should be prevented in California wines of low total acidity. In spite of this, Ingraham and Cooke (47) reported that their study of 144 California commercial wines revealed that over half had undergone malo-lactic fermentation which was 75% more common in red table wines. Likewise, Kunkee et al. (48) found that most commercial wines from southern California had undergone malo-lactic fermentation which, they suggest, may have been caused by bacteria associated with the grapes rather than from winery contamination. Metabolic products of the malo-lactic fermentation, other than lactic acid, may include esters of lactic acid as well as acetoin (acetylmethylcarbinol) and its oxidation product, diacetyl(2,3-butanedione), which is considerably more odorous than acetoin (48, 49). Rankine (6) reported the level of diacetyl to be higher in wines which had undergone malo-lactic fermentation, and that too much of this by-product is undesirable since its butterlike aroma dominates and lowers the quality of the wine.

If the malo-lactic fermentation does not occur during the primary alcoholic fermentation or during subsequent storage and aging of the wine, malo-lactic bacteria in the wine after bottling will make the wine susceptible to the malo-lactic fermentation during warehousing or distribution. If this happens, the wine will develop haze or sediment along with sauerkraut odor, a decrease in acidity caused by the decarboxylation of malic acid to lactic acid, an increase in pH, a build up of pressure caused by the carbon dioxide released from the decarboxylation of malic acid. The wine will therefore be unacceptable to the consumer on the basis of aesthetic and flavor deterioration.

Control of Microbial Disorders

The most obvious method of controlling microbial wine disorders is to prevent contamination of the wine. Yeasts which have been used for the alcoholic fermentation must be removed or inactivated before bottling. The same restriction applies to acetic- or lactic acid bacteria which may have entered the wine during or after fermentation.

The judicious use of 100–125 ml per liter of sulfur dioxide at crushing time, prior to inoculation with active yeast cultures, will control the growth and proliferation of spoilage (wild) yeasts and acetic acid and lactic acid bacteria during the primary fermentation. Pasteurization of grape juice before inoculation with selected pure yeasts will accomplish similar inactivation of the grape microorganisms. This is rarely practiced because of the economic and logistic difficulties involved in heating and cooling vast quantities of grape juice. Also, heating grape juice may bring about flavor loss or introduce a burnt or cooked flavor.

Removing yeasts and other solids, soon after fermentation by racking and rough filtration will prevent further proliferation of the fermentation yeast or invitation of acetic or malo-lactic contamination and spoilage. Periodic tasting and chemical analysis of all wines in storage is mandatory to detect such problems. Refrigerating white wines during storage will prevent the growth of microorganisms. Red wines which are stored at cellar temperatures for faster aging are more prone to bacterial spoilage if infected. Centrifugation is not widely used in California for controlling microorganisms.

Prior to bottling, wines must be polished filtered followed by either pasteurization or sterile filtration. To preserve quality, wineries are abandoning pasteurization as a method of destruction of microorganisms and have adopted cold-bottling techniques which have gained popularity since the early 1960's when inert membranes for sterile filtration became available to the wine and other beverage industries. The known and uniform pore sizes of these membrane filters guarantee yeast and bacteria removal. Controlled sanitation before bottling provides a good assurance that the wine will be free from microbial disorders for its entire shelf life.

Literature Cited

1. Amerine, M. A., Berg, H. W., Cruess, W. V., "The Technology of Wine Making," 3rd ed., Avi, Westport, 1972.
2. Amerine, M. A., Joslyn, M. A., "Table Wines; The Technology of Their Production," 2nd ed., University of California, Berkeley and Los Angeles, 1970.
3. Amerine, M. A., Singleton, V. L., "Wine: An Introduction for Americans," University of California, Berkeley and Los Angeles, 1965.

4. Amerine, M. A., "The Educated Enologist," *Proc. Amer. Soc. Enol.* (1951) 1-30.
5. Berg, H. W., Akiyoshi, M., "Some Factors Involved in Browning of White Wines," *Amer. J. Enol.* (1956) **7**, 84-90.
6. Rankine, B. C., "Influence of Yeast Strain and Malo-lactic Fermentation on Composition and Quality of Table Wines," *Amer. J. Enol. Viticult.* (1972) **23**, 152-158.
7. Amerine, M. A., unpublished data.
8. U.S. Internal Revenue Service, "Wine," *IRS Publ.* **146**, 1961.
9. Berg, H. W., Keefer, R. M., "Analytical Determination of Tartrate Stability in Wine I. Potassium Bitartrate," *Amer. J. Enol.* (1958) **9**, 180-193.
10. *Ibid.*, "Analytical Determination of Tartrate Stability in Wine II. Calcium Tartrate," *Amer. J. Enol.* (1959) **10**, 105-109.
11. Yamada, H., DeSoto, R. T., "Relationship of Solubility Products to Long Range Tartrate Stability," *Amer. J. Enol. Viticult.* (1963) **14**, 43-51.
12. Martini, M., "L'acido tartarico racemico come decalcifiante nei vini," *Riv. Viticolt. Enol.* (1965) **18**, 379-386.
13. Ribéreau-Gayon, J., "Sur les matières albuminoides des vins blancs," *Ann. Fals. Fraudes* (1932) **25**, 518-524.
14. Ribéreau-Gayon, J., "Etats, réactions, équilibres et précipitations du fer dans les vins Casses ferriques," Delmas, Bordeaux, 1933.
15. Kean, C. E., Marsh, G. L., "Investigation of Copper Complexes Causing Cloudiness in Wines. I. Chemical Composition," *Food Res.* (1956) **21**, 441-447.
16. Kean, C. E., Marsh, G. L., *Amer. J. Enol.* (1957) **8**, 80-85.
17. Kean, C. E., Marsh, G. L., "Investigation of Copper Complexes Causing Cloudiness in Wine. II. Bentonite Treatment of Wines," *Food Technol.* (1956) **10**, 355-359.
18. Lukton, A., Joslyn, M. A., "Mechanism of Copper Casse Formation in White Table Wine. II. Turbidimetric and Other Physico-Chemical Considerations," *Food Res.* (1956) **21**, 456-476.
19. Krug, K., "Die Ursachen der Eiweissnachtrübungen bei Wein des Ausfall der Eiweisschönungen," *Wein-Wiss.* (1968) **23**, 8-29.
20. Kielhöfer, E., Aumann, H., "Das Verhalten von Zinn gegenüber Wein," *Mitt. Rebe U. Wein, Serie A (Klosterneuburg)* (1955) **5**, 127-135.
21. Amerine, M. A., "Composition of Wines. I. Organic Constituents," *Advan. Food Res.* (1954) **5**, 354-510.
22. Moretti, R. H., Berg, H. W., "Variability among Wines to Protein Clouding," *Amer. J. Enol. Viticult.* (1965) **16**, 69-78.
23. Holden, C., "Combined Method for the Heat and Cold Stabilization of Wine," *Amer. J. Enol.* (1955) **6**, 47-49.
24. Thoukis, G., Amerine, M. A., "The Fate of Copper and Iron During Fermentation of Grape Musts," *Amer. J. Enol.* (1956) **7**, 45-52.
25. Marsh, G. L., "Metals in Wine," *Wine Rev.* (1940) **8** (9), 12-14, 24.
26. *Ibid.*, (10), 24-26, 28-29.
27. Joslyn, M. A., Lukton, A., "Mechanism of Copper Casse Formation in White Table Wine. I. Relation of Changes in Redox Potential to Copper Casse," *Food Res.* (1956) **21**, 384-396.
28. Peterson, R. G., Joslyn, M. A., Durbin, P. W., "Mechanism of Copper Casse Formation in White Table Wine. III. Source of the Sulfur in the Sediment," *Food Res.* (1958) **23**, 518-524.
29. Ribéreau-Gayon, J., "Le cuivre des moûts et des vins," *Ann. Fals. Fraudes* (1953) **28**, 349-360.
30. Joslyn, M. A., Lukton, A., "Prevention of Copper and Iron Turbidities in Wine," *Hilgardia* (1953) **22**, 451-533.
31. Joslyn, M. A., Lukton, A., Cane, A., "The Removal of Excess Copper and Iron Ions from Wine," *Food Technol.* (1953) **7**, 20-29.

32. Ribereau-Gayon, J., "Collage bleu; traitement des vins par le ferrocyanure. Complexes du fer dans les vins," Delmas, Bordeaux (1935).

33. Berg, H. W., "Varietal Susceptibility of White Wines to Browning. I. Ultraviolet Absorption of Wines. II. Accelerated Storage Test," *Food Res.* (1953) **18**, 399-410.

34. "Nitrogen in the Wine Industry. Part I. Sparging Process," Air Reduction Co., Madison, Wisc., 1961.

35. Bayes, A. L., "Investigations on the Use of Nitrogen for the Stabilization of Perishable Food Products," *Food Technol.* (1950) **4** (4), 151-157.

36. Cant, R. R., "The Effect of Nitrogen and Carbon Dioxide Treatment of Wines on Dissolved Oxygen Levels," *Amer. J. Enol. Viticult.* (1960) **11**, 164-169.

37. Bonetti, W., "Problems Encountered in Storage of Wine under Nitrogen," Wine Institute, Technical Advisory Committee, San Francisco, Dec. 10, 1965.

38. Rankine, B. C., "The Importance of Yeasts in Determining the Composition and Quality of Wines," *Vitis* (1968) **7**, 22-49.

39. Thoukis, G., Stern, L. A., "A review and some Studies of the Effect of Sulfur on the Formation of Off-Odors in Wine," *Amer. J. Enol. Viticult.* (1962) **13**, 133-140.

40. Rankine, B. C., "Nature, Origin and Prevention of Hydrogen Sulfide Aroma in Wines," *J. Sci. Food Agr.* (1963) **14**, 79-91.

41. Acree, T. E., Sonoff, Elizabeth P., Splittstoesser, D. F., "Effect of Yeast Strain and Type of Sulfur Compound on Hydrogen Sulfide Production," *Amer. J. Enol. Viticult.* (1972) **23**, 6-9.

42. Webb, A. D., Kepner, R. E., "The Aroma of Flor Sherry," *Amer. J. Enol. Viticult.* (1962) **13**, 1-14.

43. Lüthi, H., "La rétrogradation malololactique dans les vins et les cidres," *Rev. Ferm. Ind. Aliment.* (1957) **12**, 15-21.

44. Vaughn, R. H., Tchelistcheff, A., "Studies on the Malic Acid Fermentation of California Table Wines. I. An Introduction to the Problem," *Amer. J. Enol.* (1957) **8**, 74-79.

45. Pilone, G. J., Kunkee, R. E., "Sensory Characterization of Wines Fermented with Several Malolactic Strains of Bacteria," *Amer. J. Enol. Viticult.* (1965) **16**, 224-230.

46. Amerine, M. A., "The Acids of California Grapes and Wines. I. Lactic Acid," *Food Technol.* (1950) **4**, 117-181.

47. Ingraham, J. L., Cooke, G. M., "A Survey of the Incidence of the Malo-Lactic Fermentation in California Table Wines," *Amer. J. Enol. Viticult.* (1960) **11**, 160-163.

48. Kunkee, R. E., Pilone, G. J., Combs, R. E., "The Occurrence of Malolactic Fermentation in Southern California Wines," *Amer. J. Enol. Viticult.* (1965) **16**, 219-223.

49. Radler, F., "Die Bildung von Acetoin und Diacetyl durch die Bakterien des Biologischen Säureabbaus," *Vitis* (1962) **3**, 136-143.

50. Cambitzi, A., "The Formation of Calcium Tartrate in Wines," *Analyst* (1947) **72**, 542-543.

RECEIVED July 24, 1973.

6

The Present Status of Methods for Wine Analysis and Possible Future Trends

MAYNARD A. AMERINE

Department of Viticulture and Enology, University of California, Davis, Calif. 90241

The analytical methods used in the wine industry (mainly of the United States) are summarized, and procedures required by legal and regulatory agencies are given in some detail. Ethanol, volatile acidity, and sulfur dioxide are the important constituents to detect. Methods for detecting and measuring metal, pesticide, and fungicide contaminants are also outlined. Other procedures required for rational winery operation include determination of acids, sugars, acetaldehyde, potassium, color pigments, tannins, etc. Finally, research in enology may require methods to determine compounds such as diacetyl, succinic acid, various esters, etc. Gas–liquid chromatography is being increasingly used in wine and must analyses.

Among Pasteur's many contributions to the wine industry were several simple methods of wine analysis, and his research has had a permanent influence on the wine industry. As a well-trained chemist he was certain that chemical analysis would reveal the nature of the process of alcoholic fermentation and the type and degree of spoilage related to it. In both cases he was correct.

Even prior to Pasteur, alcohol content determination was important as a basis for local, import, and export taxes. Other important applications of accurate wine analysis have been to detect and to accurately determine food additives; now there are legal reasons for analyzing wines for sulfur dioxide, organic chloride or bromide, sodium, cyanide, diglucoside pigments, various insecticides, fungicides, etc. Winery control calls for analytical determination of iron, copper, protein, total acidity, pH, tartaric, malic and lactic acids, etc. Finally, quality control

demands accurate chemical analysis to ensure uniformity within a brand
or type for color, alcohol, total acidity, reducing sugar, etc.

By 1900 many analytical procedures were available. Dujardin-
Salleron (France) not only codified these procedures but also produced
the necessary equipment for them (*1*). Official methods of wine analysis
were soon developed in France and many other countries, and our own
Association of Official Agricultural Chemists began developing "tentative
and official" methods of wine analysis as early as 1916 (*2*); these continue
to the present (*3*). Official methods of analysis, both for reference and
routine purposes, are given on the international field by the Office International de la Vigne et du Vin (*4*) and by Amerine and Ough (*5*). For
current American practices *see* Refs. *3, 5, 6, 7, 8, 9, 10, 11, 12, 12a*. For
European procedures *see* Refs. *4, 9, 13, 14, 15, 16, 17, 18, 19, 20, 21, 22, 23*.

In many cases simple and only relatively accurate procedures are
adequate for winery control purposes. Recently a number of 5-minute
methods were developed in Europe (*23a, 24, 25*), and Schmitt (*26*) notes
that in these semimicro procedures (1 ml) care must be used to avoid
loss of volatile material, such as alcohol, and to make accurate volumetric
measurements with wines of varying viscosity. Losses during distillation
must also be minimized.

A small winery making a few determinations per year may prefer a
simple but time-consuming procedure, while a large winery (100+
determinations/day) may save money without loss of accuracy by using
a rapid procedure with expensive equipment. In some cases, particularly
in Europe, governments require a particular method even if it's not the
most modern.

Methods Required by Legal or Regulatory Standards

International limits for contaminants (*4*) are: arsenic 0.2 mg/liter,
volatile acidity 20 meq/liter for 10 vol % ethanol and 1 meq more for
each per cent alcohol above 10%, lead 0.6 mg/liter, boron 80 mg/liter
(as boric acid), bromine (total) 1 mg/liter (may be higher for wines
from grapes of certain areas), bromine (organic) 0.0, fluorine 5 mg/liter,
malvidin diglucoside 15 mg/liter, sodium 60 mg/liter (may be higher for
wines from grapes of certain areas), and sulfate 1.5 grams/liter (as
potassium sulfate).

Ethanol. The accurate, reasonably rapid, and relatively inexpensive
ethanol determination is the most important analytical procedure for the
winery analyst. Not only is it required for legal reasons but also for
winery control and research investigations.

In spite of their inefficiency (and even inaccuracy), boiling point
(ebulliometric) procedures still are used commonly by small wineries

and by home wine makers. Specific gravity methods based on hydrometry are more popular and more accurate although they are time consuming. In both cases the simple equipment makes them ideal for operators who make relatively few determinations.

For large-scale operations dichromate oxidation procedures provide precision. Microdistillation and careful temperature and acidity control make dichromate oxidation procedures rapid, accurate, time saving, and well-adapted to a multi-scale operation and automation (27, 28).

GLC is used increasingly as a rapid procedure for multi-determination laboratories. While the equipment is relatively expensive, careful operators get good results rapidly. No doubt it will be used more in the future because many other determinations can be performed nearly simultaneously, particularly in research work; Lie et al. (29) used a GLC procedure requiring only three minutes per sample.

The pycnometric procedure for determining ethanol is still the reference method for many countries although few laboratories use it routinely. In cases involving commercial transactions it is preferred, especially since the calculations can be simplified.

Ethanol also may be determined using alcohol dehydrogenase and measuring the change in nicotinamide adenine dinucleotide (NAD^+) to the reduced form (NADH) at 340 nm.

$$\text{CH}_3\text{CH}_2\text{OH} + \text{NAD}^+ \xrightleftharpoons{\text{alcohol dehydrogenase}} \text{CH}_3\text{CHO} + \text{NADH} + \text{H}^+$$

The solution is alkaline buffered, and semicarbazide is added to remove the acetaldehyde, forcing the reaction to the right. This method is particularly useful at low alcohol concentrations (30).

Methanol. In grape wines and brandies methanol determination is rarely necessary except where excessive amounts of pomace are used; Italian regulations limit the methanol content for this reason. With fruit wines and fruit brandies the legal maximum can easily be exceeded, and therefore an accurate determination of methanol is required.

Methanol may be oxidized to formaldehyde, and the color developed with formaldehyde and roseaniline or chromotropic acid as indicator is used for methanol determination. GLC has also been used (31), and Martin et al. (32) found good agreement between the two procedures at higher methanol levels.

Volatile Acidity. Acetic acid is the primary acid formed during wine spoilage. Legal limits for it exist in all wine-producing countries, varying from 0.10 to 0.25% exclusive of sulfur dioxide and sorbic acid. The United States and California State limits are among the lowest. Good winery practice demands that accurate determinations of the volatile acidity be made during processing and aging before marketing.

The current method (*3, 4, 6, 22*) involves steam distillation to separate the volatile (primarily acetic) acids from the non-volatile (fixed) acids. Special equipment has been devised for this separation (*6*). Sulfurous and sorbic acid content can be corrected, or the sulfurous acid may be removed (*33*). Carbon dioxide must be removed so that it does not interfere with the test (*6, 33*). An automated procedure is also available (*34*) which measures the volatile acids in the distillate at 450 nm using bromophenol blue.

It is probable that GLC determination of acetic acid, as distinguished from other volatile acids (propionic, lactic, etc.), will be used more often. A specific enzymatic procedure for acetate (*35*) reveals that only about two-thirds of the volatile acidity is acetic acid. However, the volatile acidity of commercial wines in modern wineries is usually well below the legal limits.

Sulfur Dioxide. The legal limits for total sulfur dioxide in wines varies from 200 to 350 mg/liter. In addition, the limit for sulfur dioxide not bound to aldehydes, polyphenolic compounds, etc. may be from 30 to 100 mg/liter. Winery control requires that the amount of sulfur dioxide present during processing and aging be carefully controlled, and increasing concerns for public health reinforce this.

For total sulfur dioxide determination, distillation procedures are preferred (*3, 4, 6, 22*). Careful attention to details is necessary since the methods are empirical (*16, 36, 37*). Automated analysis (*38*) also can be used (up to 20 samples per hr). For a summary of investigations on sulfur dioxide determination *see* Wucherpfennig (*39*).

The problem of accurately determining the non-bound (free) sulfur dioxide has not been satisfactorily solved (*6, 7, 40*). The best approach is to distill the sample in the absence of air and to recognize that the usual procedure overestimates the non-bound sulfur dioxide (*41, 42, 43*).

For this reason some countries (—*e.g.*, United States) do not specify a limit for non-bound sulfur dioxide. Nevertheless, in winery practice a limit is needed, and in countries where a legal limit exists, the determination is required.

The direct Ripper iodometric titration is still used, but it is subject to error. In its place, direct iodate–iodide titration is used (*44*). This is followed by fixing the sulfur dioxide with glyoxal in a second sample and retitrating. The difference represents the free sulfur dioxide. The second titration roughly represents the amount of reduction and the amount of ascorbic acid present. Formulas for calculating the amount of sulfur dioxide to add in order to produce a predetermined level of free sulfur dioxide have been given by Stanescu (*45*).

Metal Contaminants. There is increasing interest in the lead, sodium, and other metallic constituents of wine. The present lead limit is 0.6 mg/liter. While sodium is a normal constituent of wines, cation exchange resins may unduly increase its level. Suggested limits for sodium vary from 60 to 300 mg/liter in different countries. Wine as a low-sodium high-potassium beverage in diets for hypertension patients makes it apparent that stricter limits on sodium may be applied.

The suggested international limit for arsenic of 0.2 mg/liter should not be difficult to meet. Arsenate insecticides are not used in the United States and are seldom used abroad. While simple tests for the presence of arsenic are available, its quantitative determination is time consuming, involving ashing, special equipment, and meticulous technique and cleanliness.

Lead likewise should seldom if ever be found in wines approaching the suggested international limit of 0.6 mg/liter. The procedure by ashing and color development with dithiozone (or other reagent) is also time consuming and meticulous. Even with atomic absorption spectrophotometry there are losses in ashing and laboratory lead pick-up. Probably the best new attachments are those where the wine can be ashed directly in the apparatus.

Boron (as boric acid) has a suggested international limit of 80 mg/liter. The probable source is soil or irrigation water, and several procedures are available for its determination.

Fluorine (or fluorides) has a suggested international limit of 5 mg/liter. Fluoride pickup occurs only in the rare case where concrete tanks may have been treated with fluosilicate. With the disappearance of concrete tanks and prohibition of fluosilicate which may contaminate food, this determination should be superfluous. It is done on an ashed sample distillation of fluosilicic acid with super-heated steam and titration with thorium nitrate in the presence of sodium alizarinsulfonate.

Bromine (as bromide) may occasionally exceed the suggested 1 mg/liter limit in areas where grapes are grown on brackish soils. Here a higher limit can be applied since bromide is not toxic at this level. The determination involves ashing. Addition of Chloramine T in the presence of phenolsulfonphthalein leads to the formation of tetrabromophenolsulfonphthalein which is colored.

Organic bromine has a zero tolerance because of the reported toxicity of monobromacetic acid. It can be extracted quantitatively with ether at pH 1. The determination is then the same as with total bromide (4).

Sulfate (as potassium sulfate) has long been limited to reduce addition of calcium sulfate (plastering). The practice lowers the pH and is limited to warm climatic regions where the acidity is very low (pH high). The suggested limit of 1.5 grams/liter is somewhat lower than the 2 and 3 grams/liter limit of various countries.

The precipitation as barium sulfate requires boiling in the absence of air to remove sulfur dioxide. (The residue from the distillation procedure for sulfur dioxide may be used.) Some skill in handling the precipitate is required for accurate results. Complexometric titration after precipitation as lead sulfate (4) is possible. By using three solutions containing different amounts of barium chloride, it is possible to rapidly determine from the clouding whether a wine contains more or less than 0.7, 1.0, or 2.0 grams/liter of potassium sulfate.

Nitrate is of interest as a possible indication of watering. Many potable water supplies contain 30 mg/liter of nitrate whereas normal wines contain about 4-5 mg/liter on the average (46, 47, 48). The present recommended procedure (49) is reduction of nitrate to nitrite followed by colorimetric determination with sulfanilic acid and α-naphthylamine.

Other metals are rarely present in amounts approaching public health limits (*see* below for their determination in normal winery practice).

Atomic absorption spectrophotometry is the method of preference (50, 51). For recent reviews of this procedure as applied to wines, *see* Refs. 52 and 53. X-ray fluorescence spectrophotometry has also been used (54).

Diglucoside Anthocyan Pigments. No limits on the amounts of diglucoside anthocyan pigments present in wines exist in the United States. In Europe a limit of 4 mg/liter of malvin has been suggested (22). The limit has more economic significance than any public health hazard. Paper and thin-layer chromatographic procedures are available when this determination is required (primarily in Germany) (4, 22, 55, 56, 57, 58).

Carbon Dioxide. The U.S. limit for carbon dioxide in non-sparkling wine is 2.77 grams/liter. Above this value the tax rises from 17¢ per gallon to $2.40 or $3.40. Other countries have less stringent limits. Obviously an accurate method is required, and several are available (4, 5, 6). At present a simple enzymatic reaction using carbonic anhydrase is preferred. The bicarbonate ion is titrated with standard acid between pH 8.6 and 4.0. Carbonic anhydrase ensures that the carbonic acid is all in the bicarbonate form. A non-fading endpoint is thus obtained.

Sugar and Extract. Glucose and fructose are the main reducing sugars present in musts and wines. Even when sucrose is present, it is hydrolyzed in a few days at the relatively low pH of the wine or by sucrase. In addition to legal restrictions on reducing sugars and sucrose in certain but not all countries, good winery practice requires accurate analyses in wines and proximate analyses in musts.

The classical gravimetric copper procedures have been used for many years and are still the standard. Though they are accurate, they are time consuming. American enologists generally prefer the titrimetric Lane and Eynon (3, 4, 6) procedure.

The elegant enzymatic procedure of Drawert and Kupper (59) permits determination of glucose, fructose, and (when necessary) sucrose. Glucose reacts with adenosintriphosphate (ATP) in the presence of hexokinase (HK) to produce glucose-6-phosphate.

$$\text{glucose} + \text{ATP} \xrightleftharpoons{\quad \text{HK} \quad} \text{fructose-6-phosphate}$$

Adding glucose-6-phosphate dehydrogenase (G-6-PDH) in the presence of oxidized nicotinamide adenine trinucleotide (NADH) yields gluconic acid-6-phosphate and an equivalent amount of the reduced form, NADPH. The amount of NADPH can be measured from its absorbance at 340 nm. For fructose the first reaction is:

$$\text{fructose} + \text{ATP} \xrightleftharpoons{\quad \text{HK} \quad} \text{fructose-6-phosphate}$$

Fructose-6-phosphate is converted to glucose-6-phosphate in the presence of phosphoglucoseisomerase (PGI).

Trimethylsilyl derivatives of sugars can be made and separated by GLC (60). For musts, where 90% or more of the soluble solids are glucose and fructose, proximate procedures are usually sufficient—hydrometry or refractometry. Since only the sugar content up to 1% is often needed, sugar test pills are useful (5, 6, 61, 62, 63). Rapid spectrophotometric procedures are sometimes convenient, especially if they do not require prior filtration or decoloration (62).

Fraudulent addition of sucrose is very difficult to detect (22) because some sucrose is normally present, and added sucrose is rapidly hydrolyzed. Since commercial sucrose contains small amounts of unfermentable impurities, it has been suggested (60) that GLC would detect these in wines. Synthetic sweetening agents can be detected by thin-layer chromatography (64, 65).

The non-sugar soluble solids remaining in a table wine after dealcoholization are known as the extract. Low sugar musts or watered and sugared musts have a low extract content. The minimum extract in the United States is 1.6 and 1.8 grams/100 ml for white and red table wines respectively. These limits are so low that wines from moderately watered musts will meet them.

While the concept of extract is relatively unambiguous, its precise definition is difficult. The problem is how to dealcoholize the wine without loss of non-sugar soluble solids (lactic acid, glycerol, etc.). The empirical definition (4) is to specify a temperature (70°C), degree of vacuum (20-25 torr), period of time (1 hr plus cooling over sulfuric acid), and other conditions.

The Taberié formula is usually used: $d_r = d_w - d_a + 1.000$ where d_r is the density of the residual, d_w is that of the wine (less the volatile acidity and sulfur dioxide), and d_a that of the alcohol (both at 20°C). Tables for converting d_r to grams of extract per 100 ml are available, but the proper table to use is the Plato sucrose table, not the Ackermann empirical table (66).

Pesticides and Fungicides. Modern pure food regulations require that the food processor be responsible for their finished products. Since so many pesticides and fungicides are used in agriculture, their detection and quantitative analysis are difficult (5, 22). Organophosphorus and chlorinated hydrocarbons are the most common pesticides. When GLC is used for halogens, electron capture or microcoulometric detectors are used; for phosphorus, a thermionic flame photometric detector is required.

When the specific additive is known, simpler procedures can be devised (67). Some insecticides can be detected from their effect on cholinesterase activity (68).

Prohibited Additives. Diethylpyrocarbonate (DEPC) is now prohibited in the United States and other countries. Since a small but regular amount of diethyl carbonate is a constant byproduct of its use and is easily detectable with GLC, wines to which DEPC has been added are easily detected (69, 70, 71).

Potassium ferrocyanide has long been used to remove excess copper and iron from wines. When not used in excess it appears effective and harmless, but if any ferrocyanide residue remains, cyanide may form. While the amounts produced by a slight excess would pose little danger, the blue precipitate and distinctive odor would be undesirable. Special equipment has been devised to detect free cyanide and ferrocyanide (4, 5, 6) as Prussian blue. The suggested limit is 1 mg/liter as cyanide. Hoppe and Romminger (72) devised a rapid procedure for free and bound cyanide, and Bates (73) gives a qualitative screening method sensitive to 0.05 mg/liter.

Rarely are added benzoic or salicylic acids found in musts or wines although salicylic acid was widely used at the turn of the century. Sensitive color tests are available for their detection (3, 5, 6, 22, 74).

Just before World War II monochlor- and monobromacetic acids were used both in the United States and in Europe, but they are now prohibited. To detect organic bromide or chloride, liquid–liquid extraction is commonly used followed by destruction of the organic matter and use of the Volhard or other classical procedures. Ion-specific electrodes can also be used as well as thin-layer chromatography (75). GLC, particularly of trimethylsilyl derivatives, can detect some additives—hydroxybenzoic acid, for example (60).

Methods Required in Winery Operations

The procedures used in winery operations vary greatly, depending on the types of products produced and their market. A small winery producing only one type of red wine may need only a few different analyses. A winery producing grape juice, grape concentrate, table wines, dessert wines, special natural (flavored) wines, vermouth, fruit wines, high-proof spirits, and commercial brandy will require many different types of analyses.

Total Acid. Simple titration procedures are used to determine total acidity. Problems arise because of the widely varying amounts of different acids in wines: tartaric, malic, citric, lactic, succinic, acetic, etc. Different pK_a values for these acids make it impossible to predetermine easily the correct pH of the endpoint. Since a strong base is being used to titrate relatively weak acids, the endpoint will be greater than pH 7. In this country phenolphthalein (8.3) or cresol red (7.7) endpoints or a pH meter to 9.0 have been used (3, 6, 12, 76, 77); and the results are expressed as tartaric acid. The result at pH 7.7 $\times$ 1.05 approximately equals the result of titrating to pH 8.4. In Europe pH 7 is usually the endpoint, in France the results are expressed as sulfuric acid, and in Germany as tartaric or in milliequivalents (78).

The possibility of determining the acids in wines from the titration curve using special equations has been extensively investigated in Portugal by Pato and coworkers (79). To keep the ionic force constant, appropriate dilution is needed. Tartaric, malic, lactic, and succinic acid were determined in musts and wines.

Fixed Acid. The total acid (as tartaric) less the volatile acidity (as tartaric) is the fixed acidity. It is useful to make this calculation when one suspects activity of acid-reducing bacteria, as in the malo–lactic fermentation.

Malic Acid. This is seldom determined quantitatively in winery practice. However, qualitative paper chromatography is often done to follow malo-lactic fermentation. Using n-butyl alcohol and formic acid (80), the R_f values are: tartaric 0.28, citric 0.45, malic 0.51, ethyl acid tartrate 0.59, lactic acid 0.78, succinic 0.78, and ethyl acid malate 0.80.

For quantitative results no completely satisfactory procedure is available. Enzymatic procedures (81) using L-malic dehydrogenase suffer from possible interference of tartaric acid (82). Malic acid can be fermented by *Schizosaccharomyces pombe* (*83*). The colorimetric procedure with 2,7-naphthalene disulfonic acid (*84*) is perhaps the most practicable.

Tartaric Acid. Quantitative measures of total tartrate are useful in determining the amount of acid reduction required for high acid musts and in predicting the tartrate stability of finished wines. Three procedures may be used. Precipitation as calcium racemate is accurate (*85*), but the cost and unavailability of L-tartaric acid are prohibitive. Precipitation of tartaric acid as potassium bitartrate is the oldest procedure but is somewhat empirical because of the appreciable solubility of potassium bitartrate. Nevertheless, it is still an official AOAC method (*3*). The colorimetric metavanadate procedure is widely used (*4, 6, 86, 87*). Tanner and Sandoz (*88*) reported good correlation between their bitartrate procedure and Rebelein's rapid colorimetric method (*87*). Potentiometric titration in Me_2CO after ion exchange was specific for tartaric acid (*89*).

Lactic Acid. Qualitative and even semiquantitative data are obtained by paper chromatography. Quantitative procedures where lactic acid is oxidized to acetaldehyde and the acetaldehyde determined colorimetrically are available (*4, 13, 22, 90*).

Citric Acid. This acid is rarely determined in wines since only a small amount is present. However, should it be used to modify the acid taste, limits would be imposed, and its determination would be necessary. Enzymatic (*22, 91, 92*) and colorimetric (*93, 94*) procedures are available.

Simultaneous Determination of Acids. With the advent of trimethylsilyl derivatives of the organic acids, GLC determination has been developed (*60, 95*) as a method for their detection and quantitative determination.

pH. Because it affects the growth of microorganisms, color, taste, the ratio of free-to-total sulfur dioxide, and susceptibility to iron phosphate cloudiness, the pH is commonly measured as a guide to winery practice. Ordinary pH meters are used.

Acetaldehyde. In routine winery operation acetaldehyde is seldom measured. However, in the production of sherry, either by the film yeast or submerged culture processes, regular acetaldehyde determination is necessary.

The principles of the classical Jaulmes and Espezel (*3, 96, 97*) procedure are still used. To prevent copper-catalyzed oxidative changes, EDTA is added to remove copper (*98*). Isopropyl alcohol has the same effect (*99*). Colorimetric (*78, 94, 100, 101, 102*) and GLC (*103, 104*) procedures are becoming popular.

Hydroxymethylfurfural. When must or wines containing fructose are heated, hydroxymethylfurfural is produced. It is easily detected qualitatively, and limits are placed on it in Germany to prevent overheating of grape and fruit juices and in Portugal to prevent artificial aging of port wine by heating. Paper and thin-layer chromatography and spectrophotometric procedures have been used (6, 23, 89, 131).

Sorbic Acid and Sorbates. In addition to the correction of the volatile acidity for sorbic acid already mentioned, both sorbic acid and sorbates must be determined directly. The colorimetric procedure of Jaulmes *et al.* (10) is appropriate: oxidation of sorbic acid to malonic dialdehyde and a red color developed by reaction with 2-thiobarbituric acid (4).

Sorbitol and Mannitol. Sorbitol is present in fruits but not in grapes. A method for its determination is required to detect illegal blending of fruit wines with grape wines. Mannitol is produced by bacterial spoilage. Sorbitol dehydrogenase and thin-layer chromatography have been used for their simultaneous determination (5).

Glycerol. Glycerol is seldom determined today except in research work. GLC appears to be the method of choice (105) although the enzymatic procedure is direct and accurate (22, 106).

Copper. In the presence of sulfur dioxide, copper–protein cloudiness may develop in white wines. Only small amounts of copper (about 0.3 to 0.5 mg/liter) cause cloudiness. Widespread use of stainless steel in modern wineries has reduced copper pickup, but many wineries routinely test their wines for copper. Atomic absorption spectrophotometry is the method of choice (51) although reducing sugars and ethanol interfere, and correction tables must be used (107). To reduce this interference, chelating and extracting with ketone is recommended (108). Lacking this equipment colorimetric procedures can be used, especially with diethyldithiocarbamate (3, 4, 6, 9, 10, 22, 109). Neutron activation analysis has been used for determining copper in musts (110).

Iron. Excess iron in wines causes cloudiness, interferes with the color, and can impair flavor. The mechanism of ferric phosphate precipitation has been intensively studied, and numerous colorimetric methods have been developed. For routine purposes the color developed with thiocyanate is adequate (6, 9), but many enologists prefer the orthophenanthroline procedures (3, 4, 6, 22). Meredith *et al.* (111) obtained essentially the same results for iron using 2,4,6-tripyridyl-s-triazine (TPTZ) to develop the color. Atomic absorption spectrophotometry can be used but, as with copper, corrections for reducing sugar and ethanol are necessary (51).

Potassium. Quality standards for bottled wines now require a high degree of clarity. Even slight precipitates of potassium acid tartrate are considered detrimental. Whether wines are stabilized by cold treatment, long aging, or ion exchange, determination of their potassium content may be necessary. Precipitation as the acid tartrate (6) is widely used. However, precipitation as potassium tetraphenylborate is used in Europe (4, 22). Flame photometry and atomic absorption spectrophotometry are used by those with the necessary equipment (*5, 22, 112*). Sodium has been discussed in connection with legal restrictions.

Calcium. Excess calcium can occur in wines stored in concrete tanks or otherwise exposed to calcium (filter aids, calcium bentonite, etc.). After fortified wines are bottled, calcium tartrate may slowly precipitate.

Complexometric titration with EDTA is the usual winery procedure for determining calcium (*4, 6, 22, 113*), but atomic absorption spectrophotometry (*51, 53, 112*) and flame photometry and a rapid micro method based on oxalate precipitation (*114*) have been used successfully.

Tannin. The phenolic compounds are of increasing importance in winery operation to predict the proper fining agent and amount and method of fining as a measure of the age and flavor of the wine, as an indication of the degree of oxidation the wine will tolerate, etc. The present standard methods (*3, 4, 5, 6, 22*) use the Folin-Ciocalteu reagent (*115, 116*). Gallic acid is used to prepare the standard curve at 765 nm. For recent methods for total phenolics and for separating the phenolic fractions, *see* Singleton and Esau (*117*) and P. Ribéreau-Gayon (*118, 119*).

Histamine. The presence of small amounts of histamine in wines is now well established. These amounts are usually well below those that may have physiological effects.

A method for determining histidine and histamine simultaneously is now available (*120*). The precision and sensitivity were 0.05 mg/liter and 0.025 μg/ml and 0.7 mg/liter and 0.75 μg/ml, respectively.

Color. Precise specification of color has become more important as many producers sell large volumes of specific wines to a national clientele. There is the additional problem of low color white wines (*5*).

The usual procedure is to prepare an absorption curve in the visible spectrum. From this, the trichromatic coefficients are determined and the requisite three color parameters—luminence, dominant wavelength, and purity are obtained. The procedure is somewhat laborious, and shorter methods are often used. For white wines, changes in absorption at 420 or 430 nm are adequate to detect browning changes. For red wines the absorbance at 420 and 520 nm is measured. The ratio of these corresponds roughly with changes in tint or hue (dominant wavelength). The sum of the two absorbances is a rough measure of luminence (brightness). This method detects color changes in most red wines (*121*) even though the human eye can detect even smaller differences (*122*).

Methods Required in Research

The research enologist requires many different procedures depending on the requirements of the experiment. Research on flavor constituents requires the most sophisticated techniques of modern organic chemistry.

Among the compounds commonly determined in research laboratories are diacetyl, 2,3-butandiol, glycerol, citramalic acid, amino acids (especially proline), histamine, ammonia, succinic acid, phosphate, ash, alkalinity of the ash, ethyl, acetate, methyl anthranilate, total volatile esters, higher alcohols (both total and individually) phenolic compounds, etc. An elegant method for determining ethyl esters, capronate, caprylate, caprinate, and laurate using carbon disulfide extraction and GLC has been published (123).

Quality standards for some flavor constituents will eventually be developed—linaloöl for muscats, for example, and perhaps phenethanol for certain types of wine. Kahn and Conner (124) have published a rapid GLC method for phenethanol. It has been suggested (60) that detection of bacterial activity from the presence and amount of minor bacterial byproducts (arabitol, erythritol, and mannitol) may be useful. Based on GLC determination of carbonyls, esters, and higher alcohols, beers were accurately classified into three categories (125). Anthocyanin content has been determined quantitatively by using molar absorbance values for five anthocyanin pigments (126).

Future Trends

GLC, atomic absorption and mass spectrophotometry, enzymatic, and specific colorimetric procedures seem to be the likely candidates for routine use in the future. Automation will certainly be common. GLC is now used to detect imitation muscat wines (127). Characteristic flavor byproducts of yeasts may be detected and measured. Multiple correlation of the amounts of the more influential major and minor constituents with wine quality is the goal of such research. A simple apparatus for the simultaneous determination of the redox potential (two platinum electrodes), pH, specific conductivity, oxygen, and carbon dioxide (ion-specific electrode) has been devised (128). Molecular oxygen in wines has been determined by several procedures—polarography (129) and GLC being the latest.

Enologists are now using a variety of analytical procedures. It is difficult to predict sources of new methods, but of those currently used the trend seems to be toward rapid, automated procedures, some based on colorimetry and others based on refractometry. GLC is already used widely, but, when coupled with special integrating printout systems, this method becomes even more attractive. High pressure liquid chromatography is useful for compounds not suitable for GLC. It is recommended for separating and accurately measuring phenolic acids and flavanoids in red wines (130). Thin-layer chromatography has also been used in research and, along with rapid scanning techniques, it may find wider application. Ion-specific electrodes also appear to be especially useful. Fluorometric analysis has been used for research but could find wider use if controls develop on certain constituents, such as histamine, where it is a method of choice.

In the future new constituents may be analyzed for in an ecologically oriented society. Only a single paper on the presence of traces of urethane in diethylpyrocarbamate-treated wine was sufficient to ban its use. Under the Delaney amendment the zero-tolerance rule may require many new analytical procedures of very small amounts of certain components of wines derived directly or indirectly from additives. Obviously the future enologist will have to be as sophisticated in the chemical analytical laboratory as in sensory examination.

Literature Cited

1. Dujardin, J., Dujardin, L., Dujardin, R., "Notice sur les instruments de prècision appliqués à l'oenologie," Dujardin-Salleron, Paris, 1928.
2. Amerine, M. A., *J. Ass. Off. Agr. Chem.* (1961) **44**, 380.
3. "Official Methods of Analysis," 11th ed., Association of Official Agricultural Chemist, Washington, 1970.
4. "Recueil des Méthodes Internationales d'Analyse des Vins," Office International de la Vigne et du Vin, Paris, 1962-1972.
5. Amerine, M. A., Ough, C. S., "Wine and Must Analysis," Wiley, New York, 1974. (See also *Encycl. Ind. Chem. Anal.* **18**, in press.)
6. Amerine, M. A., "Laboratory Procedures for Enologists," Associated Students Bookstore, Davis, 1970.
7. Amerine, M. A., *Wine Inst. Technol. Advis. Comm.*, June 14, 1971.
8. Amerine, M. A., Berg, H. W., Cruess, W. V., "The Technology of Wine Making," 3rd ed., Avi, Westport, 1972.
9. Amerine, M. A., Joslyn, M. A., "Table Wines. The Technology of Their Production," University of California, Berkeley, Los Angeles, 1970.
10. Jaulmes, P., Mestres, R., Mandrou, B., *Ann. Fals. Expert. Chim.* (1964) **57**, 119.
11. Joslyn, M. A., "Methods in Food Analysis. Physical, Chemical and Instrumental Methods of Analyses," 2nd ed., Academic, New York, London, 1970.
12. "Uniform Methods of Analyses for Wines and Spirits," American Society of Enologists, Davis, 1972.
12a. Joslyn, M. A., Amerine, M. A., "Dessert, Appetizer and Related Flavored Wines. The Technology of Their Production," University of California, Berkeley, 1964.
13. Franck, R., Junge, C., "Weinanalytik. Untersuchung von Wein und ähnlichen älkoholischen Erzeugnissen sowie von Fruchtsäften," Carl Heymanns Verlag, Koln, 1970.
14. Hennig, K., Jakob, L., "Chemische Untersuchungsmethoden für Weinbereiter und Sussmosthersteller," 6th ed., Verlag Ulmer, Stuttgart, 1972.
15. Hess, D., Koppe, F., in "Handbuch der Lebensmittel Chemie," vol. VII, "Alkoholische Genussmittel," Springer-Verlag, Berlin, 1968.
16. Jaulmes, P., "Analyse des Vins," 2nd ed., Libr. Coulet, Dubois et Poulain, Montpellier, 1951.
17. Kourakou-Dragon, S., "Diethneis Methodoi Analyseos ton Glefkon ke Oinon," Ektyposis Institoutou Georgikis Mechanologias, Athens, 1971.

18. Mori, L., "Metodi Ragionali di Analisi nella Moderna Technica Enologica," 2nd ed., Luigi Scialpi Editore, Rome, 1967.
19. Nilov, V. I., Skurikhin, I. M., "Khimiya Vinodeliya i Kon'yachnogo Proizvodstva," 2nd ed., Pishchepromizdat, Moscow, 1967.
20. Rankine, B. C., "Principles and Pitfalls of Winery Analyses," Gawler Adult Education Centre, Oenology Course for Winemakers, 1961.
21. Rentschler, H., Tanner, H., "Anleitung für die Getränke-Analyse," 6th ed., Eidg. Forschungsanstalt, Wädenswil, 1971.
22. Ribéreau-Gayon, J., Peynaud, E., Sudraud P., Ribéreau-Gayon, P., "Analyse et Contrôle des Vins," Dunod, Paris, 1972.
23. Vogt, E., Bieber, H., "Weinchemie und Weinalyse," 3rd ed., Verlag E. Ulmer, Stuttgart, 1970.
23a. Jakob, L., All. Deut. Weinfachztg (1971) 107, 1163.
24. Rebelein, H., Allg. Deut. Weinfachztg. (1971) 107, 590.
25. Schmitt, A., Allgem. Deut. Weinfachztg. (1971) 107, 1962.
26. Schmitt, A., Deut. Weinbau. (1972) 27, 57.
27. Sarris, J., Morfaux, J. N., Dupuy, P., Hertzog, D., Ind. Aliment. Agr. (1969) 86, 1241.
28. Wanger, O., Mitt. Geb. Lebensmunters. Hyg. (1969) 60, 271.
29. Lie, S., Haukeli, A. D., Gether, J. J., Brygmesteren. (1970) 27, 281.
30. Tanner, H., Brunner, E. M., Mitt. Geb. Lebensmittelunters. Hyg. (1964) 55, 480.
31. Dyer, R. H., J. Ass. Off. Anal. Chem. (1972) 55, 564.
32. Martin, G. E., Caggiano, G., Beck, J. E., J. Ass. Off. Agr. Chem. (1963) 46, 297.
33. Pilone, G. J., Rankine, B. C., Hatcher, C. J., Aust. J. Wine, Brew. Spirit Rev. (1972) 91, 62
34. Jakob, L., Rebe Wein. (1971) 25, 44.
35. Postel, W., Drawert, F., Maccagnan, G., Chem., Mikrobiol., Technol. Lebensm. (1971) 1, 11.
36. Rebelein, H., Steinert, H., Ger. Offen. (1971) 2, 126.
37. Wucherpfennig, K., Bretthauer, G., Wein-Wiss. (1971) 26, 405.
38. Sarris, J., Morfaux, J. N., Dervin, L., Connais. Vigne Vin. (1970) 4, 431.
39. Wucherpfennig, K., Jahresber. Hess. Forschungsanstalt Wein. Obst. Gartenbau. Geisenheim. (1972) 1971, 46.
40. Deibner, L., Bernard, P., Chim. Anal. Paris (1970) 54, 412.
41. Jakob, L., Weinblatt. (1970) 64, 461.
42. Lay, A., Mitt. Rebe Wein, Obst. Fruchteverwert. (1970) 20, 85.
43. Rebelein, H., Mitt.-Bl. GDCH-Fachgr. Lebensmittelchem. gerichtl. Chem. (1969) 23, 107.
44. Tanner, H., Sandoz, M., Schweiz. Z. Obst. Weinbau. (1972) 108, 251.
45. Stanescu, C., Bull. O.I.V. (Off. Intern. Vigne Vin) (1972) 45, 785.
46. Junge, C., Deut. Lebensm. Rundsch. (1970) 66, 421.
47. Lotti, G., Baldacci, P. V., Riv. Viticolt, Enol. (1970) 23, 262.
48. Rebelein, H., Bull. O.I.V. (Off. Int. Vigne Vin) (1968) 41, 344.
49. Rebelein, H., Deut. Lebensm. Rundsch. (1967) 63, 233.
50. Mack, D., Berg, H., Deut. Lebensm. Rundsch. (1972) 68, 262.
51. Varjú, M., Z. Lebensm. Unters. Forsch. (1972) 148, 268.
52. Brunn, S., Bonnemaire, J. P., Rev. Franç. Oenol. (1971) 43 (3), 12.
53. Polo, M. C., Garrido, M. D., Llaguno, C., Garrido, J., Rev. Agroquím. Technol. Aliment. (1969) 9, 600.
54. Raik, S. Ya., Kryzhanovskaya, E. Kh., Sadovod. Vinograd. Vinodel. Mold. (1970) 25 (3), 36.
55. Barros, M. H. B.,v. de, Estud., Notas Relatorios. Porto (1971) 7, 59.
56. Hadorn, H., Zürcher, K., Ragnarson, V., Mitt. Geb. Lebensmittelunters. Hyg. (1967) 58, 1.

57. Hrazdina, G., *J. Agr. Food Chem.* (1970) **18**, 243.
58. Schmidt-Hebbel, H., Michelson, W., Masson, L., Steltzer, H., *Z. Lebensm. Unters. Forsch.* (1968) **137**, 169.
59. Drawert, F., Kupfer, G., *Fresenius Z. anal. Chem.* (1965) **211**, 89.
60. Ribéreau-Gayon, P., Bertrand, A., *Vitis.* (1972) **10**, 318.
61. Ough, C. S., Cooke, G. M., *Wines Vines.* (1966) **27** (8), 27.
62. Parenthoen, A., *Ann. Fals. Expert. Chim.* (1972) **65**, 279.
63. Schopfer, J. F., Regamey, R., *Rev. Suisse Viticult. Arboricult.* (1971) **3**, 107.
64. Rotolo, A., *Riv. Viticolt. Enol.* (1972) **25**, 301.
65. Woidich, H., Gnauer, H., Tunka, J., *Z. Lebensm. Unters. Forsch.* (1971) **147**, 284.
66. Matthey, E., Rentschler, H., Schopfer, J. F., *Mitt. Geb. Lebensmittelunters. Hyg.* (1971) **62**, 101.
67. Lemperle, E., Kerner, E., *Fresenius Z. anal. Chem.* (1969) **247**, 49.
68. Vitali, G., *Ind. Aliment. Pinerolo, Italy* (1970) **9**, 71.
69. Bandion, F., *Mitt. Rebe Wein, Obstbau Fruchteverwert. Klosterneuburg* (1969) **19**, 37.
70. Hara, S., Murakami, H., Marukāwa, K., Omata, Y., Sakakura, I., *J. Ferment. Technol.* (1970) **48**, 616.
71. Wunderlich, H., *J. Ass. Off. Anal. Chem.* (1972) **55**, 557.
72. Hoppe, H., Romminger, K., *Nahrung* (1969) **13**, 227.
73. Bates, B. L., *J. Ass. Off. Anal. Chem.* (1970) **53**, 775.
74. English, E., *Analyst* (1959) **84**, 465.
75. Haller, H. E., Junge, J., *Deut. Lebensm. Rundsch.* (1971) **67**, 231.
76. Guymon, J. F., Ough, C. S., *Amer. J. Enol. Viticult.* (1962) **13**, 40.
77. Wong, G., Caputi, Jr., A., *Amer. J. Enol. Viticult.* (1966) **17**, 174.
78. Owades, J. L., Dono, J. M., *J. Ass. Off. Anal. Chem.* (1968) **51**, 148.
79. Pato, M. A. da S., Pato, M. H. M. L. da S., *De Vinea et Vino Port. Doc.* (1972) **6** (3), 1.
80. Kunkee, R. E., *Wines Vines* (1968) **49** (3), 23.
81. Mayer, K., Busch, I., *Mitt. Geb. Lebensmittelunters. Hyg.* (1963) **54**, 60.
82. Poux, G., *Ann. Technol. Agr.* (1969) **18**, 359.
83. Peynaud, E., Lafon-Lafourcade, S., *Ann. Technol. Agr.* (1965) **14**, 49.
84. Tarantola, C., Castino, M., *Ann. Fac. Sci. Agr. Univ. Torino.* (1962) **1**, 137.
85. Martiniere, P., Sudraud, P., *Connais. Vigne Vin.* (1968) **2**, 41.
86. Hill, G., Caputi, Jr., A., *Amer. J. Enol. Viticult.* (1970) **21**, 153.
87. Rebelein, H., *Mitt.-Bl. GDCH-Fachgr. Lebensmittelchem. gerichtl. Chem.* (1970) **24**, 14.
88. Tanner, H., Sandoz, M., *Schweiz, Z. Obst. Weinbau.* (1972) **108**, 251.
89. Fal'kovich, Y. E., Gun'ko, G. P., Avenes'yants, R. V., *Vinodel. Vinograd. SSSR* (1972) **32** (4), 38.
90. Pilone, G. J., Kunkee, R. E., *Amer. J. Enol. Viticult.* (1970) **21**, 12.
91. Mayer, K., Pause, G., *Mitt. Geb. Lebensmittelunters. Hyg.* (1965) **56**, 454.
92. Mayer, K., Pause, G., *Lebensm.-Wiss. Technol.* (1969) **2**, 143.
93. Addeo, F., *Sci. Technol. Aliment.* (1972) **2**, 87.
94. Rebelein, H., *Deut. Lebensm. Rundsch.* (1967) **63**, 337.
95. Martin, G. E., Sullo, J. G., Schoeneman, R. L., *J. Agr. Food Chem.* (1971) **19**, 995.
96. Guymon, J. F., Wright, D. L., *J. Ass. Off. Anal. Chem.* (1967) **50**, 305.
97. Jaulmes, P., Espezel, P., *Ann. Fals. Fraudes.* (1935) **28**, 325.
98. Jaulmes, P., Hamelle, G., *Ann. Fals. Expert. Chim.* (1961) **54**, 338.
99. Burroughs, L. F., Sparks, A. H., *Analyst (London)* (1961) **86**, 381.
100. Heintze, K., Braun, F., *Z. Lebensm. Unters. Forsch.* (1970) **142**, 40.
101. Rebelein, H., *Deut. Lebensm. Rundsch.* (1970) **66**, 6.

102. Then, R., Radler, F., Z. Lebensm. Unters. Forsch. (1968) 138, 163.
103. Andre, L., Ann. Technol. Agr. (1966) 15, 159.
104. Morrison, R. L., Amer. J. Enol. Viticult. (1962) 13, 159.
105. Ough, C. S., Fong, D., Amerine, M. A., Amer. J. Enol. Viticult. (1972) 23, 1.
106. Mayer, K., Busch, I., Mitt. Geb. Lebensmittelunters. Hyg. (1963) 54, 297.
107. Caputi, Jr., A., Ueda, M., Amer. J. Enol. Viticult. (1967) 18, 66.
108. Cameron, A. G., Hackett, D. R., J. Sci. Food Agr. (1970) 21, 535.
109. Strunk, D. H., Andreasen, A. A., J. Ass. Off. Anal. Chem. (1967) 50, 334, 338.
110. Baraldi, D., Nikolai, K., Stehlik, G., Altmann, H., Kaindl, K., Inst. Biol. Agr. Seibersdorf Reactor Centre (1968) Spr. No. 17.
111. Meredith, M. K., Baldwin, S., Andreasen, A. A., J. Ass. Off. Anal. Chem. (1970) 53, 12.
112. Bergner, K. G., Lang, B., Deut. Lebensm. Rundsch. (1971) 67, 121.
113. Blouin, J., Llorca, L., Leon, P., Connais. Vigne Vin. (1971) 5, 99.
114. Weger, B., Riv. Viticolt. Enol. (1968) 21, 441.
115. Seider, A. I., Datunashvili, E. N., Vinodel. Vinograd. SSSR (1972) 32 (6), 31.
116. Singleton, V. L., Rossi, Jr., J. A., Amer. J. Enol. Viticult. (1965) 16, 144.
117. Singleton, V. L., Esau, P., "Phenolic Substances in Grapes and Wines, and Their Significance," Academic, New York, London, 1969.
118. Ribéreau-Gayon, P., Rev. Franç. Oenol. (1971) 11, 25.
119. Ribéreau-Gayon, P., Chim. Anal. (1970) 52, 627.
120. Plumas, B., Sautier, C., Ann. Fals. Expert. Chim. (1972) 65, 322.
121. Ough, C. S., Berg, H. W., Chichester, C. O., Amer. J. Enol. Viticult. (1962) 13, 32.
122. Berg, H. W., Ough, C. S., Chichester, C. O., J. Food Sci. (1964) 29, 661.
123. Koch, J., Hess, D., Gruss, R., Z. Lebensm. Unters. Fosch. (1971) 147, 207.
124. Kahn, J. H., Conner, H. A., J. Ass. Off. Anal. Chem. (1972) 55, 1155.
125. Postel, W., Drawert, F., Adam, L., Chem., Mikrobiol., Technol. Lebensm. (1972) 1, 169.
126. Niketić-Aleksić, G. K., Hrazdina, G., Lebensm. Wissen. Technol. (1972) 5, 163.
127. Spanyár, P., Kavei, E., Blazovich, M., Fr. Viticole (Montpellier) (1972) 4, 131.
128. Litchev, V., Goranov, N., Albert, H., Yanakiev, M., Bull. O.I.V. (Off. Intern. Vigne Vin.) (1972) 45, 1059.
129. Tsuyboul'kova, L. P., Balanoutse, A. P., Khramov, A. V., Nilov, B. I., Datounaschvili, E. N., Vinodel. Vinograd. SSSR (1972) 32 (4), 30.
130. Charalambous, G., Bruckner, K. J., Hardwick, W. A., Linnebach, A., Master Brew. Ass. Amer. Tech. Quart. (1973) 10, 74.
131. Izard-Verchére, C., Viel, C., Bull. Soc. Chim. Fr. (1972) 5, 2089.

RECEIVED May 29, 1973.

Malo–Lactic Fermentation and Winemaking

RALPH E. KUNKEE

Department of Viticulture and Enology, University of California, Davis, Calif. 95616

Practical and fundamental aspects of malo–lactic fermentation are given. Conditions which winemakers can use for better control of the fermentation, including detailed procedures for inoculation with Leuconostoc oenos *ML 34 and for inhibition with fumaric acid, are presented. New information on the role of malic acid decarboxylation in bacterial metabolism and on the enzymatics of malic acid decarboxylation are reviewed. The malic acid decarboxylation seems to involve two pathways: a direct decarboxylation of malic to lactic acid with NAD as a coenzyme and a concurrent but small oxidative decarboxylation to pyruvic acid and NADH. How these pathways can bring about the marked stimulation of bacterial growth rate by the malo–lactic reaction and their negligible effect on growth yield are discussed.*

Malo–lactic fermentation is a fermentation caused by growth of certain lactic acid bacteria during storage of new wine. Coincidentally with the growth of the bacteria, there is a decarboxylation of malic acid (one of the major acid components of wine) to lactic acid—hence the name. The mechanism of this decarboxylation, or release of carbon dioxide, seems simple; but it is deceptively so. From the winemaker's point of view, the importance in the loss of a carboxyl or acid group from malic acid is the resulting decrease in acidity of the wine. Historically, this aspect of the fermentation has probably been the most important aspect because it was a means of decreasing the acidity in wines made in cool climatic regions. This deacidification is not, however, such an important consideration for the winemaker in California where the wines generally are not acidic. More important for this winemaker are two other aspects of the fermentation: bacteriological stability and flavor complexity.

Malo–lactic fermentation generally manifests itself in the new wine after several months of storage as an effervescence from the escaping carbon dioxide. Sometimes an increase in turbidity from bacterial growth can be noted, and in red wines there is usually a slight decrease in color. Even with a desirable bacterial fermentation, winemakers often notice some unpleasant odors, but these dissipate within a few weeks. Acidity measurements before and after the fermentation will, of course, show a decrease. Further general descriptions of the fermentation may be found in standard enological texts and in the following selected articles: Vaughn and Tchelistcheff (1), Peynaud and Domercq (2), Radler (3), Kunkee (4), Gandini (5), and Rankine (6, 7). Malo–lactic fermentation is also a frequent occurrence in cider (see Ref. 8). For a short history of the research from Pasteur's observation of loss of acidity during wine storage and the findings of the early wine bacteriologists to the recent attempts to understand control of the fermentation, see Ref. 4.

Malo–Lactic Fermentation in the Winery

Geography of Malo–Lactic Fermentation. Malo–lactic fermentation occurs in all wine producing countries (cf. Ref. 4). It is usually thought of as being important in cold climactic areas to decrease the acidity of otherwise unpalatable wines and in warm areas where the pH of the wine is so high that it is difficult to prevent the fermentation. To assess the importance of the fermentation in any specific location, current and complete surveys of the wines of the region must be made. For example, it is often suggested that malo–lactic fermentation is of paramount importance for deacidification of highly acid German white wines although this has not been true for these wines for several years (cf. Refs. 9, 10).

Two other examples showing a high proportion of malo–lactic fermentation in California wines are easily misinterpreted. One example is from a survey of a relatively small wine producing area where the red wines are normally stored for several years before bottling. Consequently a large proportion of them undergo the fermentation (11). In the other example, the survey was of wines from a warm region where attempts generally are not made to inhibit the fermentation (12). A survey made today of all California wines would show that a lower per cent undergo malo–lactic fermentation because a large proportion of the wines are made to have a fresh, young quality. In many of these wines, one would expect the fermentation to have been inhibited by one means or another before bottling.

San Joaquin Valley (Calif.) table wines (wines with less than 14% ethanol) represent about 80% of wine production in California. In two large wineries there, the malo–lactic fermentation is inhibited and does not occur, but in two others it does (*13*). (Of the latter two, fermentation in one was not desired because of loss of acidity; in the other it was desired for increased flavor complexity.) In other smaller wineries the malo–lactic fermentation either occurred, or its occurrence was not determined.

Organisms Involved in Malo–Lactic Fermentation. The organisms which carry out malo–lactic fermentation are all from three genera of lactic acid bacteria: *Lactobacillus, Leuconostoc,* or *Pediococcus.* Older literature cites other genera which are now included in these three. Specific names of organisms which have been isolated from wine and which carry out malo–lactic fermentation have been given (*4, 14, 15*). (Not all strains of each of the species are necessarily capable of malo–lactic fermentation.) *Pediococcus pentosacens* (*16*) and *Lactobacillus trichodes* (cottony mold) should be added to these lists. The latter is a problem in dessert wines (wine with greater than 14% ethanol), but it does not decompose malic acid (*17*). However, we have isolated a strain of this organism which is dependent on ethanol for growth and which also decomposes malic acid (*18*).

A new classification of the *Leuconostocs* has been presented by Garvie (*19, 20*). According to her classification, all the *Leuconostocs* isolated from wine are included under one appellation, *Leuconostoc oenos.* These *Leuconostocs* are distinguished from the others by their tolerance to 10% ethanol and pHs less than four. These bacteria might be subdivided into two groups on the basis of their abilities to ferment pentoses. Peynaud and Domercq (*21*) have named those strains of heterolactic cocci which ferment arabinose or xylose as *L. oinos* [sic] and those which do not ferment these sugars as *L. gracile.* Garvie's new classification has simplified immensely the problem of taxonomy for the wine microbiologist since it has often been difficult to obtain fermentation data needed for older classifications (*22*). In some very fastidious strains of *Leuconostocs,* certain growth factors are required which are found only in some natural products, such as tomato juice, which also contains carbohydrates. Thus it is difficult to obtain a carbohydrate-free basal medium for fermentation tests. This problem may also be solved by the discovery and synthesis of a potent growth factor for these organisms isolated from tomato juice: 4′-*O*-(β-D-glucopyranosyl)-D-pantothenic acid (*23, 24, 25*). Nonomura *et al.* (*26*) have also presented a new scheme for classifying malic acid-decomposing *Leuconostocs* with the separation of species being based on sugar fermentation.

There is no good agreement as to the origin of malo–lactic bacteria in wineries (*cf.* Ref. 4). Apparently microflora of definite individuality can be established in a winery, presumably in used cooperage. We have discussed (*4*) several workers' suggestions for the origin of these microbes from grape skins or from a stray leaf. Others believe they might come from the winery and that the grapes themselves have little to do with it. Schanderl (27) has suggested that microorganisms might originate in a type of spontaneous generation from the grape cellular material itself, but we must await confirmation from other laboratories of this interesting but radical proposal which conflicts with Pasteur's proofs of the origin of living material. A difficulty we have experienced in this kind of investigation is the great superficial morphological similarity between microbes and fragmented grape flesh viewed microscopically (28).

Desirability of Malo–Lactic Fermentation. Three important aspects of malo–lactic fermentation must be considered in appraising its desirability: deacidification, bacteriological stability, and increased flavor complexity (4).

DEACIDIFICATION. The change in acidity may not be as important a part of modern winemaking as it has been in the past. In some localities with high-acid wines, such as Germany, it is now felt that there are better means than malo–lactic fermentation to reduce the acidity. Part of the reason for this idea is the great difficulty in obtaining malo–lactic fermentations in very acid wines. Another important consideration is the flavor change which accompanies the fermentation and perhaps brings about a loss of freshness and increase in lactic acid taste, a taste most likely associated with other end products of fermentation (cf. Ref. 29) rather than lactic acid itself. Grapes of greater natural acidity will be used for winemaking in western North America where colder climatic areas exist and varieties better suited to their climatic requirements will be used along with virus-free plantings. Thus in the future other means for deacidification also may be desirable in California. The modern German winemaker often prefers to bring about a deacidification of the must by the double salt treatment. By the use of one method, part of the must is treated with Acidex [calcium carbonate seeded with the tartrate–malate double salt of calcium (Ch. H. Boehringer Sohn, Ingelheim)]. The researches of Münz (30) and Kielhöfer and Würdig (31, 32) have shown that this treatment brings about precipitation of tartaric and malic acids of the must as the calcium double salt. Proper blending with the untreated must then allows the desired final acidity.

In addition to malo–lactic fermentation, another biological method for deacidification of high-acid must is to use malic acid-metabolizing *Schizosaccharomyces* yeast for the alcoholic fermentation. Benda and Schmidt (33) have selected strains of these yeasts which produce wines with no off-flavors. In using some of these same strains we have also been able to make wines of sound character (18).

In certain wines where the excess acidity is only moderate, the condition may be corrected by adding grape juice as a sweetening agent to bring about a harmonious flavor balance. Naturally, this latter treatment must be done just before bottling, and the wine must be made microbiologically stable to prevent a secondary yeast fermentation in the bottle.

We have been considering only the desirability of deacidification. In regions where the deacidification is not desired and where a malo–lactic fermentation often occurs, the consequences of this biological deacidification are not grave if the fermentation is a clean one. The loss or lack of acidity of the grapes can be adjusted by adding acidulating agents or by ion exchange treatment. In this situation it is probably easier for the winemaker to adjust the acidity, where permitted by law, rather than attempt to inhibit the fermentation. Some secondary effects of the deacidification by malo–lactic fermentation have been given (4).

BACTERIOLOGICAL STABILITY. The bacteriological stability provided by malo–lactic fermentation is its most important attribute. Wines aged before bottling (and which are susceptible to the fermentation) will nearly always be fermented during the first or second year. With proper post-fermentation treatment, these wines can be safely bottled without fear of further bacterial attack. We have seen no instances where a second bacterial fermentation has occurred once the malo–lactic fermentation was completed unless additions had been made to the wines or they had been blended.

Using inhibitory manipulations as outlined below, the attentive winemaker can also bottle, without danger, aged wines which have not had a malo–lactic fermentation. The great difficulty comes with wines to be bottled soon after vintage—at the very time when the malo–lactic fermentation might be expected to occur. The winemaker must make the difficult decision whether to try to inhibit the fermentation or whether to attempt to encourage the complete fermentation before bottling.

FLAVOR COMPLEXITY. A more subtle effect of malo–lactic fermentation is the change in flavor complexity resulting from end products formed by the bacteria. There is a general opinion among many winemakers that flavor changes not related to the change in acidity occur, but this is difficult to evaluate. Wines sampled before and after the fermentation must be adjusted to give identical titratable acidities and pH levels before they can be compared organoleptically. Rankine (7) argues that except for flavors from diacetyl formation, one cannot reliably detect malo–lactic fermentation by taste. We have also found inconsistency among experienced tasters in their capabilities of detecting the fermentation. Nevertheless, some flavor differences do occur. We found some tasters were able to make significant preference rankings of several wines which had undergone malo–lactic fermentation by various different strains of malo–lactic bacteria (34).

The highly flavorable compound diacetyl is an important by-product of lactic acid bacterial fermentation. The mechanism of its formation has recently been unraveled (35). Diacetyl (measured as diacetyl rather than as diacetyl plus acetoin) is present in higher concentrations in wines with malo–lactic fermentation (*cf*. Ref. 36). At approximately threshold levels, this compound might contribute favorably to the flavor of wine (7) since increased complexity has been shown to enhance the quality of wine (37).

Pilone (38) has shown that the increase in volatile acidity which accompanies malo–lactic fermentation is from increased formation of acetic acid with only a small contribution from volatilization of lactic acid. Although it is our contention that the increase in flavor complexity of wine is not an overriding factor in evaluating the desirability of malo–lactic fermentation, the opposite situation of spoilage can certainly be detected and is to be avoided. Besides diacetyl, several other odors are associated with some malo–lactic fermentations—*e.g.*, the odors of Spanish olives, cabbage, and sauerkraut. However, we have seen no case where the conditions of Koch's Postulates (*cf*. Ref. 39) for medical microbiology have been met where an organism isolated from spoiled wine has been used to inoculate another wine to demonstrate that this organism was the causative agent of the spoilage (*cf*. Ref. 40).

Possibilities of Control of Malo–Lactic Fermentation. Although the factors mentioned above concerning the desirability of malo–lactic fermentation will greatly influence the winemaker's decision, of even greater importance is the susceptibility of the wine to the fermentation. It may be well for the winemaker to wish to encourage or discourage the fermentation, but this decision must be based on knowledge of the wine, *i.e.*, its potential for the fermentation. Although one can state with some assurance the qualitative effects of pH, temperature, concentrations of alcohol, free sulfur dioxide, and nutrients on the fermentation rate (*cf*. Ref. 4), the extent of interrelationship of these factors, and the synergisms involved are essentially unknown. However, we feel that a preliminary assessment of susceptibility can be obtained from the pH of the wine. As a general rule, we have found that in a wine with a pH less than 3.3, special efforts must be made to obtain malo–lactic fermentation, and with a pH greater than 3.3, special efforts must be taken to inhibit it. The borderline of pH 3.3 must be considered only in context of California conditions since the relationship between pH and nitrogenous content of the grape at maturity or other factors may vary from location to location. Rice and Mattick (41) reported that malo–lactic fermentation occurred regularly in New York State wines with initial pHs of 3.0 and occasionally in wines below pH 3.0.

Inhibition of Malo–Lactic Fermentation. Some important actions a winemaker may take to inhibit malo–lactic fermentation include early racking at the end of alcoholic fermentation to prevent autolysis of yeast and release of micronutrients; continual surveillance to maintain free sulfur dioxide concentrations at a reasonable level dependent upon pH (perhaps 30 mg/liter); maintenance of storage temperature at less than 18°C; adjustment of acidity to lower the pH to at least below 3.3 or some other empirically established safe pH; and storing the wine in new cooperage or other containers known to be devoid of malo–lactic bacteria. (We know of no way to sanitize successfully wooden cooperage heavily infested with bacteria.) If pH is not lowered by ion exchange, caution should be used in choosing acidulating agents. If the wine is not bacteriologically stable, adding citric acid may lead to inordinate amounts of diacetyl; addition of malic acid (usually available commercially as the DL isomer) may lead to stimulation of growth of bacteria, although the D isomer would be biologically inactive; and fumaric acid is difficult to solubilize. Perhaps tartaric acid addition is the best choice even though much of it will be lost by precipitation as potassium bitartrate. Radler and Yannissis (42) have shown decomposition of tartaric acid by lactic acid bacteria, even to some extent at pH less than 4, but we have seen only one modern report (6) of bacterial spoilage of wine resulting from tartrate decomposition. We have previously suggested (4) that tartaric acid can generally be considered microbiologically inert in wine.

More definite steps to stabilize wine against malo–lactic fermentation include killing the bacteria or removing them from the wine. Although simple pasteurization has been used to prevent bacterial spoilage of wine, we are hesitant to recommend it for modern winemakers because of its detrimental effect on wine quality. However, we would not rule out a future practice of very careful HTST treatment (high temperature, short time—98°C for one second, with rapid cooling). This treatment is claimed to bring about no deterioration of white grape juice, but we are not in a position to comment on its effect on wine. Obviously, bacteria can be removed by sterile filtration, for example, with the use of membrane filters of 0.45 μm. If the post-filtration treatments are also done sterilely, no malo–lactic fermentation can occur. However, with membrane filters which are at present commonly available, the extent of silting of those of 0.45 μm porosity is too great with most wines to be of practical use. Membranes of 0.65 μm would give better flow rates although they would not necessarily remove all bacteria. The sterilizing type of depth filters will also satisfactorily remove bacteria, but as with membrane filters, the wine should first be pre-filtered (43, 44).

Absolute stabilization may not be required. We have stated that the factors which influence the malo–lactic fermentation are qualitative, and stabilization may possibly be realized if all of the factors are at borderline conditions. However, the unfavorable condition of any one factor might support the commencement of reasonable rates of bacterial growth and malo–lactic fermentation.

FUMARIC ACID INHIBITION. Another means of preventing malo–lactic fermentation is to add fumaric acid after alcoholic fermentation is complete (45, 46, 47, 48). The inhibition is relative and its extent is dependent on the amount added. The susceptibility to fumaric acid is also dependent on the strain of malo–lactic bacteria tested (49). However, we know of no case where fumaric acid addition at the levels suggested by Cofran and Meyer (45) (about 0.05%) did not delay malo–lactic fermentation under normal winemaking conditions. This includes several experiments from our pilot winery (50). Nevertheless, we have not been hasty to recommend the use of fumaric acid as an inhibitor because: 1) of the difficulty in solubilizing the acid in wine; 2) we do not know the mechanism of action of its inhibition [Pilone (47, 48) has shown that the bacteria metabolize low levels of fumaric acid to lactic acid but, at inhibitory levels at wine pH, the acid is bactericidal]; and 3) of the desirability of minimizing the use of chemical additives.

Stimulation of Malo–Lactic Fermentation. Winemakers may wish to encourage malo–lactic fermentation to bring about rapid bacteriological stability to wines so that they may either be bottled or stored for aging with no danger of spoiling. One may also wish to bring about a controlled fermentation with a known malo–lactic organism. To encourage malo–lactic fermentation, practices opposite to those for inhibition are used: the first racking off of the yeast lees should be delayed a short time, no addition of sulfur dioxide should be made after dryness is reached, the wine should be stored no lower than 18°–22°C, and no adjustment of acidity (lowering of pH) should be made. Of course each of these practices tend to bring about the conditions which also encourage bacterial spoilage, thus the wine must be faithfully attended. An additional stimulatory measure and one tending to lower the danger of spoilage is to store the wine in cooperage harboring microflora from a previous acceptable malo–lactic fermentation or to induce a rapid and clean fermentation by inoculating the wine with a known strain of malo–lactic bacteria. Peynaud and coworkers (51, 52) also have discussed conditions for encouraging malo–lactic fermentation, including the use of starter cultures. Ardin (53) has given detailed methods for induction of malo–lactic fermentation by use of commercial starter cultures.

The efficiency of using starter cultures is in dispute. Rankine and Pilone (54) have found no consistency in the rate of completion of malo–lactic fermentation in relation to the amount of inoculum used or the time of addition. Indeed we, and others, also have found from time to time a more rapid malo–lactic fermentation in control uninoculated wines than in the inoculated (*18*). These results might be explained by the complete utilization in the inoculated wine of a specific micronutrient (present in low concentrations) needed for growth of the bacteria used as inoculum and thus preventing complete growth of this added organism. At the same time there would be partial utilization of other nonspecific nutrients in this wine. That is, the inoculated wine would become more depleted of nonspecific nutrients than the control wine and less able to support the growth of any wild malo–lactic organisms which may already be present.

To show correlation between occurrence of malo–lactic fermentation and bacterial inoculation, and to show consistency of results among the control lots and inoculated lots, it is imperative that the division of lots be made before alcoholic fermentation. For red wines, great care must be taken to obtain equitable distribution of the crushed grapes in each of the lots. The inconsistency in the results mentioned above may have been caused by variation in amounts of grape skins during the alcoholic fermentation with the accompanying effect on fermentation temperature, sulfur dioxide concentration, pH after pressing, and concentration of nutrients (*55, 56*).

We have found a high degree of correlation between induction of rapid malo–lactic fermentation and bacteria addition in starter cultures when the wines were treated as suggested above to encourage the fermentation. Peynaud and Domercq (*51*) obtained malo–lactic fermentation by inoculating commercial wines of a locality which had not had a natural fermentation for many years.

PROCEDURE FOR BACTERIAL INOCULATION. We are convinced of the potential usefulness of bacterial inoculation, but we do not urge its use unless microbiologically trained personnel and microbiological facilities are available.

Organism. For bacterial inoculation, we advocate the use of *Leuconostoc oenos* ML 34 (UCD Enology No. ML 34). This organism was formerly named *L. citrovorum* ML 34 and is called *L. gracile* Cf 34 by Bordeaux workers (*57*). It was isolated from a red wine from a Napa Valley, Calif. winery by Ingraham and Cooke (*58*). Although this is a very fastidious organism, and it will not grow after serial transfer in defined medium without the factor from tomato juice, it seems to have greater capabilities than other organisms for growth in the hostile environment of wine with its high alcohol and sulfur dioxide concentrations and low pH, storage temperature, and nutrient supply. Further characterization of the organism has been given (*59*).

Preparation of Cultures. *L. oenos* ML 34 can be maintained as a stab culture (with transfer every four months) on a freshly prepared modified

Rogosa-type medium (60) made as follows: 2% Tryptone (Difco), 0.5% yeast extract (Difco), 0.5% peptone (Difco), 0.5% glucose, 0.005% Tween 80 (Nutritional Biochemicals Corp.), and 2% agar (Difco) in a filtered or centrifuged fourfold dilution (with water) of tomato juice (containing no preservatives). The medium is adjusted to pH 5.5 with HCl before adding agar.

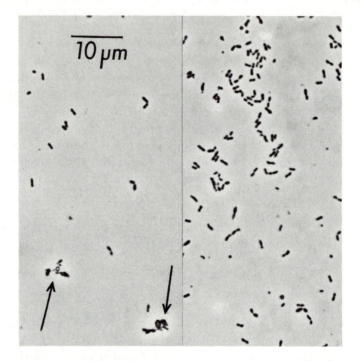

Figure 1. Photomicrographs of Leuconostoc oenos *ML 34 grown on a grape juice medium for use as starter cultures.*

Photographs were made with oil-immersion, phase-contrast optics. Arrows indicate nonbacterial debris from grapes.

A grape juice medium can be made as follows: 1 volume grape juice, 1 volume water, 0.05% yeast extract; titrate to pH 4.5 with NaOH. With some grape juices a supplemental addition of 0.005 to 0.05% Tween 80 is helpful. The media are sterilized by autoclave for 15 min at 15 psi.

Starter cultures of *L. oenos* ML 34 for inoculating wine are obtained by inoculating the grape juice medium with 1 vol % of a subculture (or another starter culture). The subculture is prepared by inoculating 5 ml of the grape juice medium from a stab culture. The cultures are incubated at room temperature until turbidity is seen, approximately one week. [Temperatures greater than 30°C should be avoided; ML 34 can be differentiated from many other strains of *L. oenos* by its slow growth at 37°C (*61*).] The cell concentration at this time can be determined by microscopic count with a Petroff-Hausser counting chamber. The extent of maximal growth of *L. oenos* ML 34 on the grape juice medium is variable depending on the grape juice. The maximal concentration is usually 1–5 $\times$ 10^8 chains of cells per ml. In this stage of growth on this medium, the chains are mostly two cells in length, but some chains of four or more cells are also evident (Figure 1). Because of the great difficulty in differentiating bacteria microscopically from some grape flesh material, it is essential to make the cell counts with a high quality microscope equipped with an oil-immersion objective and phase-contrast optics. Maximum growth can be assumed about two days after the complete disappearance of malic acid (as detected by paper chromatography). Evidence of maximum growth is also obtained by noting the settling of the cells in undisturbed cultures several days later. Viability of *L. oenos* ML 34 in the grape juice medium, as contrasted to wine, is remarkably stable, showing a loss of only about 50% over a two-month period at refrigerator temperature. The nutritional requirements of many strains of malo–lactic bacteria have been published (*16, 62*). Weiller and Radler (*16*) found nicotinic and pantothenic acids required for all strains, and several other common vitamins for many. However, all of these growth factors should be present naturally at sufficient concentrations in grape juice to have little effect on growth in the starter culture.

Bacteria Addition. A convenient time for inoculating red wine is at pressing where the red wine-must is pressed before the end of the alcoholic fermenation. The amount of starter culture to use is about 0.1 vol %. Depending on availability of starter and its bacterial concentration, less might be used (0.01 vol %). There are essentially no data to show what is the best time to add this amount of inoculum (*63*). Bordeaux workers have added the bacteria at the same time as the yeast (*51*), but this practice might lead to excessive end products by metabolism of the high concentration of sugars at this stage (*64*). We suggest the inoculation be made when the wine must is about 5° Brix because at this time there is no inhibitory effect of free sulfur dioxide, and the alcohol concentration will be lower than at dryness. Furthermore, the inoculation is more easily performed after the wine must is pressed. We have had less experience inoculating white wine, but we can state that the malo–lactic fermentation is greatly encouraged by a delay of pressing of up to one day after crushing (*55*).

Malo–lactic strains of bacteria are commercially available from Equilait (38, avenue de la République, 15-Aurillac, France). The cultures come as 1-gram packages of freeze-dried bacteria with 500 ml of growth medium and directions for use. This firm has also published a more detailed description of necessary conditions and procedures for greater assurance of induction by this inoculation (53). We have examined one of their strains of *Lactobacillus* and found it to be a hearty grower, but we have not tested its effectiveness in malo–lactic fermentation.

Post Malo–Lactic Fermentation Treatment. Soon after malo–lactic fermentation is complete, good winery practice demands that the wine receive special cellar treatment even though we have not seen any cases of a second bacterial fermentation. This treatment includes addition of sulfur dioxide, lowering the storage temperature if it had been elevated to encourage the fermentation, and adjusting the acidity where needed. Some winemakers at this time carry out aeration, fining, and high quality filtration. If the wine were to receive sterile filtration and storage treatment, the sulfur dioxide addition for bacteriological stabilization would not be necessary.

Detection of Malo–Lactic Fermentation. It is imperative that the winemaker, to control malo–lactic fermentation, has a satisfactory method for its detection. Disappearance of malic acid is the indication of the fermentation, but the formation of lactic acid is not sufficient evidence since it might also be formed by yeast and by bacteria from other carbohydrate sources. The rate of conversion of malic acid is expected to reflect bacterial metabolism and growth. In New York State wines, Rice and Mattick (41) showed bacterial growth (as measured by viable count) to be more or less exponential to 10^6–10^7 cells/ml, preceding disappearance of malic acid. The rate of loss of malic acid is probably also exponential. Malic acid seems to disappear so slowly that its loss is not detected until a bacterial population of about 10^6–10^7 cells/ml is reached; then it seems to disappear so rapidly that its complete loss is detected within a few days (41). Rice and Mattick (41) also showed a slight increase in bacterial population for a few days following this.

An easy procedure for detecting malic acid is a simple paper chromatographic method which is a combination of several previously described procedures (65): chromatographic grade filter paper, such as S&S 2043b (Schleicher & Schuell, Keene, N.H.) or Whatman 1 Chr is cut into 20 $\times$ 30 cm rectangles. The wines are spotted on a pencil line approximately 2.5 cm parallel to the long edge. The spots are placed along this line about 2.5 cm apart. Each spot is made four times (and allowed to dry in between) from micropipettes 1.2 $\times$ 75 mm to give approximately 10 μl volume. Spotting may also be made from Pasteur pipettes, the spots being somewhat larger but still satisfactory. A cylinder is formed from the paper by stapling the short ends. Care should be taken not to overlap the edges. Wide-mouth, one-gallon mayonnaise jars serve well as chromatography jars: 14 cm in diameter and 25 cm in height. The solvent is prepared by shaking in a separatory funnel 100 ml H_2O, 100 ml n-butyl alcohol, 10.7 ml. concentrated formic acid, and 15 ml 1% bromocresol green (water soluble: Matheson, Coleman, and Bell). After about 20 min, the lower (aqueous) phase is drawn off and discarded. Seventy ml of the upper layer are placed in the jar, the chromatogram is inserted (the spotted edge down), and the jar lid is attached. The developing time is approximately 6 hrs; this may be safely extended to overnight even if the solvent reaches the upper edge. After the yellow chromatogram is removed, it should be stored in a well ventilated area until dry and until the formic acid has volutilized, leaving a blue-green background with yellow spots of acids having the approximate Rf values: tartaric acid 0.28, citric acid 0.45, malic acid 0.51, lactic and succinic acids 0.78, and fumaric acid 0.91. Standard solutions (0.2%) of these acids should also be run as controls. The solvent may be used repeatedly if care is taken to remove any aqueous layer which may have separated after each run.

For faster results, thin layer chromatography has been used (*66*), but we are not confident that lower levels of malic acid (0.06%) (*cf.* Ref. *11*), sometimes found in California wine before malo–lactic fermentation, are easily detected by this means. Malic acid can be more precisely measured by using the quantitative enzymatic method (*67*). Only the L isomer, the natural form present in grapes and wine, is detected by this method.

Measuring the deacidification does not seem to be a reliable means for detecting malo–lactic fermentation, even though the change in titratable acidity is theoretically equal to one-half the amount of malic acid originally present. The change in acidity is complicated by the possibility of lactic acid formation by the bacteria from other carbohydrate sources [even though this might be small (*41*)] and by loss of acidity by precipitation of potassium bitartrate (*4*). Changes in pH resulting from malo–lactic fermentation can be as great as 0.5 unit (*41*), nonetheless pH change is not a reliable means of following fermentation. Under California conditions the change might be quite small (less than 0.1 unit), and the extent of pH change is dependent on the initial pH (*68*). Wejnar (*69, 70*) has made a thorough study of the interactions between the acid and salt forms of tartaric and malic acids and the effect of malo–lactic fermentation on them. He has calculated that, because of the differences between the dissociation constants of the two acids and of the low solubility of potassium bitartrate, it is theoretically possible to obtain a decrease in pH resulting from malo–lactic fermentation (*70*).

An increase in bacterial population would indicate a fermentation of some sort. However, measuring bacteria by plate count is cumbersome, and direct microscopic count in wine is difficult because of the similar appearances of bacteria and grape debris.

Malo–Lactic Fermentation in the Laboratory

The role that the malo–lactic reaction—the decarboxylation of malic acid—plays in the overall physiology of the cell, and the enzymatic pathway of the reaction are two questions which have plagued enologists since the discovery of the reaction. Morenzoni (71) has described this discovery by Ochoa and coworkers (72, 73, 74), and he has related how their early representation of the reaction led to some confusion. We are now convinced, from studies by Radler and coworkers (75, 76) with partly purified enzymes from *Lactobacillus* and *Leuconostoc,* that the reaction is catalyzed by an inducible malate carboxy lyase with NAD (nicotinamide adenine dinucleotide) and Mn^{2+} as cofactors:

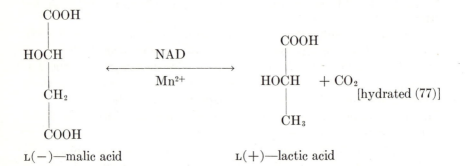

$$L(-)\text{—malic acid} \qquad\qquad L(+)\text{—lactic acid}$$

The reaction is shown as reversible because calculations of the thermodynamics show the release of free energy to be only -7.1 to -2.6 kcal/mole (4, 77). The actual reversibility of the reaction with this enzyme has not been shown. The activity is generally measured as carbon dioxide formation in a respirometer in the presence of L-malic acid and the cofactors. Lonvaud and Ribéreau-Gayon (78) have simplified the method with the use of a carbon dioxide specific electrode.

The free energy released from the above reaction would not be biochemically available to the organism in the reaction as shown; there is no change in the redox state of the coenzyme NAD, and there is no provision for formation of high energy phosphate. Results from carefully collected data show essentially no influence of malic acid on cell yield. Pilone and Kunkee (59) found a molar growth yield for malic acid of 1.34 grams dry weight/mole L-malic acid added in complex medium with no added carbohydrate. Furthermore, there was no statistically significant effect of malic acid on molar growth yield for glucose: we found Y(glucose) yields of 9 to 10 with *L. oenos* ML 34, and Kandler *et al.* (79) found 10.5 in *L. mesenteroides* 39—with and without malic acid.

A slight stimulatory effect on cell yield might have evolutionary significance in establishing primeval strains; however, from gross observation, it is not very interesting. Of greater interest is a rather spectacular effect of malic acid on end product formation.

Role of the Malo–Lactic Reaction. For the *Leuconostocs*, D-lactic acid is formed from glucose, but L-lactic acid is formed from L-malic acid. Looking at end product formation, several workers have noted an increase in formation of D-lactic acid in the presence of malic acid— where malic acid itself was very nearly stoichiometrically converted to L-lactic acid. An early stimulation of lactic acid production (*80*) was found in several malo–lactic bacteria after a stimulated initial use of glucose (*81*) in the presence of malic acid. Doelle (*82*) and Kandler *et al.* (*79*) found increases in D-lactic acid in the presence of malic acid in *L. mesenteroides;* however, the latter authors attributed the increased formation to the smaller decrease in pH allowed by the added malic acid. We (*59*) found (in *L. oenos* ML 34) the formation of D-lactic acid in the presence of malic acid (in a complex medium with limited glucose) to be approximately three times that theoretically expected from glucose for a heterolactic fermentation. It seems that this effect did not result from any accompanying pH change. These experiments were carried out at pH 5.5, the optimal pH of growth (*47*). Without added malic acid, the final pH dropped to 5.3; with malic acid, the final pH rose to 6.1—both final pHs being away from the optimal (*59*).

STIMULATION OF GROWTH RATE. The striking effect which malic acid has on end product formation led us to examine more closely its effect on growth. Instead of measuring growth yields, we looked at growth rates. Other workers have found stimulation of growth rate but this has been attributed to the more favorable pH which accompanied the decarboxylation of malic acid (*3*). By taking measurements at the optimal pH of the bacteria, any deviation from this pH because of malo–lactic fermentation would give a decreased growth rate. Nevertheless, even at optimal pH, a great increase in specific growth rate in the presence of malic acid was found: 0.18 hr^{-1} to 0.26 hr^{-1} for *L. oenos* ML 34 (*47, 83*).

At the low pH of wine the stimulation was also noteworthy (Figure 2). In this case, the utilization of malic acid is brought about by an increase in pH of the medium from 3.65 to 3.77, yet the growth was faster than in the control experiment with an initial pH of 3.85. The stimulation in cell yield, however, seems to result from a pH effect; it is found in both the sample with malic acid and in the sample with increased pH (*see* Figure 2). Thus, we feel confident in answering the first question we raised. For the first time, a biological function for the malo–lactic reaction has been shown—it stimulates the growth rate. This brings up the question of how the malic acid utilization stimulates the growth rate.

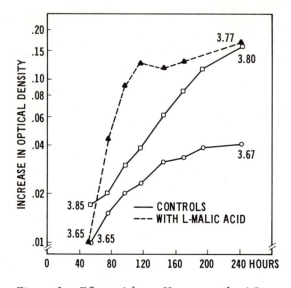

Figure 2. Effect of low pH on growth of Leu-conostoc oenos ML 34 without malic acid (○, □) and with 0.2% L-malic acid (▲).

One optical density unit change is equivalent to 311.5 μg/ml cellular dry weight. Initial and final pHs are given at the beginning and end of each curve.

Enzymatic Pathway of Conversion of Malic Acid to Lactic Acid. If the total reaction is a simple decarboxylation of malic acid to lactic acid, it is difficult to understand how the utilization of malic acid can influence cell growth. Carbon dioxide also has been shown to stimulate growth of *L. oenos* ML 34 (*84*), yet it would seem the carbon dioxide formed from the reaction would be superfluous in new wine already saturated with it. In the original conception of the pathway, the reaction was considered as two stages with pyruvic acid and reduced NAD (NADH) as intermediates (*72*). Morenzoni (*71*) discussed the research and thinking (*4, 85, 86, 87*) which lead to the belief that there was only one step in the reaction and no formation of NADH or pyruvic acid as intermediates. However, any evidence to show a real formation of pyruvic acid would help explain the stimulatory effect of the reaction on cellular metabolism.

EVIDENCE OF TWO ENZYMATIC ACTIVITIES. We found no formation of NADH spectrophotometrically during decomposition of malic acid by cell-free extracts of *L. oenos* ML 34. However, using a more sensitive fluorometric technique to search for NADH (*88*), we found a small, continual formation of NADH (about 5 nmoles NADH/min/mg extract protein) as an end product (*89, 90*). Thus, in addition to the decarboxyla-tion of malic acid to lactic acid, there is another reaction that gives

NADH as an end product. This implicates oxidized malic acid, either pyruvic or oxaloacetic acid, as another end product. By adding commercial preparations of L-lactic dehydrogenase or malic dehydrogenase to the reaction mixture, Morenzoni (90) concluded that the end product was pyruvic acid. Attempts were then made to show whether two enzymes—malate carboxy lyase and the classic "malic" enzyme, malate oxidoreductase (decarboxylating), were involved or if the two activities were on the same enzyme. The preponderance of evidence indicated that only one enzyme is involved. This evidence came from temperature inactivation studies, heavy-metal inhibition studies, and ratio measurements of the two activities of partially purified preparations of Schütz and Radler's malo-lactic enzyme (76, 90). This is not the first case of a single enzyme having two different activities (91).

NADH-FORMING ACTIVITY. The NADH-forming activity (from malic acid) of extracts of *L. oenos* ML 34 was very low, less than 1% of the main malic decarboxylation reaction. This is consistent with the results of the first studies of the reaction where only 0.2% of malic acid was converted to pyruvic acid (73). The formation of NADH is also consistent with the recent work of Alizade and Simon (92) with radioactive substrates. They found some of the hydrogen of malic acid in the D-lactic acid end product. This substantiates the intermediate formation of NADH. Although the NADH producing activity of the enzyme is low, at least *in vitro*, perhaps it is important enough to effect the metabolism of the cell—more so than the malic-to-lactic decarboxylation reaction. Stamer and Stoyla (84) and Meyrath and Lüthi (93) have shown that the presence of hydrogen acceptors, *e.g.*, fructose, stimulate the early growth of *Leuconostocs*. Thus, the pyruvic acid formed in the above reaction might be used to form acetyl phosphate which could in turn act either as a source of high energy phosphate or as a hydrogen acceptor. A similar situation is found in *Streptococcus diacetilactis* where the growth rate is stimulated by citric acid because of a rapid formation of acetyl coenzyme A from the pyruvic acid formed in the metabolism (94).

These speculations of the intermediary metabolism of malic acid are consistent with the physiology of *L. oenos* ML 34. The small amount of pyruvic acid formed as end product can account for the stimulation of growth rate, yet the amount of high energy formed, if any, would be too small to bring about an increase in growth yield.

The NADH-forming activity described here is different from the classical malic enzyme activity found by London *et al.* (95) in *Lactobacillus casei*. In their system, NADH is a major end product and detectable by spectrophotometry while lactic acid is only a minor product. *L. casei* uses malic acid as an energy source with carbon dioxide, acetate, and ethanol as the main fermentation products. The optimal pH of this malic enzyme is about 8, but the malic acid decomposing activities in *L. oenos* ML 34 are optimum at pH 5.6.

For discussion of three aspects of malo–lactic fermentation not presented here, *see* Refs. 4 and 96. These are: the relationship between malo–lactic fermentation and the yeast strain used in the alcoholic fermentation; inducibility of the malic decomposing enzyme(s); and the role of oxaloacetic acid in the pathway of conversion of malic acid to lactic acid (*cf.* Ref. 76).

Conclusion

To control malo–lactic fermentation, the modern winemaker should first consider the susceptibility of the wine to this bacterial attack and expected storage time of the wine before it is to be bottled. Of less importance in the decision to attempt to control the fermentation are the changes in acidity and flavor brought about by the fermentation. Standard winemaking techniques can be applied to encourage or inhibit the fermentation with a reasonable degree of success. Helpful in implementing this decision is the process of inoculation with bacterial cultures such as *Leuconostoc oenos* ML 34 as well as the new information regarding stimulation by malic acid and inhibition by fumaric acid. Other new fundamental information from several laboratories helps answer some perplexing questions as to the role of the malo–lactic fermentation in bacterial metabolism and in the enzymatic pathway of the conversion of malic acid to lactic acid. The major pathway seems to be a direct decarboxylation to lactic acid with concomitant oxidative decarboxylation to form a small amount of pyruvic acid, both activities being apparently on the same enzyme. The second activity of the enzyme can explain the practically negligible increase in growth yield since only small amounts of pyruvic acid are formed. However, since pyruvic acid can serve as a hydrogen acceptor, its formation could explain the marked stimulation of growth rate which occurs in the presence of malic acid.

Literature Cited

1. Vaughn, R. H., Tchelistcheff, A., *Amer. J. Enol.* (1957) **8**, 74–79.
2. Peynaud, E., Domercq, S., *Ann. Technol. Agr.* (1961) **10**, 43–60.
3. Radler, F., *Zentr. Bakteriol. Parasitenk. Abt. II* (1966) **120**, 237–287.
4. Kunkee, R. E., *Advan. Appl. Microbiol.* (1967) **9**, 235–279.
5. Gandini, A., *Vini d'Italia* (1969) **11**, 125–134, 227–233.
6. Rankine, B. C., Fornachon, J. C. M., Bridson, D. A., Cellier, K. M., *J. Sci. Food Agr.* (1970) **21**, 471–476.
7. Rankine, B. C., *Amer. J. Enol. Viticult.* (1972) **23**, 152–158.
8. Beech, F. W., *Rept. Long Ashton Res. Stn. for 1971* (1972) 166–197.
9. Radler, F., *Z. Lebensm. Untersuch. Forsch.* (1968) **138**, 35–39.
10. Radler, F., *Wein Wiss.* (1970) **25**, 418–424.

11. Ingraham, J. L., Cooke, G. M., *Amer. J. Enol. Viticult.* (1960) **11**, 160–163.
12. Kunkee, R. E., Pilone, G. J., Combs, R. E., *Amer. J. Enol. Viticult.* (1965) **16**, 219–223.
13. Cooke, G. M., private communication.
14. Radler, F., *Vitis* (1962) **3**, 144–176.
15. Peynaud, E., Lafon-Lafourcade, S., Guimberteau, G., *Rev. Ferment. Ind. Aliment.* (1967) **22**, 61–66.
16. Weiller, H. G., Radler, F., *Mitt. (Klosterneuburg) Rebe Wein, Ser. A* (1972) **22**, 4–18.
17. Fornachon, J. C. M., Douglas, H. C., Vaughn, R. H., *Hilgardia* (1949) **19**, 129–132.
18. Kunkee, R. E., unpublished data.
19. Garvie, E. I., *J. Gen. Microbiol.* (1967) **48**, 431–438.
20. Garvie, E. I., *J. Gen. Microbiol.* (1967) **48**, 439–447.
21. Peynaud, E., Domercq, S., *Ann. Inst. Pasteur-Lille* (1968) **19**, 159–169.
22. Breed, R. S., Murray, E. G. D., Smith, N. R., "Bergey's Manual of Determinative Bacteriology," 7th ed., Williams & Wilkins, Baltimore, 1957.
23. Amachi, T., Yoshizumi, H., *Agr. Biol. Chem.* (1969) **33**, 139–146.
24. Amachi, T., Imamoto, S., Yoshizumi, H., *Agr. Biol. Chem.* (1971) **35**, 1222–1230.
25. Imamoto, S., Amachi, T., Yoshizumi, H., *Agr. Biol. Chem.* (1973) **37**, 545–551.
26. Nonomura, H., Yamazaki, T., Ohara, Y., *Mitt. (Klosterneuburg) Rebe Wein, Ser. A* (1965) **15**, 241–254.
27. Schanderl, H., *Weinberg Keller* (1969) **16**, 279–298.
28. Nelson, W. G., Kunkee, R. E., unpublished data.
29. Dittrich, H. H., Kerner, E., *Wein Wiss.* (1964) **19**, 528–535.
30. Münz, T., *Weinberg Keller* (1960) **7**, 239–247.
31. Kielhöfer, E., Würdig, G., *Duet. Wein. Zt.* (1963) **99**, 1022–1028.
32. Kielhöfer, E., Würdig, G., *Wein Wiss.* (1964) **19**, 159–168.
33. Benda, I., Schmidt, A., *Weinburg Keller* (1969) **16**, 71–83.
34. Pilone, G. J., Kunkee, R. E., *Amer. J. Enol. Viticult.* (1965) **16**, 224–230.
35. Collins, E. B., *J. Dairy Sci.* (1972) **55**, 1022–1028.
36. Rankine, B. C., Fornachon, J. C. M., Bridson, D. A., *Vitis* (1969) **8**, 129–134.
37. Singleton, V. L., Ough, C. S., *J. Food Sci.* (1962) **27**, 189–196.
38. Pilone, G. J., *Amer. J. Enol. Viticult.* (1967) **18**, 149–156.
39. Stanier, R. Y., Doudoroff, M., Adelberg, E. H., "The Microbial World," 3rd ed., Prentice-Hall, Englewood Cliffs, 1970.
40. Lüthi, H., *Amer. J. Enol.* (1957) **8**, 176–181.
41. Rice, A. C., Mattick, L. R., *Amer. J. Enol. Viticult.* (1970) **21**, 145–152.
42. Radler, F., Yannissis, C., *Arch. Mikrobiol.* (1972) **82**, 219–239.
43. Neradt, F., private communication.
44. Neradt, F., *Weinberg Keller* (1970) **17**, 403–412.
45. Cofran, D. R., Meyer, J., *Amer. J. Enol. Viticult.* (1970) **21**, 189–192.
46. Tchlistcheff, A., Peterson, R. G., van Gelderen, M., *Amer. J. Enol. Viticult.* (1971) **22**, 1–5.
47. Pilone, G. J., Ph.D. Thesis, University of California, Davis, 1971.
48. Pilone, G. J., Long Ashton Symp., 4th, 1973, "Lactic Acid Bacteria in Foods and Beverages," in press.
49. Radler, F., private communication.
50. Ough, C. S., Kunkee, R. E., unpublished data.
51. Peynaud, E., Domercq, S., *C.R. Acad. Agr. Fr.* (1959) **45**, 355–358.
52. Domercq, S., Sudraud, P., Cassignard, R., *C.R. Congr. Soc. Saventes Paris Dept., Sect. Sci.* (1960) 1959, 239–245.
53. Ardin, F., *Rev. Fr. Oenol.* (1972) no. 46, 66–68.
54. Rankine, B., Pilone, G. J., private communication.

55. Kunkee, R. E., *Amer. J. Enol. Viticult.* (1967) **18**, 71–77.
56. Beelman, R. B., Gallander, J. F., *Amer. J. Enol. Viticult.* (1970) **21**, 193–200.
57. Peynaud, E., private communication.
58. Ingraham, J. L., Vaughn, R. H., Cooke, G. M., *Amer. J. Enol. Viticult.* (1960) **11**, 1–4.
59. Pilone, G. J., Kunkee, R. E., *Amer. J. Enol. Viticult.* (1972) **23**, 61–70.
60. Rogosa, M., Wiseman, R. F., Mitchell, J. A., Disraely, M. N., Beaman, A. J., *J. Bacteriol.* (1953) **65**, 681–699.
61. Pilone, G. P., M.S. Thesis, University of California, Davis, 1965.
62. Peynaud, E., Lafon-Lafourcade, S., Domercq, S., *Bull. Off. Int. Vin* (1965) **38**, 945–958.
63. Kunkee, R. E., Ough, C. S., Amerine, M. A., *Amer. J. Enol. Viticult.* (1965) **15**, 178–183.
64. Barre, B., *Rev. Fr. Oenol.* (1972) no. 45, 37–41.
65. Kunkee, R. E., *Wines Vines* (1968) **49** (3), 23–24.
66. Tanner, H., Sandoz, M., *Schweiz. Z. Obst. Wein.* (1972) **108**, 182–186.
67. Mayer, K., Busch, J., *Mitt. Geb. Lebensmittelunters. Hyg.* (1963) **54**, 60–65.
68. Bousbouras, G. E., Kunkee, R. E., *Amer. J. Enol. Viticult.* (1971) **22**, 121–126.
69. Wejnar, R., *Mitt. (Klosterneuberg) Rebe Wein, Ser.* A (1969) **19**, 193–200.
70. Wejnar, R., *Mitt. (Klosterneuberg) Rebe Wein, Ser.* A (1972) **22**, 19–37.
71. Morenzoni, R. A., ADVAN. CHEM. SER. (1974) **137**, 171.
72. Korkes, S., Ochoa, S., *J. Biol. Chem.* (1948) **176**, 463–464.
73. Korkes, S., del Campillo, A., Ochoa, S., *J. Biol. Chem.* (1950) **187**, 891–905.
74. Kaufman, S., Korkes, S., del Campillo, A., *J. Biol. Chem.* (1951) **192**, 301–312.
75. Radler, F., Schütz, M., Doelle, H. W., *Naturwissenschaften* (1970) **12**, 672.
76. Schütz, M., Radler, F., *Arch. Mikrobiol.* (1973) **91**, 183–202.
77. Pilone, G. J., Kunkee, R. E., *J. Bacteriol.* (1970) **103**, 404–409.
78. Lonvaud, M., Ribéreau-Gayon, P., *C.R. Acad. Sci. Ser. D* (1973) **276**, 2329–2331.
79. Kandler, O., Winter, J., Stetter, K. O., *Arch. Mikrobiol.* (1973) **90**, 65–75.
80. Flesch, P., private communication.
81. Flesch, P., *Arch. Mikrobiol.* (1968) **60**, 285–302.
82. Doelle, H. W., *J. Bacteriol.* (1971) **108**, 1290–1295.
83. Pilone, G. J., Kunkee, R. E., unpublished data.
84. Stamer, J. R., Stoyla, B. O., *Appl. Microbiol.* (1970) **20**, 672–676.
85. Brechot, P. J., Chauvet, L., Croson, M., Irmann, R., *C.R. Acad. Sci. Ser. C* (1966) **262**, 1605–1607.
86. Peynaud, E., Lafon-Lafourcade, S., Guimberteau, G., *Amer. J. Enol. Viticult.* (1966) **17**, 302–307.
87. Peynaud, E., Lafon-Lafourcade, S., Guimberteau, G., *Mitt. (Klosterneuburg) Rebe Wein, Ser.* A (1968) **18**, 343–348.
88. Storey, B., private communication.
89. Morenzoni, R. A., Kunkee, R. E., unpublished data.
90. Morenzoni, R. A., Ph.D. Thesis, University of California, Davis, 1973.
91. Houston, L. L., *J. Biol. Chem.* (1973) **248**, 4144–4149.
92. Alizade, M. A., Simon, H., *Z. Physiol. Chem.* (1973) **354**, 163–168.
93. Meyrath, J., Lüthi, H. R., *Lebensm. Wiss. Technol.* (1969) **2**, 21–27.
94. Collins, E. B., Bruhn, J. C., *J. Bacteriol.* (1970) **103**, 541–546.
95. London, J., Meyer, E. Y., Kulczyk, S. R., *J. Bacteriol.* (1971) **108**, 196–201.
96. Amerine, M. A., Kunkee, R. E., *Ann. Rev. Microbiol.* (1968) **22**, 323–358.

RECEIVED August 20, 1973.

8

The Enzymology of Malo–Lactic Fermentation

RICHARD MORENZONI[1]

Department of Viticulture and Enology, University of California,
Davis, California 95616

The literature concerning malo–lactic fermentation—bacterial conversion of L-malic acid to L-lactic acid and carbon dioxide in wine—is reviewed, and the current concept of its mechanism is presented. The previously accepted mechanism of this reaction was proposed from work performed a number of years ago; subsequently, several workers have presented data which tend to discount it. Currently, it is believed that during malo–lactic fermentation, the major portion of malic acid is directly decarboxylated to lactic acid while a small amount of pyruvic acid (and reduced coenzyme) is formed as an end product, rather than as an intermediate. It is suspected that this small amount of pyruvic acid has extremely important consequences on the intermediary metabolism of the bacteria.

Malo–lactic fermentation can be defined as the bacterial conversion of L-malic acid to L-lactic acid and carbon dioxide during storage of new wine. Malic acid is dicarboxylic, but lactic acid is monocarboxylic; therefore, the net result of malo–lactic fermentation in wine, aside from the production of carbon dioxide, is a loss in total acidity. In commercial practice, this fermentation is not well understood, and better methods of controlling it are sought.

Considering the malo–lactic fermentation microbiologically, several factors are apparent. For example, the enzyme cofactor nicotinamide-adenine dinucleotide (NAD) is required for completion of the reaction, although there is no net oxidation–reduction change in proceeding from L-malic acid to L-lactic acid. Classically, the involvement of NAD in an

[1] Present address: E. & J. Gallo Winery, Modesto, California 95353.

171

enzymatic reaction indicated that an oxidation–reduction reaction was occurring and, because the malic acid–lactic acid reaction showed no change in redox state, it was inferred that there must be an intermediate between malic acid and lactic acid. By considering the structures of malic and lactic acids, it was felt that the intermediate must be pyruvic acid although its presence was never proved. This concept has led to a large degree of confusion. The pertinent literature on the enzymology of malo–lactic fermentation will be reviewed here, and recent evidence will be cited which supports the current concepts of its pathway.

History of the Malo–Lactic Fermentation

Kunkee (1) has reviewed the history of malo–lactic fermentation. The first observations of a loss in total acidity in wines which was higher than could be accounted for by known mechanisms was recorded as early as 1864 by Berthelot and de Fleurieu (2). Alfred Koch, in the early 1900's, was able to isolate malo–lactic bacteria and induce malo–lactic fermentation by reinoculation of the organisms. Subsequently, Möslinger (3) in 1901 first presented an overall equation describing malo–lactic fermentation as the conversion of malic acid to lactic acid and carbon dioxide. In 1913, Müller-Thurgau and Osterwalder (4) published a description of bacteria which were able to carry out malo–lactic fermentation along with a taxonomic key for their classification. In 1943, Cruess (5) published a review of microorganisms involved in winemaking in which he noted that some bacteria indigenous to wine were capable of utilizing malic acid; however, it was not until the late 1940's that the biochemistry of the malo–lactic fermentation first came under serious scrutiny in work initiated by Ochoa and his co-workers, Korkes, del Campillo, Blanchard, and Kaufman. Although their work is not directly concerned with malo–lactic fermentation per se, it has been taken as the accepted basic reference in this field. Their work will be discussed in detail, and, in order to interpret some of the findings, reference will be made to recent work (6, 7, 8). Malo–lactic fermentation has been studied by several investigators, and the more recent work has been reviewed thoroughly by Schanderl (9), Radler (10), and Kunkee (1).

Importance of Malo–Lactic Fermentation to the Organism

In the Korkes and Ochoa (11) mechanism proposed for the malo–lactic reaction (see top of next page), pyruvic acid is either a short-lived, fleeting intermediate, or it is bound to malic enzyme so that as soon as it is formed by the enzyme, it is converted to lactic acid by lactate dehydrogenase. [Both malic enzyme ("malic") and malate dehydrogenase (de-

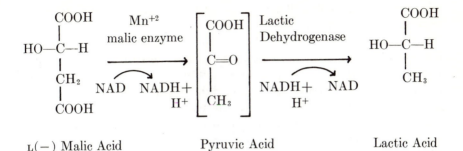

L(−) Malic Acid	Pyruvic Acid	Lactic Acid

carboxylating) are trivial names for the enzyme described as L-malate: NAD oxidoreductase (decarboxylating) E.C. 1.1.1.38 (*12*)].

According to this mechanism the malo–lactic reaction, if stoichiometrically complete, cannot yield any net amount of reduced cofactor (NADH) because in proceeding from pyruvic acid to lactic acid, the NADH generated in the preceding step would immediately be reoxidized to NAD. This poses a problem because energetically the organism would derive no benefit from the fermentation. Concerning the energetics of the reaction, Kunkee (*1*) has calculated the conversion of malic acid to lactic acid and anhydrous carbon dioxide yields a standard free-energy of −6 kcal/mole. Pilone and Kunkee (*13*) subsequently reported that anhydrous carbon dioxide is not produced directly by malo–lactic fermentation; instead one of its hydrated species is formed. That is, carbonic acid, bicarbonate ion, or carbonate ion is the primary decarboxylation product, and subsequent chemical dehydration at low pH produces anhydrous carbon dioxide which is observed as gas evolution. Their calculations showed that if carbonic acid or bicarbonate ion were produced, the standard free-energy would be −6.2 and −7.1 kcal/mole, respectively. If carbonate ion were produced, the standard free-energy change decreases to −2.6 kcal/mole. Thermodynamically, then, the malo–lactic fermentation is capable of producing some energy, but the reaction for converting malic acid to lactic acid does not yield enough energy for the production of ATP.

The possibility now arises that if, in fact, there is an intermediate involved in the conversion of malic acid to lactic acid, the cell may, in some way, be capable of deriving energy from it. In 1950, Korkes *et al.* (*14*), working with the malo–lactic system of *Lactobacillus plantarum*, demonstrated the production of a very small amount of pyruvic acid (0.2%) from malic acid. However, 98% of the malic acid was recovered in lactic acid, and the recovery of carbon dioxide was consistent with conversion of 90% of the malic acid. The pyruvic acid recovery was attributed to spillage from the enzyme surface. We will see below that a small amount of pyruvic acid and NADH are indeed produced during malo–lactic fermentation but not by spillage, and that they play an important part in the metabolism of the malo–lactic bacteria.

When considering the mechanism of the malo–lactic fermentation, the possibility that malic acid may be converted first to oxaloacetic acid (by malic dehydrogenase) must be recognized. This acid could then be decarboxylated to pyruvic acid, and subsequent reaction would yield lactic acid. However, if this were the case, there then should be no situation where malic acid would be decarboxylated faster than oxaloacetic acid. This, however, was shown to occur at pH 6 (14). Similarly, Flesch and Holbach (15) report that malic dehydrogenase has an optimal pH of 10, but that the malo–lactic reaction proceeds at pH 5.6. Therefore, it would not seem likely that the cell would degrade malic acid by this mechanism; hence, the oxaloacetic acid intermediate would not be available to the organism.

What we now have is confusing at best. We have seen that from the biochemical as well as the energetic point of view, there is no logical reason for the malo–lactic fermentation to proceed. Nevertheless, Pilone (16) and Pilone and Kunkee (17) have shown that as a result of malo–lactic fermentation, cell yield is slightly increased and growth rate is greatly stimulated. What, then, can be the explanation for this? It must be either that something (in addition to lactic acid) is being produced which the organism is able to use for growth or that malic acid is toxic to the cell, and malo–lactic fermentation represents a detoxification mechanism. Such a detoxification mechanism has been reported for acetoin formation from excess pyruvic acid by *Streptococcus* (18).

Meyrath and Lüthi (19) reported enhanced growth of leuconostocs in the presence of citric, pyruvic, and oxaloacetic acids, compounds which probably function as hydrogen acceptors. Similarly, Stamer and Stoyla (20) reported a fructose stimulation of the growth of *Leuconostoc citrovorum (oenos)* ML 34, and they suggest that this carbohydrate may serve as an essential reductant or catalyst to initiate the onset of vital growth processes. Schmidt (21), using isotopic techniques, has verified that lactic acid and carbon dioxide are the end products of the malo–lactic reaction, and Morenzoni (6) has shown that pyruvic acid and NADH are produced as a corollary to malic acid decomposition and may account in part for the stimulation of growth rate and cell yield.

Previous Work on the Malo–Lactic System

Mechanism. Ochoa and his co-workers were the first to delve into the biochemistry of malo–lactic fermentation. In the late 1940's and the early 1950's, Ochoa became interested in mechanisms of carbon dioxide fixation and dicarboxylic acid synthesis in higher organisms. This work necessitated the development of an accurate, sensitive method for the quantitative determination of dicarboxylic acids, of which malic acid was one of the most important. To do this, Ochoa began working with the malic acid decomposing system of *Lactobacillus arbinosus (plantarum)*, a system which is analogous to the malo–lactic system of *Leuconostoc oenos* ML 34, an organism reportedly used by some California wineries. Unfortunately, at this point in the literature, the understanding of the enzymology of malo–lactic fermentation becomes somewhat confused and difficult to follow. This area will be discussed in light of recent work (6, 7), and the current views concerning the enzymology of malic acid decomposition by *L. oenos* ML 34 will be presented.

Previously, a proposed mechanism for the malo–lactic fermentation was shown (11), and this proposal seems to be responsible for much of the confusion. In the same work, the authors reported a strain of *Lactobacillus arabinosus (plantarum)* which was capable of adaptively carrying out the following two reactions:

$$\text{L-Malate} \rightarrow \text{Lactate} + CO_2 \tag{1}$$

$$\text{Oxaloacetate} \rightarrow \text{Pyruvate} + CO_2 \tag{2}$$

These reactions were demonstrated by manometric measurement of the carbon dioxide evolved when resting cells were placed in contact with the substrates. The auhors (11) state: "The extracts contain lactic dehydrogenase but little or no malic dehydrogenase as tested spectrophotometrically with reduced DPN and pyruvate or oxaloacetate respectively." It was also reported (11, 14) that Reaction 1 results from the combination of the following two reactions:

$$\text{L-Malate} + \text{DPN}_{ox} \rightarrow \text{Pyruvate} + CO_2 + \text{DPN}_{red} \text{ (malic enzyme)} \tag{3}$$

$$\text{Pyruvate} + \text{DPN}_{red} \rightarrow \text{Lactate} + \text{DPN}_{ox} \text{ (lactate dehydrogenase)} \tag{4}$$

They stated further that, "the new adaptive enzyme catalyzing Reaction 3 appears to be similar to the malic enzyme of pigeon liver, although strictly DPN (instead of TPN)-specific. The coenzyme specificity explains the ready occurrence of Reaction 1." Therefore, the authors showed that exogenous NAD was required for the overall reaction (malic acid → lactic acid), but because this activity was measured manometrically, they never demonstrated the formation of reduced NAD. Similarly, they did not attempt to show that pyruvic acid was the intermediate between L-malic acid and lactic acid. Instead, the formation of pyruvic acid was inferred from the NAD requirement and because the malic acid dissimilation activity remained constant during purification while the lactate dehydrogenase activity decreased (14). In fact, attempts to show any appreciable amounts of pyruvic acid intermediate failed (22).

It is not surprising that the pyruvic acid intermediate seemed plausible because in a paper earlier in that same year (23), the authors described a malic enzyme from pigeon liver. This enzyme was shown to form appreciable amounts of pyruvic acid from malic acid, but it was NADP instead of NAD specific. The end product was shown to be pyruvic acid by spectrophotometric assay involving lactate dehydrogenase.

Using isotopic techniques, Korkes *et al.* did further work on the *Lactobacillus* system. They report (14) that over a range of 30-fold purification, the ratio of activities with malate and oxaloacetate as substrates remains constant while the ratio of lactic dehydrogenase to malic activity falls from about 5 to 1. They also showed that for each mole of malic acid disappearing, one mole of lactic acid and carbon dioxide was formed. Interestingly, they were able to demonstrate the recovery of 0.2% of added malic acid carbon in pyruvic acid carbon. From the purification ratios which they reported, it may be stated that one enzyme (or two enzymes being purified simultaneously) was responsible for the activity with malic and oxaloacetic acids as substrates and that lactic dehydrogenase was probably not involved in the reaction unless it was incorporated into a bound, coupled system.

The reasons for the confusion surrounding the mechanism of the malo–lactic fermentation are now apparent. In the malate system from *Lactobaccillus plantarum*, Korkes *et al.* (14) demonstrated carbon dioxide and lactic acid production from malic acid, but they were unable to show a large amount of pyruvic acid production. However, the cofactor requirement for the system indicated the need for an intermediate between malic acid and lactic acid, and pyruvic acid was the logical choice. At this time, the occurrence of enzymes requiring NAD in a function other than reduction–oxidation was not realized, so it was logical to conclude that the malic acid to lactic acid conversion involved a redox reaction. The later information, however, indicates that this is probably not the case.

In discussing the studies of Brechot *et al.* (24) and Peynaud *et al.* (25), Kunkee (1) found it odd that bacteria which ordinarily produce D or DL-lactic acid from glucose produce L-lactic acid in wine as a result of malo–lactic fermentation. Peynaud *et al.* (26) reported that organisms which produced only D-lactic acid from glucose produced only L-lactic acid from L-malic acid. He postulated further that the malo–lactic fermentation pathway has no free pyruvic acid as an intermediate because the optical nature of L-malic acid would be lost when it was converted to pyruvic acid since pyruvic acid has no asymmetric carbon atom. Therefore, if pyruvic acid were the intermediate, one would expect D, L, or DL-lactic acid as the end product whereas L-lactic acid is always obtained. These results lend considerable support to the hypothesis that free pyruvic acid is not an intermediate in the malo–lactic fermentation, but it does not discount the possibility that L-lactate dehydrogenase is complexed with malic enzyme, and together they catalyze the conversion of L-malic acid to L-lactic acid. However, Radler (27) described a malic enzyme preparation from *Lactobacillus plantarum* which was capable of catalyzing the conversion of L-malic acid to L-lactic acid and carbon dioxide but exhibited no L-lactic dehydrogenase activity. Similarly, Schütz (28) attempted to disrupt chemically this postulated complex but was unsuccessful.

The most confusing aspect of the pathway proposed by Ochoa and his group now rests with the NAD requirement. In proceeding from L-malic acid to L-lactic acid, there is no net change in oxidation state. Yet in whole cells or cell-free extracts, the malo–lactic fermentation will not proceed in the absence of NAD. Therefore, by the proposed mechanism, one is unable to demonstrate the appearance of reduced cofactor, and the NAD specificity cannot be explained as a redox requirement. However, in the time since this mechanism was proposed, an NAD dependent enzyme (glyceraldehyde-3-phosphate dehydrogenase) has been described which requires NAD in a non-redox capacity (29), and it is possible that the same is true for the enzyme causing the malic acid–lactic acid transformation.

Oxaloacetic Acid Decarboxylase Activity

It was reported (14) that the adaptive enzyme from *Lactobacillus plantarum* could decarboxylate oxaloacetic acid as well as malic acid. However, in the same organism, Nathan (30) carried this work further and showed that the oxaloacetate decarboxylase activity is not related at all to the malic acid–lactic acid transformation activity. She based this conclusion on the ability of malic and oxaloacetic acids to induce oxaloacetate decarboxylase activity as well as malic enzyme activity. In her words,

Whole organisms of *Lactobacillus plantarum* grown with L-malic acid were unable to decarboxylate oxaloacetate whereas acetone-dried preparations from the same batch of organisms decarboxylated oxaloacetate readily. That is, L-malate induced the synthesis of malic enzyme and oxaloacetate decarboxylase but not the permease which permits the penetration of oxaloacetate into the intact organism. Organisms grown with oxaloacetate did not decarboxylate malate nor did acetone-dried preparations made from these organisms. Thus oxaloacetate induced oxaloacetate permease and oxaloacetate decarboxylase but not malic enzyme, and malate induced oxaloacetate decarboxylase and malic enzyme but not oxaloacetate permease. It therefore seems highly unlikely that the malic enzyme and oxaloacetate decarboxylase are part of one "bifunctional protein."

Also, Flesch and Holbach (15) showed that malic enzyme activity was inhibited by parachloromercuribenzoate, but oxaloacetic acid decarboxylation was not.

In *Leuconostoc oenos* ML 34, we have shown oxaloacetic acid decarboxylation manometrically (6, 7, 8). We were also able to demonstate fluorometrically the enzymatic production of reduced NAD with malic acid as a substrate, but, of course, were unable to do so with oxaloacetic acid since no NADH could be formed from this substrate. It is likely that this oxaloacetic acid decarboxylation activity, as in *Lactobacillus plantarum*, is distinct from the activity causing the malic–lactic transition. It is also possible that oxaloacetic acid decarboxylation is caused by a malic enzyme. However, there is no verified NAD dependent malic oxidoreductase (decarboxylating) enzyme which does so (12). For example, Macrae (31) isolated a malic enzyme from cauliflower bud mitochondria which showed no activity with oxaloacetic acid. Similarly, Saz (32) isolated a malic enzyme from *Ascaris lumbricoides* which is also inactive toward oxaloacetic acid. True, the Enzyme Commission (12) lists an enzyme described as L-malate: NAD oxidoreductase (decarboxylating) (E.C. 1.1.1.38) which is said to be capable of decarboxylating oxaloacetic acid, but its description dates back to the studies of Ochoa and his group, and we now feel this listing may be improper.

Cofactor and Ion Specificity

Ochoa's group reported that in their malic enzyme, Co^{2+} could replace the Mn^{2+} requirement, but that Mg^{2+} was considerably less effective. Macrae (31) reported that malic enzyme from cauliflower bud mitochondria has an absolute requirement for either Mn^{2+}, Co^{2+}, or Mg^{2+}. Morenzoni (6) has shown that the NADH producing activity of *Leuconostoc oenos* exhibited an absolute specificity for Mn^{2+}; Co^{2+} and Mg^{2+} could not substitute, nor could Fe^{3+}, Zn^{2+}, or Cu^+. Furthermore, Cu^{2+} inhibits this activity as well as the malo–lactic acivity.

Ochoa reported that malic enzyme from *L. plantarum* was NAD and not NADP specific. The malic enzyme of cauliflower bud mitochondria (31) is NAD and NADP specific, with NAD being the preferred cofactor. Both the malo–lactic activity and NADH producing activity of the *Leuconostoc oenos* system (6, 7, 8) was strictly NAD specific. Nicotinamideadenine dinucleotide phosphate, flavin adenine dinucleotide, and flavin mononucleotide could not substitute in either of these activities.

Enzyme Nomenclature

In wine microbiology, it has been common since the early work of the 1950's to refer to the malo–lactic fermentation and malo–lactic reaction as being caused by the malic enzyme. As we have seen from this review, this usage, which has led to great confusion, should have been questioned from the first. In the accepted enzyme listing (*12*), malic enzyme from lactic acid bacteria has been given the Enzyme Commission number E.C. 1.1.1.38. [Since this enzyme does not seem to decarboxylate oxaloacetate (*30*), the correct number would seem to be E.C. 1.1.1.39, as previously proposed by Pilone and Kunkee (*13*).] In this listing, the enzyme is described as catalyzing the reaction,

$$\text{L-Malate} + \text{NAD} \rightarrow \text{Pyruvate} + CO_2 + \text{Reduced NAD},$$

and is classified as L-malate: NAD oxidoreductase (decarboxylating). This description was taken from the work of Ochoa's group in which they measured carbon dioxide evolution from malic acid. As was pointed out previously, these workers did not show the production of pyruvic acid from malic acid. Instead, their work described the decarboxylation of malic acid to lactic acid (the malo–lactic fermentation). The rationale of using either the trivial names malic enzyme, or malate dehydrogenase (decarboxylating) or the accepted name L-malate: NAD oxidoreductase (decarboxylating) was based on the suspicion of the involvement of two enzymes (this one and lactic dehydrogenase). Nevertheless, all of the activity results were measurements of the malo–lactic reaction, the formation only of carbon dioxide, and confusingly referred to as a unit of malic enzyme activity.

Our work (*6, 7, 8*) has shown that the same protein which causes the malic acid–lactic acid transformation will also cause the production of a small amount of pyruvic acid from malic acid. However, the pyruvic acid produced is not involved with lactic acid production. Apparently, one protein is producing two end products from the same substrate. The malic acid to lactic acid activity is not an oxidoreductase whereas the malic acid to pyruvic acid activity is. Since the pyruvic acid producing activity is only a small per cent (about 0.2%) of the malic acid to lactic acid activity, and since the enzyme should be classified according to the major end product, the enzyme has been given the trivial name of malo–lactic enzyme (*6, 7, 8*). Schütz (*28*) has speculated that if this enzyme were crystallized, it should be called malate-carboxy-lyase. In either case, use of either the trivial or accepted terms for the malic enzyme is not recommended.

London and Meyer (*33*) have demonstrated the presence of a malic enzyme, E.C. 1.1.1.39, in *Streptococcus faecalis*. In their system, the organism grows at the expense of L-malic acid, producing carbon dioxide, acetate, and ethyl alcohol and NADH as major end products, and a small amount of lactic acid as a minor end product. The authors speculate that the major function of the malic enzyme is to provide energy, presumably *via* a phosphoroclastic cleavage of pyruvate. This system is most likely not operational in *Leuconostoc oenos* as it is glucose repressible in *Streptococcus faecalis* but not in *Leuconostoc oenos*.

In discussion of nomenclature of malic acid decomposing enzymes, mention should be made of malate–lactate transhydrogenase. This enzyme, isolated from *Micrococcus lactiyticus (Viellonella alcalescens)*, a bacterium found in vertibrates, can catalyze reversibly the conversion of L-malic and pyruvic acids to L-lactic and oxaloacetic acids with NAD as coenzyme (36).

Antohi (34) has reported several possible isozyme structures of a malic acid decarboxylating enzyme in *Bacillus subtilis*, and Peak (35) has reported the same in *Euglena gracilis;* this must be kept in mind for the *Leuconostoc oenos* system. It is possible that the enzymatic activities that we have reported (6, 7, 8) may be the result of isozyme interactions of the same protein.

Current Concepts

In a previous section, we have reported the presence of two separate and distinct end products produced by one enzyme, the malo–lactic enzyme. One product is, naturally, lactic acid, and the other is a small but real amount of pyruvic acid, as well as reduced NAD.

The results of Alizade and Simon (37) also indicate the formation of NADH during the decarboxylation of L-malic acid in *Leuconostoc mesenteroides*. During growth on glucose and in the presence of L-malic acid tritiated at the C_2 position, tritium was recovered in the C_2 position of L-lactic acid, as would be expected from a first decarboxylation. Surprisingly, a significant amount of label was recovered in D-lactic acid which indicated the intermediate formation of tritiated NADH from malic acid.

Our results, then, allow for a ready explanation of the pyruvic acid spillage which Korkes observed in the *L. plantarum* system (14). The pyruvic acid possibly was not spilling off the malic acid–lactic acid enzyme, but it was being produced directly by the malic acid–pyruvic acid activity (6, 7, 8). Because of this new activity, it is possible that the NAD requirement for the malo–lactic activity (the main activity) does not indicate the presence of an oxidation–reduction reaction. (On the molecular level, it might be stated that during the internal rearrangement of the malic acid–enzyme–cofactor complex to form lactic acid, a transitory production of bound, reduced cofactor may occur. This bound NADH may have no fluorescent properties. This reasoning is not germane to the biological arguments, but it might be the basis for an investigation of the biochemical nature of the decarboxylation.) As stated previously, the NAD is possibly used catalytically in a non-redox function. A speculative point must be made in that the formation of pyruvic acid and NADH by this second reaction might account for the stimulatory effect of malic acid on the growth of these bacteria (*16, 17*), possibly by utilization of pyruvic acid as an energy source or a hydrogen acceptor.

In summary, the current concept (*6, 7, 8*) of malic acid utilization in the *Leuconostoc oenos* ML 34 system involves two separate enzyme activities (located on the same protein) which act simultaneously on malic acid. One activity, which we call the malo–lactic activity, catalyzes the reaction,

$$\text{L-Malic acid} \xrightarrow[\text{Mn}^{+2}]{\text{NAD}} \text{L-Lactic acid} + CO_2$$

and the other activity catalyzes the reaction

$$\text{L-Malic acid} + \text{NAD} \longrightarrow \text{Pyruvic acid} + \text{NADH}_2 + CO_2$$

This system appears to be somewhat novel, and it is expected that malic acid utilization in malo–lactic bacteria other than *Leuconostoc oenos* may occur through this mechanism.

Literature Cited

1. Kunkee, R. E., "Malo–Lactic Fermentation," *Advan. Appl. Microbiol.* (1967) **9**, 235–279.
2. Berthelot, M., de Fleurieu, A., *C.R. Acad. Sci.* (1864) **58**, 720–723.
3. Möslinger, D., *Z. Unters. Nahr.* (1901) 673, 1121.
4. Müller-Thurgau, H., Osterwalder, A., "Die Bakterien In Wein Und Obstwein Und Die Dadurch Verursachten Veranderrungen," *Zentralbl. Bakteriol. Parasitenk. Abt.* (1913) II, **36**, 129–338.
5. Cruess, W. V., "The Role of Microorganisms and Enzymes in Wine Making," *Advan. Enzymol.* (1943) **3**, 349–386.
6. Morenzoni, R. A., "A Second Enzymatic Malic Acid Decomposing Activity in *Leuconostoc oenos*," Doctoral Thesis, University of California, Davis, 1973.
7. Morenzoni, R. A., Kunkee, R. E., unpublished data.
8. Kunkee, R. E., "A Second Enzymatic Activity for Decomposition of Malic Acid by Malo–Lactic Bacteria," Symposium on Lactic Acid Bacteria in Beverages and Foods, Long Ashton Research Institute, University of Bristol, September 19–21, 1973.
9. Schanderl, H., "Mikrobiologie Des Mostes Und Weins," "Handbuch der Kellerwirtschaft, II," 2nd ed., 157–172, Ulmer, Stuttgart, 1959.
10. Radler, F., "Die Mikrobiologischen Grundlagen Des Saureabbaus Im Wein," *Zentralbl. Bakteriol.* (1966) **120**, 237–287.
11. Korkes, S., Ochoa, S., "Adaptive Conversion of Malate to Lactate and Carbon Dioxide by *Lactobacillus arabinosus*," *J. Biol. Chem.* (1948) **176**, 463–464.
12. Barman, Thomas E., "Enzyme Handbook," Springer-Verlag, New York, 1969.
13. Pilone, G. J., Kunkee, R. E., "Carbonic Acid From Decarboxylation by 'Malic' Enzyme in Lactic Acid Bacteria," *J. Bacteriol.* (1970) **103**, 404–409.

14. Korkes, S., Del Campillo, A., Ochoa, S., "Biosynthesis of Dicarboxylic Acids by Carbon Dioxide Fixation, IV. Isolation and Properties of an Adaptive 'Malic' Enzyme From *Lactobacillus arabinosus*," *J. Biol. Chem.* (1950) **187**, 891–905.

15. Flesch, P., Holbach, B., "Zum Abbau Del L-Apfelsäure Durch Milchsäurebacterien. I. Uber Die Malat-Abbauenden Enzyme Der Bakterium 'L' Unter Besanderer Berücksichtigung Der Oxalessigsäure–Decarboxylase," *Arch. Mikrobiol.* (1965) **51**, 401–413.

16. Pilone, G. J., "Energetics and Control of Malo–Lactic Fermentation," Doctoral Thesis, University of California, Davis, 1971.

17. Pilone, G. J., Kunkee, R. E., unpublished data.

18. Harvey, R. J., Collins, E. B., "Roles of Citrate and Acetoin in the Metabolism of *Streptococcus diacetilactis*," *J. Bateriol.* (1963) **86**, 1301–1307.

19. Meyrath, J., Lüthi, H. R., "On the Metabolism of Pentoses by *Leuconostoc* Isolated From Wines and Fruit Juices," *Levensm. Wiss. U. Technol.* (1969) **2**, 21–27.

20. Stamer, J. R., Stoyla, B. O., "Growth Stimulants in Plant Extracts for *Leuconostoc citrovorum*," *Appl. Microbiol.* (1970) **20**, 672–676.

21. Schmidt, H. J., Huskens, G., Jerchel, D., "Zum Abbau Von ^{14}C-Markierter Apfelsäure Durch *Bacterium gracile*," *Arch. Mikrobiol.* (1962) **43**, 162–171.

22. Kaufman, S., Korkes, S., Del Campillo, A., "Biosynthesis of Dicarboxylic Acids by Carbon Dioxide Fixation. V. Further Study of the 'Malic' Enzyme of *Lactobacillus arabinosus*," *J. Biol. Chem.* (1951) **192**, 301–312.

23. Ochoa, S., Mehler, A. H., Kornberg, A., "Biosynthesis of Dicarboxylic Acids by Carbon Dioxide Fixation. I. Isolation and Properties of an Enzyme From Pigeon Liver Catalyzing the Reversible Oxidative Decarboxylation of L-Malic Acid," *J. Biol. Chem.* (1948) **174**, 979–1000.

24. Brechot, P. J., Chauvet, L., Croson, M., Irmann, R., "Configuration Optique de l'acide Lactique Apparu Au Cours De La Fermentation Malo–lactique Pendant La Vinification," *C.R. Acad. Sci.* (1966) **C262**, 1605–1607.

25. Peynaud, E., Lafon-Lafourcade, S., Guimberteau, G., "L(+) Lactic Acid and D(−) Lactic Avid in Wines," *Amer. J. Enol. Viticult.* (1966) **17**, 302–307.

26. Peynaud, E. J., Lafon-Lafourcade, S., Guimberteau, G., "Uber Den Mechanismus Der Apfelsäure-Milchsäure-Garung," *Mitt. Klosterneuburg* (1968) **18**, 343–348.

27. Radler, F., Schütz, M., Doelle, H. W., "Die Beim Abbau Von L-Apfelsäure Durch Milchsäurebakterien Entstehenden Isomeren Der Milchsaure," *Naturwissenschaft.* (1970) **57**(12), 672.

28. Schütz, M., "Uber Das Malatenzym Von *Lactobacillus plantarum* und *Leuconostoc mesenteroides*," Doctoral Thesis, Johannes Gutenberg-Universitat, Mainz, 1972.

29. Harting, J., Velick, S. F., "Transfer Reactions of Acetyl Phosphate Catalyzed by Glyceraldehyde-3-Phosphate Dehydrogenase," *J. Biol. Chem.* (1954) **207**, 867–878.

30. Nathan, H. A., "Induction of Malic Enzyme and Oxaloacetate Decarboxylase in Three Lactic Acid Bacteria," *J. Gen. Microbiol.* (1961) **25**, 415–420.

31. Macrae, A. R., "Isolation and Properties of a 'Malic' Enzyme From Cauliflower Bud Mitochondria," *Biochem. J.* (1971) **122**, 495–501.

32. Saz, H. J., Hubbard, J. A., "The Oxidative Decarboxylation of Malate by *Ascaris lumbricoides*," *J. Biol. Chem.* (1957) **225**, 921–933.

33. London, J., Meyer, E. Y., "Malate Utilization by Group D *Streptococcus*: Physiological Properties and Purification of an Inducible Malic Enzyme," *J. Bacteriol.* (1969) **98**, 705–711.

34. Antohi, S., Mararu, I., Cotae, D., "Possible Isozyme Structures in Malate Dehydrogenase From *Bacillus subtilis*," *Biochem. Biophys. Res. Commun.* (1970) **39**, 226–230.
35. Peak, M. J., Peak, J. G., Ting, I. P., "Isozymes of Malate Dehydrogenase and Their Regulation in *Euglena gracilis* Z.," *Biochem. Biophys. Acta* (1972) **284**, 1–15.
36. Allen, S. H. G., Patil, J. R., "Studies on the Structure and Mechanism of Action of Malate–Lactate Transhydrogenase," *J. Biol. Chem.* (1972) **247**, 909–916.
37. Alizade, M. A., Simon, H., "Zum Mechanismus und zur Kompartmentierung der L- und D-Lactatbildung aus L-Malat bzw. D-Glucose in *Leuconostoc mesenteroides*," *Z. Physiol. Chem.* (1973) **354**, 163–168.

RECEIVED July 24, 1973.

9

Analytical Fractionation of the Phenolic Substances of Grapes and Wine and Some Practical Uses of Such Analyses

VERNON L. SINGLETON

Department of Viticulture and Enology, University of California, Davis, Calif. 95616

The molar color yield under recommended conditions with a molybdotungstophosphoric heteropoly anion (Folin-Ciocalteu) reagent was determined for 150 phenol derivatives and eight potentially interfering substances. Determination of total phenol content was shown to be reliable, consistent, and stoichiometrically predictable from the reactions of known phenols. The expression of results as mg gallic acid equivalent per kg (grapes) or per liter (wine) is recommended. Potentially interfering substances are either very low in wines or grapes, or they can be corrected for by separate analysis. Among grape and wine constituents, only flavonoids readily precipitated when allowed to react with acidic formaldehyde. Total phenol analysis before and after such precipitation estimates flavonoid and nonflavonoid content. Practical analyses of wines, effects of barrel aging, and browning tendency are described.

The importance of phenolic substances in wines cannot be over-emphasized (*1*). Although generally present in smaller amounts than alcohols, acids, and carbohydrates, they are among the next most abundant substances with a typical range of a few tenths of a gram to a few grams per liter of wine and roughly 1 to 7 grams/kg for whole fresh grape berries (*1*). Wines would be much less interesting beverages without phenols. They form the red, golden, and brown pigments of most wines. Astringent and bitter flavors in wines involve phenols. The major reservoir of readily autoxidizable substances in a wine is its phenols, and they are crucial in aging changes, browning reactions, and oxidation changes. The phenolic substances of grapes appear in wine made by various methods of vinification to varying degrees. In wine, the grape's phenols are modified by fermentation and storage, and additional phenols may be contributed to wine from storage in wooden containers.

Because they are important, the phenolics and their roles in wine have been studied considerably (1), but more needs to be known. Since different classes of phenols behave quite differently in the various roles as pigments, oxidation substrates, flavors, etc., separate consideration and analysis is essential to completely understand them. Every individual phenol should be separately determined precisely and quantitatively. Chromatography may solve these problems, but for now, chromatography and other detailed separatory methods have drawbacks, particularly for economical application to many individual wine or grape samples.

Methods of analysis are needed to determine total phenolic content and the relative content of phenolic fractions by means of their different characteristics. Many analytical methods used for phenols have been empirical and not easily reproduced or rationalized (1). Procedures that are based on sound chemical principles and that are sufficiently verified deserve wider application. We are concerned here with recent work on such analyses for phenols in wines. Application of these results may help solve a major problem in phenol research—the many different, too empirical, unrelatable values (ml $KMnO_4$, vanillin-to-leucoanthocyanin ratio, etc.) obtained in different ways by different researchers. Uniform use of verified methods and uniform standards and methods of expressing results will aid in developing an understanding in this field.

Total Phenol Content of Wine

Singleton and Esau (1) reviewed the methods for phenol analysis of wine. They pointed out that study would be greatly advanced if one could determine the total content of phenolic substances and express it in such a way that analysis of subclasses of phenols could be related to the original total and a balance sheet could be obtained. One could then say, for example, "this wine has a total phenolic content of 1200 mg/liter calculated as gallic acid, and of that total, cinnamic acid derivatives account for 200 mg/liter, anthocyanins for 300 mg/liter, other small flavonoids for 200 mg/liter, and condensed tannins complete the total with 500 mg/liter of gallic acid equivalent." To accomplish this, the total phenol analysis not only must meet ordinary criteria of reproducibility and precision, but it also must be based on chemical relationships such that fractions determined separately can be converted to units of the total. Of course when clearcut fractionation can be accomplished by other means, analysis for total phenols before and after treatment would then yield the desired information.

The colorimetric method based on the reagents of Folin and Denis or of Folin and Ciocalteu has been generally preferred over other methods to determine total phenols in complex natural materials such as wines and fruits (1, 2, 3, 4, 5). This method is relatively simple, convenient, reliable, generally applicable, and it is accepted as an official analysis in several countries for total phenols in wines and a number of other products. Although it is a preferred method, it can be even better than is commonly recognized.

Swain and Goldstein (6, 7) noted a rather large difference in the molar color yield from different phenols with the Folin-Denis reagent. They attributed this to differences in relative oxidation–reduction potentials of the different phenols, but under their conditions pyrogallol gave about half the color of catechol and more than resorcinol. However, they also reported that the molar absorptivity produced by a flavonoid was approximately equal to the sum of the values for the separate phenolic moieties which it contained.

The Folin-Ciocalteu Assay for Total Phenols

The Folin-Ciocalteu (F-C) reagent is superior to the Folin-Denis formulation (8, 9), and improved conditions for the assay have been developed (9). With the improved F-C procedure (compared with Folin-Denis analysis), the molar color yield was higher and more consistent, particularly with less reactive phenols. The improved method gave less deviation among replicate analyses, better recovery of added gallic acid, and was affected less by nonphenolic reductants. Based on limited numbers of known phenol structures, it appeared that with this procedure (9) monophenols reacted similarly and polyphenols gave additional reductive color generation depending on their quinoidal possibilities, e.g., phloroglucinol reacted as a monophenol and pyrogallol as a diphenol (1). Further studies have now been made with a wider group of phenols.

Method. The method of Singleton and Rossi (9) was used. Commercial samples of the compounds to be tested, generally without further purification, were accurately weighed (about 100–200 mg unless the supply was very limited), dissolved in ethanol, and diluted with water so that the final solution was 10 vol % ethanol and had a known concentration of phenol which yielded an absorbance of about 0.3 in the analysis. For incompletely soluble substances the suspension was kept dispersed, fine, and homogeneous. The analysis was essentially as published (9): 1.00 ml sample, gallic acid standard, or blank solution was mixed with 10.00 ml distilled water, 1.00 ml F-C reagent, 3.0 ml 2% Na_2CO_3 solution, and 5.00 ml distilled water. The absorbance was determined in a

Zeiss PMQII spectrophotometer, 1-cm cells at 765 nm after either 5 min at 50°C in a water bath or 2 hrs at room temperature (24°C).

The Beer-Lambert plots for this method are linear, and it is sensitive —approximately $5 \times 10^{-5}M$ monophenol in the final assay solution (*e.g.*, 7 mg/liter 4-hydroxybenzoic acid) gives about 0.6 absorbance at 765 nm in 1-cm cells. Five min at 50°C gave maximal absorbance at 765 nm with gallic acid. The absorbance produced was nearly the same as that obtained with 2 hrs at room temperature whereas more rapid fading tended to produce lower maximum absorbance at 60°C. For 150 different phenols, the absorbance produced by 2 hrs at room tempreature averaged 100.6% of the absorbance produced in 5 min at 50°C with a standard deviation of ±0.5%. In the few instances that the hotter incubation gave appreciably higher readings than did the room temperature one, the phenols were suppressed in activity and produced lower than usual molar absorptivities (3,5-diisopropylcatechol, ellagic acid, ethyl *p*-hydroxybenzoate, *p*-hydroxybenzaldehyde, *p*-hydroxyacetophenone, and pentachlorophenol).

The gallic acid mean molar absorptivity and standard deviation of 21 replicates which were weighed separately, diluted, and allowed to react with the assay reagents for 5 min at 50°C was 24967 ± 847. The standard deviation thus was 3.4% for replicate assays and would be expected to be about 5% for a single analysis compared with a single standard and about 7% for analysis by difference between two such assays. The reproducibility of a given value would be better if more standard levels, duplicate unknown samples, etc., were used. Nearly identical mean molar absorptivity and standard deviation were obtained for the samples held 2 hrs at room temperature and, since the warmer temperature had advantages including slightly greater reaction with difficult-to-oxidize phenols, only the 5 min at 50°C values will be reported further.

Monophenol Derivatives. The molar absorptivities at 765 nm, produced in the F-C assay (*9*) with 150 phenols or phenolic derivatives and eight other substances after color development for 5 min at 50°C, are listed in Tables I through IX. The values shown are mostly from a single assay, but any result which seemed questionable was repeated one or more times with the same chemical from another source if possible. Table I lists 28 monophenols and one biphenol. The maximum molar absorptivity for a monophenol is about 16,000 in this assay. The typical value is *ca.* 12,500–13,000 for phenol itself and electronically equivalent structures. Phenols producing extinction coefficients higher than 13,000 generally have electron-repelling substituent groups, and those producing lower than this have electron-attracting substituents. However, substituents which strongly suppress proton loss to form the phenolate ion (*e.g.*,

Table I. Molar Absorptivity in the Total Phenol Assay and
Precipitability with Acidic Formaldehyde: Monophenols

Substance	ε_{765} ($\div 1000$)	% Pptd by HCHO
2-Hydroxycinnamic acid	15.9	0
4-Hydroxystilbene	15.8	insol.
Tyrosine	15.7	0
3-Hydroxycinnamic acid	15.6	0
4-Hydroxycinnamic acid	15.6	0
4-Hydroxybenzoic acid	15.6	0
3,5-Dimethylphenol	13.6	83
Phenol	12.7	0
4-Chlorophenol	12.1	0
2,4,6-Trichlorophenol	11.4	0
3,4-Dimethylphenol	11.1	40
2,4-Dichlorophenol	10.6	0
Salicylamide	10.2	0
4-tert-Butylphenol	10.2	0
2,2'-Biphenol	(20.2 ÷ 2) 10.1	0
Salicylic acid	8.0	0
2,6-Dimethylphenol	8.0	0
2,6-Di-tert-butylphenol	7.2	insol.
2,4,5-Trichlorophenol	4.6	0
Phenyl salicylate	3.4	insol.
Methyl 4-hydroxybenzoate	2.4	0
4-Hydroxyacetophenone	1.9	0
Ethyl 4-hydroxybenzoate	1.7	0
4-Hydroxybenzaldehyde	0.6	0
Pentachlorophenol	0.4	insol.
4-Chloro-2-nitrophenol	0.2	0
4-Nitrophenol	0.1	0
2,4-Dinitrophenol	0.1	0
Picric acid	0	0

2,6-dimethylphenol) tend to lower the reducing power toward the F-C
reagent. Conversely, substituents which make the phenol very acidic but
suppress the oxidative removal of electrons (e.g., 2,4,5-trichlorophenol)
suppress or prevent oxidation of the phenol by the F-C reagent. The
only phenols almost completely unreactive in this assay are the highly
acidic nitrophenols (Table I). Aldehyde, ketone, and ester groups suppressed the phenol's oxidizability and therefore the color yield as might
be expected. However, in the alkaline assay solution (about pH 8.8),
carboxyl ions enhanced rather than suppressed the removal of electrons
by the heteropoly anions of the F-C reagent (cf. 4-hydroxybenzoic acid
with its esters, Table I).

Catechol and Guaiacol Derivatives. The only diphenol in Table I
behaved as two separate monophenols. Catechols also generally react as
diphenols, i.e., they produce twice the 765 nm absorbing color with F-C
reagent as does a typical monophenol. Table II shows molar extinc-

tions from 14 catechol and eight guaiacol derivatives. The catechols show much the same effects as, and perhaps more clearly than, the monophenols. Molecular configuration conveying ready ionization and also ready oxidative removal of the electron promotes reduction of the F-C reagent to produce the blue pigment. A few highly inhibited catechols behaved as monophenols, or at least their color yield was less than the maximum expected from a monophenol.

Table II. Molar Absorptivity in the Total Phenol Assay and Precipitability with Acidic Formaldehyde: Catechol and Guaiacol Derivatives

Substance	ε_{765} ($\div 1000$)	Phenolic Hydroxyls A Free	B Reactive	ε/B	% Pptd by HCHO
Pyrocatechol violet	52.7	3	3	17.6	0
Chlorogenic acid	28.9	2	2	14.4	0
3,4-Dihydroxyphenylalanine	24.6	2	2	12.3	0
Caffeic acid	22.9	2	2	11.4	0
3-Methoxycatechol	22.7	2	2	11.4	7
Catechol	22.5	2	2	11.2	0
4-Methylcatechol	21.6	2	2	10.8	0
3-Isopropylcatechol	20.9	2	2	10.4	7
3-Methylcatechol	20.4	2	2	10.2	0
2,3-Dihydroxybenzoic acid	17.8	2	2	8.9	0
3,4-Dihydroxybenzoic acid	17.3	2	2	8.6	0
3,5-Diisopropylcatechol	12.5	2	1	12.5	7
4-*tert*-Butylcatechol	11.1	2	1	11.1	0
Tetrabromocatechol	9.9	2	1	9.9	insol.
3-Hydroxy-4-methoxy-cinnamic acid	19.2	1	1 (2)	19.2 (9.6)	0
Ferulic acid	19.2	1	1 (2)	19.2 (9.6)	0
Vanillic acid	18.5	1	1 (2)	18.5 (9.2)	0
3-Ethoxy-2-hydroxy-benzaldehyde	16.3	1	1	16.3	0
Vanillin	14.9	1	1	14.9	0
Zingerone	14.3	1	1	14.3	0
Acetovanillone	12.8	1	1	12.8	0
Ethyl vanillate	10.1	1	1	10.1	0

Guaiacol derivatives (Table II) mostly react as monophenols as expected. The three most reactive guaiacol analogs gave more than the maximum color expected from monophenols, and the rest tended to be slightly higher than might be predicted from Table I. There may be some tendency for phenol regeneration from the methoxy groups. Veratric acid did not give a small color yield from F-C assay but 2,4-dimethoxycinnamic acid did (Table IX).

Pyrogallol Derivatives. Table III lists molar absorptivities for 11 pyrogallol derivatives. The result with sinapic acid was uniform with samples from three different commercial lots, and it is interpreted as clear evidence for conversion of methoxyl to hydroxyl by the assay conditions. Furthermore, recrystallization of one sample twice from 95% ethanol did not change its melting point (193–195°C, dec, uncorrected) or color yield. The demethylation does not appear to be the result of excessive acid contact during assay, because the color yield from sinapic acid did not increase as the time between F-C reagent and sodium carbonate addition was increased from 2 to 16 min. Such demethylation has been reported from other studies [sinapyl alcohol readily loses a methyl group during storage in solution (10)] to be relatively facile, but apparently it occurred only to a smaller degree with the carbonyl analogs and not with syringic acid. Otherwise, the data in Table III agree well with predictions which could be made from Tables I and II.

Table III.   Molar Absorptivity in the Total Phenol Assay and
Precipitability with Acidic Formaldehyde:
Pyrogallol Derivatives

Substance	ε_{765} ($\div 1000$)	Phenolic Hydroxyls		ε/B	% Pptd by HCHO
		A Free	B Reactive		
Bromopyrogallol red	47.4	5	3	15.8	0
Purpurogallin	33.7	4	3	11.2	84
Methyl gallate	30.0	3	2	15.0	0
Gallic acid	25.0	3	2	12.5	0
Pyrogallol	24.8	3	2	12.4	20
Ellagic acid	23.5	4	2	11.8	insol.
2,3,4-Trihydroxybenzoic acid	18.3	3	2	9.1	0
Syringic acid	11.9	1	1	11.9	0
Syringaldehyde	15.9	1	1	15.9	0
3,5-Dimethoxy-4-hydroxyacetophenone	19.6	1	2	9.8	—
Sinapic acid	33.3	1	2	16.6	0

Namely, these series show that monophenols react according to their oxidizability, catechols react similarly except twice as much (presumably *via* o-quinone production), and pyrogallol derivatives generally react as catechols if vicinal hydroxyls are free and as monophenols if not. Ellagic acid appears perhaps anomalous, but it is known that two of the four hydroxyls are considerably more acidic than the other two (11).

Table IV.   Molar Absorptivity in the Total Phenol Assay and Precipitability with Acidic Formaldehyde: *m*-Polyphenols

Substance	ε_{765} ($\div 1000$)	Phenolic Hydroxyls A Free	B Reactive	ε/B	% Pptd by HCHO Alone	With Phloroglucinol
Phloroglucinol Derivatives						
Rottlerin	24.4	5	2	12.2	97	—
2,4,6-Trihydroxy-benzoic acid	13.6	3	1	13.6	95	—
Phloroglucinol	13.3	3	1	13.3	99	—
2,4,6-Trihydroxy-acetophenone	12.7	3	1	12.7	97	—
Resorcinol Derivatives						
2,2',4,4'-Tetra-hydroxybiphenyl	26.5	4	2	13.2	91	—
2,2',4,4'-Tetra-hydroxybenzophenone	22.0	4	2	11.0	31	52
Resorcinol	19.8	2	2	9.9	98	—
Resorcinol monoacetate	10.2	1	1	10.2	84	—
3,5-Dihydroxybenzoic acid	16.2	2	1	16.2	0	13
Orcinol	15.4	2	1	15.4	97	—
2,4-Dihydroxybenzoic acid	14.6	2	1	14.6	0	17
2,6-Dihydroxybenzoic acid	14.6	2	1	14.6	94	—
Sesamol	11.7	1	1	11.7	23	77

Phloroglucinol and Resorcinol Derivatives. Table IV lists absorptivities for four phloroglucinol and nine resorcinol derivatives. Each phloroglucinol ring reacts as a monophenol. However, resorcinol and certain of its analogs react as diphenols. That this is not merely a difference in oxidizability or oxidation–reduction potential is shown by the fact that resorcinol monoacetate gives almost exactly half the molar yield of blue pigment as does resorcinol, yet the dipole moments of the hydroxyl and acetate groups are nearly identical. Further, this result indicates that under the conditions of the assay the acetate group was not appreciably hydrolyzed. Other resorcinol derivatives, depending on their substitution, apparently react as either monophenols (especially with electron-repelling substituents) or diphenols (especially with electron-attracting substituents). Reaction of sesamol as a monophenol shows that the methylene dioxy group is not destroyed in the conditions of the assay.

Hydroquinone Derivatives. The nine hydroquinone derivatives listed in Table V all appear to act as monophenols. It appears from the quin-

Table V. Molar Absorptivities in the Total Phenol Assay and
Precipitability with Acidic Formaldehyde:
Hydroquinone Derivatives

| Substance | ε_{765} ($\div 1000$) | Phenolic Hydroxyls | | ε/B | % Pptd by HCHO |
		A Free	B Reactive		
2,5-Dihydroxybenzoic acid	16.1	2	1	16.1	0
Tetrahydroxy-1,4-benzoquinone	14.7	4	1	14.7	0
Quinhydrone	14.7	2	1	14.7	0
Hydroquinone	12.8	2	1	12.8	0
4-n-Butoxyphenol	12.1	1	1	12.1	0
4-Methoxyphenol	11.6	1	1	11.6	0
2-$tert$-Butylhydroquinone	10.0	2	1	10.0	0
2,5-Di-$tert$-butyl-hydroquinone	8.7	2	1	8.7	insol.
2,5-Dihydroxy-1,4-benzoquinone	8.4	2	1	8.4	87

hydrone complex that phenols oxidized to the quinone form are not
further oxidizable by the F-C reagent as would be expected. However,
oxidations expected to give 1,4-benzoquinones give half the molar color
yield with F-C reagent than when 1,2-benzoquinones are expected. This
is noteworthy especially since the oxidation–reduction potential is about
0.8 for o-quinone and 0.7 for p-quinone (*i.e.*, hydroquinone is a stronger
reductant than catechol). Furthermore, it appears that the second
phenol equivalent of a catechol can be produced by the F-C reagent
acting on a hydroxyl substituent on a 1,4-benzoquinone (Table V).
Since phenol oxidation by the F-C reagent is demonstrated to be very
reproducible and quantitative, a study of the nature of the oxidation
products with different classes of phenol would be worthwhile.

Naphthalene and Anthracene Derivatives. In Table VI the molar
color yield in the F-C assay with 11 naphthalene and 6 anthracene-
derived phenols is given. In general, the results seem consistent with
what one would predict from the preceding tests and general chemical
knowldge. Anthraquinone (Table IX) does not give appreciable color
formation and, presumably owing to strong hydrogen bonding of the
peri OH with the quinone carbonyl, the 1- and 8-hydroxyanthraquinones
are strongly suppressed in reducing power. Apparently analogous to
1,4-benzoquinone, 1,4,9,10-tetrahydroxyanthracene, and 1,4-dihydroxy-
naphthalene react as monophenol equivalents. Further hydroxyls in the
anthraquinone series apparently contribute up to two more monophenol
equivalents depending on their position in the different rings. The meta
substitution in 1,3-dihydroxynaphthalene parallels to reactivity of orcinol

(Table IV) and the other naphthols react according to the number (one or two) of their hydroxyls.

Flavonoid and Coumarin Derivatives. Table VII lists the results with 17 aglycones and nine glycosides in the flavonoid and chalcone series and five coumarins including one glycoside. In general the results agree excellently with predictions from preceding tables. The phloroglucinol (or resorcinol) ring reacts as a monophenol, and the other ring reacts according to its substitution. Malvin, like sinapic acid, behaves as a diphenolic B-ring through loss of a methyl substituent to regenerate a catechol structure. Similarly, the color yield from biochanin A, hesperetin, and neohesperidin dihydrochalcone is higher than would be predicted from the free phenolic groups only, and conversion from the 4'-methoxy to the relatively acidic 4'-hydroxyl is indicated.

Table VI.   Molar Absorptivity in the Total Phenol Assay and Precipitability with Acidic Formaldehyde: Naphthalene and Anthracene Derivatives

Substance	ε_{765} ($\div 1000$)	Phenolic Hydroxyls A Free	Phenolic Hydroxyls B Reactive	ε/B	% Pptd by HCHO
Naphthalenes					
2,6-Dihydroxynaphthalene	27.8	2	2	13.9	98
2,3-Dihydroxynaphthalene	26.0	2	2	13.0	92
2,7-Dihydroxynaphthalene	24.0	2	2	12.0	94
1,7-Dihydroxynaphthalene	21.8	2	2	10.9	99
1,5-Dihydroxynaphthalene	18.5	2	2	9.2	96
1,3-Dihydroxynaphthalene	15.9	2	1	15.9	99
1,4-Dihydroxynaphthalene	5.6	2	1	5.6	81
1-Hydroxy-2-carboxy-naphthalene	12.9	1	1	12.9	81
1-Naphthol	12.4	1	1	12.4	96
2-Naphthol	8.8	1	1	8.8	84
Naphthochrome green	2.4	1	1	2.4	6
Anthracenes					
1,2-Dihydroxy-anthraquinone	31.8	2	2	15.9	96
1,2,5,8-Tetrahydroxy-anthraquinone	24.9	4	2	12.4	94
Emodin (1,3-8-trihydroxy-6-methyl-anthraquinone)	15.8	3	1	15.8	89
1,4,9,10-Tetrahydroxy-anthracene	15.6	4	1	15.6	85
1,8-Dihydroxy-anthraquinone	2.6	2	1	2.6	70
1-Hydroxyanthraquinone	0.7	1	0	—	42

Table VII.   Molar Absorptivity in the Total Phenol Assay and
Precipitability with Acidic Formaldehyde:
Flavonoids and Coumarins

Substance	ε_{765} ($\div 1000$)	Phenolic Hydroxyls A Free	B Reactive	ε/B	% Pptd by HCHO Alone	With Phloroglucinol
Aglycones						
Myricetin	48.5	5	3	16.1	35	80
Dihydroquercetin	49.1	4	3	16.4	90	—
Quercetin	48.3	4	3	16.1	92	—
Morin	42.5	4	3	14.2	96	—
l-Epicatechin	43.9	4	3	14.6	91	—
d-Catechin	38.7	4	3	12.9	95	—
Cyanidin	34.3	4	3	11.5	90	—
Fisetin	42.7	3	3	14.2	0	0
Hesperetin	34.9	3	2 (3)[a]	17.6 (11.6)	85	99
Naringenin	32.2	3	2	16.1	79	97
Kaempferol	29.6	3	2	14.8	91	—
Apigenin	28.6	3	2	14.3	97	—
Biochanin A	23.8	2	1 (2)[a]	23.8 (11.9)	95	—
Chrysin	10.8	2	1	10.8	96	—
Acacetin	3.0	2	1	3.0	85	—
Tectochrysin	1.8	1	1	1.8	51	—
2-Hydroxychalcone	7.0	1	1	7.0	insol.	—
Glycosides						
Quercitrin	44.8	4	3	14.9	14	38
Rutin	43.4	4	3	14.5	12	24
Neohesperidin dihydrochalcone	36.4	3	2 (3)	18.2 (12.1)	34	47
Phlorizin	25.6	3	2	12.8	43	55
Naringin dihydrochalcone	21.2	3	2	10.6	13	29
Malvin	40.5	2	2 (3)	20.2 (13.5)	28	65
Pelargonin	24.9	2	2	12.4	20	51
Robinin	20.0	2	2	10.0	17	46
Naringin	18.9	2	2	9.4	0	16
Coumarins						
4-Methylesculetin	31.1	2	2	15.6	0	11
4-Methyldaphnetin	28.0	2	2	14.0	0	—
Esculin	16.8	1	1	16.8	1	4
7-Hydroxycoumarin	10.0	1	1	10.0	0	21
4-Methyl-7-hydroxycoumarin	7.6	1	1	7.6	4	25

[a] The higher value assumes conversion of the 4′-methoxyl to a hydroxyl.

The color yield of acacetin is low in spite of a 4'-methoxyl group, and this is attributed to suppression of the A-ring activity owing to 5-hydroxyl to 4-keto hydrogen bonding, carbonyl attraction of electrons, and perhaps low solubility.

Chrysin shows a similar effect which is much enhanced in tecto-chrysin, owing presumably to methylation of the 7-hydroxyl in techto-chrysin leaving only the keto-associated 5-hydroxyl free in that flavone. Opening the flavone heterocyclic ring to produce 2-hydroxychalcone leaves monophenol activity still partly suppressed because of the remaining carbonyl effect on oxidizability and H-bonding. Since the structure is relatively free to rotate about the ring-to-carbonyl bond, the suppression is less than in tectochrysin.

The coumarins (Table VII) react as expected with some suppression of reactivity of the 7-hydroxyl which is attributable to the alkyl–ester substitution. Note that the ester function is not appreciably hydrolyzed (Tables VII, IX) to liberate the 2-hydroxyl under the conditions of the assay.

Table VIII.   Molar Absorptivity in the Total Phenol Assay and Precipitability with Acidic Formaldehydes: Amines

		Amine + Hydroxyls			% Pptd
Substance	ε_{765} ($\div 1000$)	A Free	B Reactive	ε/B	by HCHO
3-Aminophenol	24.1	2	2	12.0	46
2-Aminophenol	21.8	2	2	10.9	36[a]
1,2-Diaminobenzene	21.4	2	2	10.7	95[a]
3-Diethylaminophenol	19.1	2	2	9.6	1[a]
N-(4-Hydroxyphenyl) glycine	19.9	2	2	10.0	0
4-Aminophenol	12.5	2	1	12.5	0
p-Toluidine	11.6	1	1	11.6	0
p-Anilinophenol	17.1	2	2	8.6	46[a]
2-Amino-4-nitrophenol	10.7	2	1	10.7	0

[a] Not precipitated but lost to the assay presumably by carbonyl amine reaction.

Amine and Aminophenol Derivatives. Amines and aminophenols (Table VIII) react with the F-C reagent about as predicted considering the aromatic amino groups equivalent to phenolic hydroxyls. This would be an important interference with total phenol assay in samples with appreciable aromatic amine content. Fortunately, for this and other reasons as well, the major wine grapes and most other fruit and vegetable products are free of significant concentrations of aromatic amines which would interfere. Correction might be made for methyl anthranilate present at low levels in foxy grapes. Formaldehyde can tie up and render primary amine groups inactive in this assay (Table VII). This might be developed as a means of differentiation between the two functions in a mixture.

Table IX. Molar Absorptivity in the Total Phenol Assay and Precipitability with Acidic Formaldehyde: Nonphenolic Substances

Substance	ε_{765} ($\div 1000$)	% Pptd by HCHO
Ferrous sulfate	3.4	0
Sodium sulfite	17.1	40[a]
D-Fructose	0	0
D-Glucose	0	0
Ascorbic acid	17.5	0
Mandelic acid	0.04	0
Veratic acid	0	0
2,4-Dimethoxycinnamic	0.1	yes
Acetylsalicylic acid	0.2	0
Chalcone	0.08	0
Flavone	0.1	insol.
Flavanone	1.9	0
3-Hydroxyflavone	3.5	insol.
Rotenone	1.8	insol.
Coumarin	0.1	0
4-Hydroxycoumarin	0.09	0
Anthraquinone	0.3	12

[a] Prevented from reaction with F-C reagent, but not precipitated.

Other Substances. Table IX gives the reaction in the F-C assay of several other substances which might interfere and a number of non-phenolic derivatives of phenols. Ascorbic acid (and no doubt other reductones and ene-diols) could be an important source of interference, but apparently reducing sugars and α-hydroxy acids are not. Readily reduced inorganic substances such as ferrous ion and bisulfite ion may interfere, but bound bisulfite is less reactive (Table IX). The presence of these substances is low in grapes or wine, and it is routinely determined. Their effect could be subtracted from a total phenol analysis when necessary, but it is seldom necessary since they ordinarily make a very small contribution. The amount of ascorbic acid can be appreciable in fresh grapes, but neither it nor significant amounts of any other substance reductive enough to react rapidly with sodium 2,6-dichloroindophenol is present in ordinary wine unless it is added. Good success has been seen in separately determining the ascorbic acid content with Tillman's reagent, calculating its contribution to the F-C analysis, and subtracting to adjust the total phenol value.

Reference has already been made to the effect in F-C determinations of a number of the phenolic derivatives no longer containing free phenolic hydroxyls (Table IX). The near absence of reaction with acetylsalicylic acid reinforces the result (Table IV) with resorcinol monoacetate. These results together with those for the coumarins lacking free phenolic groups show that esters of phenolic hydroxyls generally do not hydrolyze sufficiently in the course of the assay, as used here, to cause interference.

Methoxyl groups can convert to or behave as phenolic hydroxyls in assay under some conditions of activation. However, in most cases, particularly when all phenolic hydroxyls are methylated, they are unreactive. Quinones do not react unless further hydroxylated (or reduced to hydroquinones). There is some reaction by 3-hydroxyflavone that is attributed to enol reaction involving the substitution pattern of carbons 2, 3, and 4 in the flavonol series. This may explain the general tendency of flavonols to give higher color yield, based simply on the phenolic hydroxyls, than do flavanols (*e.g.*, quercetin *vs.* catechin, Table VII). Flavanone gives (Table IX) small but measurable color formation in the F-C assay, and this is attributed to partial formation of the chalcone, a known equilibrium reaction in solution, thus producing a new free phenolic hydroxyl. This effect is used to justify the slightly high color yield from naringenin and dihydroquercetin (Table VII). If one assumes no conversion of the 2-hydroxychalcone (Table VII) to the flavanone, then 27% flavanone is converted to the chalcone (Table IX) under the assay conditions.

Phenols Alone and in Mixtures. The phenolic substances of grapes and wine have been well classified. They are predominantly common types with the expected substitution patterns and they are well covered by the model substances studied here (*1*). The phenolics are known not to contain significant amounts of polycyclic phenols like naphthols or anthraquinones; the flavonoids seem to be the common types (*e.g.*, phloroglucinol rather than resorcinol derivatives) as do the cinnamates. The content of potentially interfering nonphenols is generally small and subject to correction by separate determination. From the data presented, a good estimate of the molar absorptivity to be expected in the F-C determination for any one of the phenols known in grapes can be made by inspecting the structure and comparing it with compounds already tested. Several levels of sophistication are possible. Suppose an assay has been made and the total phenol content has been determined as so many gallic acid quivalents (*e.g.*, 340 mg/liter GAE) in a wine. To estimate the equivalent concentration of, say, *d*-catechin, the GAE value could be multiplied by the ratio of the two phenolic equivalents in the assay for gallic acid and the three equivalents for catechin, and then by the molecular weight ratio of the two substances. The result ($340 \times 2/3 \times 290/170$) indicates that 387 mg/liter of catechin would be equiva-

Table X. Behavior of Phenols Alone and in Mixtures Toward Total

Assay	Caffeic Acid, 44.8[a] 40.9[b]	Cate- chin, 47.2[a] 42.5[b]	Vanillic Acid, 74.8[a] 56.0[b]
Assay Without Formaldehyde			
Mg/liter GAE by assay (A)	43.4	42.0	55.8
% A/C	106.3	98.9	99.8
Mixture 1, mg/liter GAE	17.4	8.4	11.2
Mixture 2, mg/liter GAE	8.7	16.8	11.2
Mixture 3, mg/liter GAE	8.7	8.4	22.3
Mixture 4, mg/liter GAE	8.7	8.4	11.2
Assay With Formaldehyde			
Mg/liter GAE by assay (B)	40.0	5.8	62.4
% B/C	97.9	13.5	111.5
Apparent % pptd by HCHO (100-100 B/A)	7.8	86.3	0
Mixture 1, mg/liter GAE	16.0	5.8	12.5
Mixture 2, mg/liter GAE	8.0	5.8	12.5
Mixture 3, mg/liter GAE	8.0	5.8	25.0
Mixture 4, mg/liter GAE	8.0	5.8	12.5

[a] Mg/liter added.

lent to 340 mg/liter of gallic acid in the F-C assay. In many instances of adding phenols to wine, such estimations have proved accurate enough. By applying actual molar absorptivity ratios or other improvements culminating in direct co-analysis of the components involved, improved estimates can be made. The above figure becomes 340 mg of gallic acid which is equivalent to 374 mg of catechin by our best estimate.

To demonstrate that such calculations are valid and that the different phenols act independently when a mixture is analyzed, solutions of four phenols chosen to represent different types important in wine were analyzed alone and as mixtures (Table X). Separate assays of solutions of the four phenols gave gallic acid equivalent concentrations 98.9–106.3% of that calculated from the known concentration of each phenol corrected for the molecular weight and average molar absorptivity compared with gallic acid. A second assay (Table X) of the same solutions after treatment with acidic formaldehyde gave similar results except for slightly larger deviations and nearly quantitative precipitation of catechin. Analysis of four mixtures, each with all four phenols but with one in turn higher in concentration, showed very good agreement in both assays with the value calculated from the assays of the individual substances. Therefore, the phenols reacted independently with the F-C reagent and did not affect each other's assay. A similar conclusion can be

Phenol Assay Before and After Formaldehyde Treatment

Syringic Acid, 99.9[a] 40.8[b]	Total Phenol in Mixture, GAE mg/liter		
	By Sum (E)	By Assay (D)	Recovery % E/D
40.8			
100.0			
8.2	45.1	43.5	96.5
8.2	44.8	45.8	102.2
8.2	47.6	46.2	97.0
16.3	44.6	43.4	97.3
49.8			
122.1			
0			
10.0	44.2	45.4	102.8
10.0	36.2	38.9	107.6
10.0	48.7	47.1	96.8
19.9	46.2	44.2	95.7

[b] Mg/liter GAE by calculation: gallic acid equivalent (GAE) calculated from the differences in molecular weight and molar absorptivity in the assay (C).

reached from quantitative recovery of gallic acid added to wine (9) and other similar tests.

The exact chemical nature of the yellow heteropoly molybdo- and tungstophosphoric anions and their blue reduction products is still uncertain. Several types and intergraded forms appear to exist, but detailed discussion would be out of place here. Many of the studies on these substances have been related to their use in phosphate determination which is carried out in acidic solution. Nevertheless, a few observations from recent reports seem significant in relation to the F-C assay for total phenols. The F-C reagent is prepared as a mixture of molybdo- and tungstophosphates. The tungsten ions can substitute freely for the molybdenum ions in the complex yellow heteropoly acid formed in making the active reagent (9).

The heteropoly molybdophosphates are much more reducible than the corresponding tungstates, but the latter exhibit one-electron transfer steps at high pH whereas the molybdophosphates did not (12). Otherwise, the electronic structures and behavior of the two are very similar (13), and perhaps the F-C reagent operates well, partly because one-electron transfers are possible, yet the reagent is a good oxidant.

Alkali slowly decomposes the yellow heteropoly anions, but the reduced blue forms are more stable. Presumably alkaline conditions are required to convert the phenols to phenolate ions and this fact would explain poor reactions of cryptophenols. The reduction involves conversion of some of the molybdenum (or tungsten) ions in the complex from valence six to five without the complex structure of the yellow anion being broken down in converting to the blue form (13, 14). Therefore, any two substances, say phenols, capable of reducing the complex to the blue form behave similarly and without interference as long as excess oxidized heteropoly anion is present. Furthermore, the phenol oxidized by the F-C reagent apparently acts as an electron-contributing reductant and does not become a part of the blue pigment. The blue pigment formed seems to be the same regardless of the nature of the reductant.

The amount of blue pigment formed by reduction of a given portion of yellow heteropolyanion is relatively constant, but the rate of reduction increases as the acidity decreases (15), becoming pH independent with molybdophosphate blue at pH 4 for the transfer of the first two electrons and at pH 8 for the second pair (12). The molar absorptivity of the blue pigment is proportional to the number of electrons accepted at about ε 6000 for each electron transferred in either the molybdenum or tungsten series (14, 15, 16). Thus molar absorptivity in the F-C analysis of 12,000 as for phenol itself would indicate the transfer of a pair of electrons and gallic acid at ε 25,000 two pair. Blue pigments formed from increasing pairs of electrons transferred to the same pigment molecule have maxima shifted toward shorter wavelengths within 850–660 nm, with the 4-electron molybdophosphate blue at about 760 nm (16).

In summary, the F-C assay for total phenols (9) is stoichiometrically reproducible for a given phenol within a modest standard deviation. The color formed is reasonably predictable from the number of monophenol equivalents to be expected in a given phenolic molecule, and predictions can be improved if they are based on the activation and deactivation effects of substituents. The latter effects can best be interpreted from properly chosen model substances but they generally conform to predictions from well known chemical principles. The quantitative assay of total phenols by the F-C reagent is thus highly suitable for comparing the relative contributions of different classes of natural phenols in products like wine. The potentially interfering substances in wine, grapes, and most other foods are limited or analyses can be easily corrected for their presence.

Carbonyl Reactions as a Flavonoid Assay

The substitution of aldehydes as electrophiles into activated nucleophilic positions of aromatic rings, such as the ortho or para positions in a phenol, is well known from the commercial importance of phenol–formaldehyde resins. The reaction can be catalyzed by heat and alkali, in which case many phenols react and oxidation is likely. The reaction can also be catalyzed by strongly acidic conditions, but the more strongly nucleophilic centers produced in *m*-dihydric phenols then may be required for reaction unless the mixture is heated. The products of reaction between aromatic aldehydes, particularly vanillin, and phenols have been analyzed by colorimetry. In the strongly acidic solution used, molar equivalents of vanillin and phloroglucinol rings, such as the A-rings of flavonoids and certain resorcinol derivatives, react to produce color at 500–520 nm maximum (6, 7). The molar extinction coefficients indicate stoichiometric conditions ranging from 28,000 to 45,000 with resorcinol and phloroglucinol derivatives (6, 7). Chlorogenic acid did not react. Although reaction with vanillin has been used to determine catechins and other flavonoids in wine (1, 17), it has serious drawbacks. The strong mineral acid used converts some of the anthocyanogens to anthocyanidins which absorb at the same wavelength (1, 18) while use in red wine is virtually impossible. Furthermore, phloracetophenone and flavonoids with a 4-keto group do not react nor do phlobaphenes from acid-catalyzed polymerization of catechin (1, 5, 6). Polymeric condensed tannins react, but the reaction is decreased, apparently by loss of reactive sites to the polymer linkages and by steric coverage (6). Therefore, the vanillin/HCl reaction gives values which are very difficult to interpret (1).

Formaldehyde in the Stiasny test (with HCl) has long been used to distinguish (by their precipitation) condensed from hydrolyzable tannins, and high formaldehyde uptake is a characteristic of vegetable tannin preparations that have high flavonoid *vs.* gallic acid derivatives. Hillis and Urbach (19) showed that the reaction of formaldehyde at room temperature was confined to the phloroglucinol A-ring, and at pH 1 this remained so even if the solution was heated. Formaldehyde is both more reactive and smaller than vanillin so that A-ring sites inaccessible to vanillin are substituted by formaldehyde (20). Hillis and Urbach (19) tested formaldehyde/HCl precipitation as a gravimetric determination but found the resin too difficult to prepare and dry reproducibly. The combination of this reaction with the reliable F-C total phenol assay before and after precipitation by formaldehyde seemed to be a good prospect (1) for a meaningful flavonoid (precipitable) *vs.* nonflavonoid (nonprecipitable) determination in products such as wine.

Formaldehyde Precipitation of Phenols

Kramling and Singleton (21) developed an analytical procedure which gave reproducible values on a series of wines with a variability of about 2.5%. They found that the formaldehyde did not significantly affect the F-C total phenol assay. Provided the pH was below 0.8, catechin, phloroglucinol, and wine flavonoids were nearly quantitatively precipitated by excess formaldehyde in 24 hrs at room temperature, but pyrocatechol was essentially unaffected. Formaldehyde should be present in excess (about 10-fold molar excess was used (21)), but at very high levels the precipitate was solubilized, and it could be dissolved in ethanol (22).

The factors affecting the completeness of the precipitation of phenols reactive to formaldehyde under the assay conditions have been studied, and it is planned to report improvements in the method later (22). For example, a nomograph to correct for the moderate but real effects of temperature, ethanol content of wine samples, and sugar content of juice samples is under consideration. In this report the general applicability and specificity of the reaction is considered further. The products of the reaction of formaldehyde with flavonoids would be first a methylol substitution followed by crosslinking via a methylene bridge to a second flavonoid (or phloroglucinol) unit. The natural flavonoids of grapes, wine, and most foods are phloroglucinol rather than resorcinol derivatives and have two free nucleophilic centers at positions 6 and 8. Thus the reaction can continue to produce a polymer. The insolubilization of the flavonoid–formaldehyde polymer depends on the effects of the increased size, decreased polarity, and decreased hydration of the polymer. An additional factor may be the tendency of the relatively dense structure to roll up and self associate by hydrogen bonding (23). Also, in the presence of ethanol, methylol groups can convert to ethers, and with HCl they can convert to chloromethyl derivatives (24) which would reduce solubility. Since initial formaldehyde reaction with the phenol would still be required to produce precipitation, these side reactions would only improve the method. The phenols already discussed were also tested to determine their precipitability by reaction with formaldehyde in acid solution.

Method. Essentially the method of Kramling and Singleton (21) was used. The wine or phenol solution, 10.00 ml, was mixed with 5.00 ml HCl solution (100 ml concentrated HCl diluted to 250 ml with distilled water) and 5.0 ml aqueous formaldehyde (13 ml of 37% HCHO diluted to 100 ml with distilled water). The air in the test tube was displaced with nitrogen and the stoppered tube was left at room temperature for 24 hrs. Any precipitate was centrifuged, and the supernatant liquid was filtered through a 0.45μ membrane filter. The filtered solution was assayed in the usual (9) manner for total phenols with the F-C reagent using 5 min at 50°C and correcting for the additional dilution. When phloroglucinol was added, it was added at 4–5-fold molar equivalents of the co-precipitating phenol in such a manner that the final solution volume remained the same.

The results with different substances are shown in Tables I–IX. Among the monophenols (Table I) appreciable precipitation occurred only with 3,4- and 3,5-dimethylphenols. Since the methyl group has an electron-repelling effect similar to, but smaller than, that of a hydroxyl group, it is to be expected that these could readily react and precipitate with formaldehyde. However, the other monophenols did not react, and this type of phenol should not occur among the substances present in grapes, wines, or most foods (*1*). None or insignificant precipitation occurred with acid formaldehyde among the catechol and quaiacol derivatives (Table II).

Most pyrogallol derivatives did not precipitate either (Table III). A small precipitation occurred with pyrogallol (which may be considered as a hydroxyresorcinol) and it reacts with vanillin/HCl also (*7*). However, any additional substitution eliminates this reactivity, *e.g.*, gallic acid (Table IV). Purpurogallin did precipitate with formaldehyde. It is not very soluble in the reaction mixture, and it is a tropolone. Structures of this type might interfere, but in tea and probably in wine the known tropolone derivatives are flavonoids (*1, 25*).

The resorcinol and phloroglucinol derivatives (Table IV) all precipitated rather completely as expected except for 3,5- and 2,4-dihydroxybenzoic acids. Their precipitation was enhanced if phloroglucinol was added; this indicates that formaldehyde substitution occurred but the products were too soluble and too polar to precipitate until crosslinked with phloroglucinol. It had been shown previously that phenols which did not react with formaldehyde were not appreciably entrained in the precipitate formed with those which did (*21*) (*see* Table X). Hydroquinone derivatives, except for one which is also a resorcinol derivative, did not precipitate with formaldehyde (Table V).

Naphthalene- and anthracene-derived phenols did, however, almost uniformly precipitate (Table VI). In natural materials (not grapes or wines) which contain them they would be included in the formaldehyde precipitable group. Several primary amines capable of Schiff's base formation reacted with formaldehyde to lose their F-C oxidizability, but only the resorcinol analog, 3-aminophenol, precipitated (Table VIII). Sulfite also reacted but did not precipitate with formaldehyde, and the F-C oxidizability was suppressed (Table IX). The resorcinol derivative, 2,4-dimethoxycinnamic acid, formed a precipitate with formaldehyde, but it did not react appreciably in the F-C assay.

The most important group of compounds for our purposes, the flavonoids, are shown with coumarins in Table VII. The flavonoid aglycones precipitated nearly quantitatively except for fisetin, myricetin, and tectochrysin. Tectochrysin, 5-hydroxy-7-methoxyflavone, still preceipitated to 51% without phloroglucinol even though two of the three potential hydroxyl groups of the molecule are converted to ethers with lesser dipoles and the third is strongly hydrogen-bonded to the 4-keto group and, therefore, decreased in acidity. Myricetin apparently is too polar for complete precipitation alone, but it readily precipitates when crosslinked with phloroglucinol. Fisetin is the only flavonoid in the series tested lacking the 5-hydroxyl, a flavonoid structure uncommon among food plants. Its failure to precipitate with formaldehyde even with phloroglucinol agrees with the failure of such compounds to react with vanillin (26). Since it represents a type of flavonoid restricted to wattle and a few other natural sources, its failure to react does not present a problem in flavonoid estimation in wine and similar materials.

The coumarins did not precipitate (Table VII) without phloroglucinol and not very well with it. This also indicates the importance of the phloroglucinol substitution pattern as opposed to the resorcinol. These data show that the coumarins reacted weakly at a single center and therefore could not propagate polymers, although some linking and co-precipitation with polymers of other phenols can occur. Esculin, the 6-glucoside of esculetin, has only the 8 position available for formaldehyde substitution and, by analogy with fisetin, its apparent lack of reaction seems logical.

The glycosides of normal flavonoids (Table VII), except naringin, showed some precipitation alone, and without exception, precipitation increased considerably when they co-reacted with a few moles of phloroglucinol. The chalcones tended to precipitate more completely than the true flavonoids, probably because of the more free and reactive phloroglucinol moiety.

It is difficult for the formaldehyde product to drag the flavonoid polymer from aqueous solution if it is highly solubilized by attached sugars, as evident from these data. That the presence in the same reaction mixture of sufficient phloroglucinol will crosslink and overwhelm the solubilizing effect and precipitate the glycosides is also seen in the data. This is not as great a problem for successful flavonoid estimation in wine and grape extracts as it may first appear because these products contain a relatively large fraction of their flavonoids in non-glycoside forms, especially catechins and oligomeric condensed anthocyanogenic tannins (1). The resultant mixed precipitate renders the glycosides insoluble also.

The glycosidic anthocyanins are almost completely precipitated with red wines of a few months age and normal tannin content. However, with young wines made from the same grapes, a rosé wine with only 713 mg GAE/liter total phenol gave 311 mg GAE/liter as not-precipitated nonflavonoid, and with increasing levels of total phenol in red wines to a high of 1421, the apparent nonflavonoid content decreased until a constant level of 250 mg GAE/liter was reached (*21*). Although with grape wine the method has been quite informative, further improvement through routine addition of a glycoside-precipitating co-reactant such as phloroglucinol is under study.

Table X illustrates the successful application of formaldehyde precipitation as a means of estimating the flavonoid and nonflavonoid contents in a mixture. The mixture consisted of catechin as the flavonoid and caffeic, vanillic, and syringic acids as the nonflavonoids. The catechin was 86% precipitated (lower than usual because of the low level), but the other substances were not significantly precipitated. The slight apparent loss of caffeic acid is attributable to experimental variation since in many other experiments the lack of reaction and precipitation or co-precipitation of caffeic acid or chlorgenic acid has been demonstrated. Allowing for the same slight solubility of the catechin-formaldehyde product in the mixtures as in the single component solution, the analysis of the mixtures gave 95.7–107.6% of the calculated value. This indicates no significant co-precipitation or entrainment of the nonflavonoids as the flavonoid was removed. This result has been verified a number of times with different substances added to model solutions and wines (*21, 22*).

Viewing the data (Table X) as if it had been the usual assay of unknowns and subtracting the assay values after formaldehyde treatment from those before, the mixtures 1, 3, and 4 would apparently contain no flavonoid when in fact they contained 8.4 mg/liter GAE by separate assay. On the other hand, mixture 2 with 16.8 mg/liter GAE of flavonoid by separate assay gave 6.9 mg/liter by formaldehyde precipitation. If correction was made for 5.8 mg/liter GAE residual solubility of the catechin–formaldehyde product then mixures 1–4 would be indicated to have, respectively, 3.9, 12.7, 4.9, and 5.0 flavonoid and 39.6, 33.1, 41.3, and 38.4 mg/liter GAE nonflavonoid. These values are considered very close to the true content considering the results are based on differences between two assays with the attendant increase in variability.

In summary, formaldehyde precipitates only phloroglucinol derivatives (including flavonoids), some resorcinol analogs, and polycyclic phenols of naphthalene or higher ring systems under the conditions tested. Total phenol analysis before and after such precipitation gives a useful estimate of nonflavonoids (remaining in solution) and flavonoids (precipitating) provided that, as in grape wine, conditions are such or can be arranged so that flavonoid glycosides precipitate, and potentially interfering phenols such as 5-deoxyflavonoids, certain coumarins, and polycyclic phenols are absent or in low concentration.

Examples of Practical Application of These and Similar Analyses to Wine

Kramling and Singleton (*21*), in addition to the original development of the flavonoid–nonflavonoid analysis just described, showed that white and red table and dessert wines had very similar nonflavonoid phenol content. In the wines studied, the total phenol content ranged from 205 to 1421 mg/liter GAE, but the nonflavonoid phenols were relatively constant at 190–343 mg/liter GAE. Nearly all of the variation was in the formaldehyde-precipitable flavonoid content which ranged from 10 to 1169 mg/liter GAE. The nonflavonoids of grapes, largely caffeic acid derivatives like chlorogenic acid (*1*), were shown to be essentially confined to the vacuolar fluids of the grape which are easily expressed as juice. The flavonoids are very low in juice, but very high in grape solids, skins, and seeds. As a result, white wine phenols are almost exclusively nonflavonoid unless appreciable pomace extraction had been made, and red wines had increased flavonoid in proportion to pomace extraction. Although this has been inferred from other largely qualitative studies (*1*), it is believed that this study (*21*) represented the first direct quantitative evidence for it.

Grape varieties vary slightly in nonflavonoid content, but they vary considerably in flavonoid contribution to wines prepared in similar ways (*21, 27*); large differences in the flavonoid content caused by vinification and processing often obscure varietal differences. Wines made (*21*) from the same grapes had a flavonoid content of 132 mg/liter for the lightest rosé and 1169 mg/liter GAE for the darkest red, but the nonflavonoid content assayed 201 and 252 mg/liter, respectively, and averaged 262 mg/liter GAE for a series of five wines.

Table XI. Average Analysis of Nonflavonoid Phenol Content of Typical Woods Which Contact Wine (*27*)

Wood	Solids Extd, grams/ 100 grams Dry Wood	Total Phenols, mg GAE/gram Extd Solids	Non-flavonoid Phenols, mg GAE/gram Extd Solids	Non-flavonoids, % of Phenols
American oak	6.45	365.1	320.2	87.7
European oak	10.39	560.7	505.3	87.7
Redwood	15.94	607.5	590.8	97.2
Cork	2.44	141.4	137.9	97.5

Formaldehyde analysis has been used to detect and measure oak extract in wines aged in wood cooperage and to correlate the amount of extract with the aging effect (*27*). Tannins and phenols of oak (and redwood and cork) are predominantly nonflavonoid–hydrolyzable tannins (Table XI), and they add to the otherwise relatively low and

Table XII.   Selected Analyses of Wines with Known Treatments with Oak (21)

Wine	Wood Treatment[a]	Total Phenol, mg/liter GAE	Non-flavonoid Phenol, mg/liter GAE	Estd[b] Non-flavonoid from Oak, mg/liter GAE
Carignane	None	950	164	0
Carignane	AN 2.5 grams/liter, 4 days	1000	203	39
Carignane	AN 5 grams/liter, 4 days	1034	256	92
Carignane	AN 10 grams/liter, 4 days	1040	318	154
Carignane	AN 20 grams/liter, 4 days	1204	477	313
Carignane	AN 50 grams/liter, 4 days	1504	884	720
Cabernet Sauvignon	none	1744	353	0
Cabernet Sauvignon	EU 60 gal., 3 mo.	1798	352	0
Cabernet Sauvignon	EU 60 gal., 1 yr	1670	414	61
Cabernet Sauvignon	EU 60 gal., 3 yr	1374	458	105
Chardonnay	none	—	200	0
Chardonnay	EN 60 gal., 6 mo.	404	314	114
Chardonnay	EN 60 gal., 1 yr	518	478	278
Chardonnay	EN ———, 5 mo.	374	320	120
Chenin blanc	none	291	233	0
Chenin blanc	EU 340 gal., 2 mo.	326	251	18
Chenin blanc	EU 340 gal., 6 mo.	311	248	15
Flor sherry	AU 50 gal., 8-9 yr	324	300	100

[a] A = American oak, E = European oak; N = new, U = previously used; chips grams/liter, barrels capacity; time of contact.
[b] Nonflavonoid content of treated wine minus that of the untreated wine (or for Chardonnay and sherry values on the high end of the typical range).

constant flavonoid content of wine during aging (Table XII). Thus, increased nonflavonoid in wine is proportional to the newness of the barrel, time, and contact of wood surface per unit contents (28), and it can be followed even if total phenol varies considerably because of various reactions during aging. Results with wines having known barrel aging treatments suggest that this technique is capable of reliable detection and estimation of treatment by aging in wooden cooperage of purchased bottled wines whose history is unknown. This is feasible because of the relatively constant and low content of nonflavonoid phenols in wines from a single variety and in wines in general so that a relatively small increase in nonflavonoid phenol from the barrel or other source becomes significant (21, 27, 28).

The flavonoid fraction, especially catechin, from grapes has been reported to be the primary source of yellow and brown colors in white wines (29). In work not yet reported in detail (30, 31, 32) the flavonoid content was shown to determine and control the potential of a white table wine to brown. Wines were prepared from several white grape varieties with increasing periods of contact between the juice and the pomace, allowing various extractions of flavonoid compounds. When exposed to oxygen under standardized conditions, each wine's increase in absorbance at 420 nm correlated almost perfectly with the assayed flavonoid content. Experimental data on two of the varieties, Clairette blanche (with the greatest tendency to brown) and Colombard (with the least tendency) are shown in Figure 1 (32). These results explain the greater stability and resistance to browning of wines or beers deliberately treated with formaldehyde (33, 34).

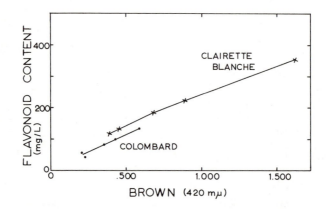

Figure 1. Flavonoid content mg/liter GAE, in wines made from the same grapes, as correlated with browning of each wine upon exposure to oxygen under standardized conditions

Peri and Pompei (35, 36) have helped verify the methods elaborated here (9, 21), and they have extended them by adapting cinchonine precipitation to separate tannins from nontannins. Thus, in a mixture of the two tannins precipitated by cinchonine they were able to redissolve them and quantitate tannic acid separately from grape seed tannin because of tannic acid's failure to precipitate with formaldehyde and by appropriate application of the F-C analysis. They applied these techniques to analysis of a rather tannic white wine containing a total of 813 mg/liter GAE phenols, 33% (267 mg/liter GAE) of which was nonflavonoids, 25% (205 mg/liter GAE) of which was nontannin flavonoids, and 42% (347 mg/liter GAE) of which was tannins. Thin layer chromatograms illustrated the validity of the separations.

Table XIII.   Total Phenol, Tannin, and Nontannin Content of Wines From the Data of Mitjavila *et al.* (*37*)

	Total Phenol by F-C Assay, mg/liter GAE	Nontannin, mg/liter GAE	*PVP/Cl₃CCOOH Pptn*	
			Tannin (Pptd), mg/liter GAE	Sum, % of Total
Wine				
Fronton red	1328	846	418	95
Rhone red	1701	1019	568	93
Burgundy red	1664	1019	627	99
Bordeaux red	2019	1210	882	104
Rioja red	1842	1064	782	100
Toulouse red	1595	1364	200	98
Piedmont red	2092	1455	746	105
Corbières rose	455	364	91	100
Sherry white	437	313	84	91
Juracon white	318	293	20	98

Other separations can easily take advantage of the F-C total phenol method in the same way that has been indicated for formaldehyde and cinchonine precipitation. In one recent example (*37*), tannins were determined by complexing with polyvinylpyrrolidinone and precipitation of the complex with trichloroacetic acid. The precipitate was redissolved and assayed by the improved F-C method. They showed quantitative absence of tannic acid and enotannin in the supernatant liquid and a similar absence of gallic acid and catechin from the precipitate by this method. The only component tested which divided into both fractions was leucocyanidin. Monomeric flavan-3,4-diols are very low if not absent in wines (*1, 38*). In a series of wines, the sum of the precipitated tannins and the nonprecipitated phenols by this method (*38*) was identical (within experimental variation) with a separate assay of the total phenol content (Table XIII). Direct comparability with values obtained by others is an advantage of this method. Their data (*37*) in Table XIII are recalculated to gallic acid equivalents, since *d*-catechin had been used as the standard, by multiplying by the conversion factor derived from the best data in this paper, 0.9094. It appears to us most convenient to express such results in terms of mg of gallic acid equivalent per liter of beverage or per kg of fruit (*1*), and it is hoped others will adopt this terminology for even easier comparisons.

Acknowledgment

The author thanks the California Wine Advisory Board for financial support and C. Kramer for her capable assistance in conducting the analyses.

Literature Cited

1. Singleton, V. L., Esau, P., "Phenolic Substances in Grapes and Wine, and Their Significance," 282 p., *Advan. Food Res.* Suppl. 1, Academic, New York, 1969.
2. Pro, M. J., *J. Ass. Off. Agr. Chem.* (1952) **35**, 255; (1954) **37**, 665; (1955) **38**, 757.
3. Ribéreau-Gayon, P., Satore, F., *Chim. Anal.* (Paris) (1970) **52**, 627.
4. Swain, T., in "Chemistry and Biochemistry of Plant Pigments," T. W. Goodwin, ed., p. 545, Academic, New York, 1965.
5. Swain, T., Hillis, W. E., *J. Sci. Food Agr.* (1959) **10**, 63.
6. Goldstein, J. L., Swain, T., *Phytochemistry* (1963) **2**, 371.
7. Swain, T., Goldstein, J. L., in "Methods in Polyphenol Chemistry," J. B. Pridham, ed., p. 131, Macmillan, New York, 1964.
8. Joslyn, M. A., Morris, M., Hugenberg, G., *Mitt. Rebe Wein Obstbau Fruchteverwert. (Klosterneuburg)* (1968) **18**, 17.
9. Singleton, V. L., Rossi, J. A., Jr., *Amer. J. Enol. Viticult.* (1965) **16**, 144.
10. Schroeder, H. A., *J. Chromatogr.* (1967) **30**, 537.
11. Singleton, V. L., Marsh, G. L., Coven, M., *J. Agr. Food Chem.* (1966) **14**, 5.
12. Papaconstantinou, E., Pope, M. T., *Inorg. Chem.* (1967) **6**, 1152.
13. Varga, G. M., Jr., Papaconstantinou, E., Pope, M. T., *Inorg. Chem.* (1970) **9**, 662.
14. Reznik, B. E., Tsyganok, L. P., *Zh. Neorg. Khim.* (trans.) (1968) **13**, 548.
15. Crouch, S. R., Malmstadt, H. V., *Anal. Chem.* (1967) **39**, 1084.
16. Papaconstantinou, E., Pope, M. T., *Inorg. Chem.* (1970) **9**, 667.
17. Pompei, C., Peri, C., *Vitis* (1971) **9**, 312.
18. King, H. G. C., White, T., in "The Chemistry of Vegetable Tannins," p. 51, Society of Leather Trades' Chemists, Croydon, 1956.
19. Hillis, W. E., Urbach, G., *J. Appl. Chem. London* (1959) **9**, 474.
20. Hillis, W. E., Urbach, G., *Nature* (1958) **182**, 657.
21. Kramling, T. E., Singleton, V. L., *Amer. J. Enol. Viticult.* (1969) **20**, 86.
22. Singleton, V. L., Kramling, T. E., Sullivan, A. R., private communication.
23. Imoto, M., Ijichi, I., Tanaka, C., Kinoshita, M., *Makromol. Chem.* (1968) **113**, 117.
24. Mironov, G. S., Farberov, M. I., Shem, V. D., Vetrova, V. V., *Kinet. Katal.* (trans.) (1968) **9**, 612.
25. Singleton, V. L., in "The Chemistry of Plant Pigments," C. O. Chichester, ed., p. 143, *Advan. Food Res.*, Suppl. 3, Academic, New York, 1972.
26. Hillis, W. E., Urbach, G., *J. Appl. Chem.* (London) (1959) **9**, 665.
27. Singleton, V. L., Sullivan, A. R., Kramer, C., *Amer. J. Enol. Viticult.* (1971) **22**, 161.
28. Singleton, V. L., ADVAN. CHEM. SER. (1974) **137**, 254.
29. Rossi, J. A., Jr., Singleton, V. L., *Amer. J. Enol. Viticult.* (1966) **17**, 231.
30. Kramling, T. E., M.S. Thesis, University Calif., Davis, 1969.
31. Singleton, V. L., *Die Wynboer* (1969) No. **455**, 13.
32. Singleton, V. L., DeWet, P., Kramling, T. E., Van Wyk, C. J., private communication.

33. Cantarelli, C., Pompei, C., Peri, C., Montedoro, G., *Amer. J. Enol. Viticult.* (1971) **22**, 59.
34. Macey, A., Stowell, K. C., White, H. B., *J. Inst. Brewing* (1965) **72**, 29.
35. Peri, C., Pompei, C., *Amer. J. Enol. Viticult.* (1971) **22**, 55.
36. Peri, C., Pompei, C., *Phytochemistry* (1971) **10**, 2187.
37. Mitjavila, S., Schiavon, M., Derache, R., *Ann. Technol. Agr.* (1971) **20**, 335.
38. Weinges, K., Piretti, M. V., *Justus Liebigs Ann. Chem.* (1971) **748**, 218.

RECEIVED May 29, 1973.

10

Wine Quality Control and Evaluation

RICHARD G. PETERSON

Beaulieu Vineyard, Rutherford, California[a]

Existing quality control programs of California wineries are discussed, and a type of quality control program is presented. This includes check lists showing individual trouble areas, arranged chronologically, with discussion of each point so that the reader could set up a winery quality control system using this paper as a foundation. Quality control of winemaking has its basis in the vineyards. One aspect of quality control may be unique to wine; wines may improve in general quality characteristics after storage in bottles under ambient conditions for specific, though often unpredictable, periods. Storage beyond the optimum time often damages quality. The paper presents one approach to the control of quality under these circumstances.

Winery quality control has not been formalized in the United States to the point where standard methods, practices, or performance criteria are established. The literature devotes little attention to the subject of wine production quality control, *per se*. The first survey of quality control in the California wine industry appeared as late as 1972 (*1*), and it is limited to sanitation aspects. One does not find chapters on quality control of winemaking in any of the major wine technology texts.

Yet quality control in this increasingly important industry does exist. Winery quality control might include all the procedures, tests, and practices which are used to guarantee that the winery production operation will produce and ship the right wine of the right quality level in the right package. Each approach to this end has generally been unique for each winery, but there are a number of similarities which apply throughout.

In most cases, quality control as actually practiced has arisen from the myriad of do's and don'ts which themselves have arisen through

[a] Present address: Monterey Vineyard, P.O. Box 650, Gonzales, Calif. 93926.

trial and error as this early art assumed more and more of a scientific foundation. Unfortunately the best methods of final product evaluation are organoleptic and therefore highly subjective. This is as true in 1974 A.D. as it was in 1974 B.C.

Acetic acid content is used as a criterion for aerobic bacterial spoilage in wines. We can easily analyze two wines and determine that one has a lower volatile acidity than the other. But by every available standard of product quality judgment, the wine with the higher level of acetic acid may be the superior product. It is no accident that Subpart ZZ, Part 240, Title 26 of the Code of Federal Regulations allows the direct addition of acetic acid "to correct natural deficiencies in grape wine." As sanitation practices have improved, the so-called natural acetic acid content diminished, and this has been correlated with lower consumer acceptance in certain cases (2).

Likewise, a number of materials such as SO_2, charcoal, gelatin, pectic enzymes, and ion exchange processes which are normally used in wine-making may greatly improve a given wine while greatly downgrading a second one. There is no substitute for trial and error in wine fining, stabilization, and clarification. For this reason, no winemaking recipe can be precise, and many of the most respected and successful wine-makers never think of using a recipe. Conversely, where product conti-nuity from year to year is desired, as is the case in many of the larger volume producers, a recipe approach is mandatory to continued economic success. Up to 0.5 ppm of copper sulfate may be added to wine provided the residual level in the bottled product contains less than 0.2 ppm. The usual purpose of adding copper is to remove H_2S formed as a by-product of yeast fermentation under certain conditions (3). Prior to copper's at-taining legal status, it was common practice for wineries to exaggerate the use of brass or bronze fittings, valves, and pumps because it made cleaner smelling wine. Many commercial wines exist in which copper levels of 0.5 ppm or more would go unnoticed, both to the winemaker or quality control chemist and to the wine consumer. However, other commercial wines exist which can develop a whitish copper–protein cloud, copper *casse*, with levels of copper as low as 0.2 ppm (4).

Even when no additive is used in winemaking, the necessity for small-lot trial before production scale operation is apparent. Because a high percentage of wine is consumed only after chilling, and because chilling may accelerate the precipitation of potassium acid tartrate, ill-defined colloids, anthocyanin–tannin polymers, proteins, etc., simple cold stabilization by refrigeration in the winery may irreversibly alter the product and its eventually-perceived quality level. It often happens, especially in heavy-bodied red varietal wines, that a dark, amorphous precipitate may form in the bottle over several years. Usually tannoid, this material may have either a distinctly bitter taste or no apparent taste at all, but the clear wine above usually becomes softer, lighter in character, and more pleasing for having lost the insoluble polymer. A chemically similar, dark, amorphous precipitate usually could have been separated from the same wine if that wine had been held at temperatures near freezing for several days or weeks prior to the original bottling. However, the result is rarely the same. The cold treatment of the relativly young wine often immediately strips the wine of its body and changes the character undesirably, and usually that wine never recovers, regardless of subsequent aging. It is this general type of recurring phenomenon and man's inability to define it precisely in scientific terms that slows the transformation of winemaking from an art into an exact science. All these things mean to the quality control chemist in the winery that nothing can be taken for granted. Each wine must be judged on its own merits and few, if any, standard checks can be safely bypassed. The winery quality control manager must assume that every wine is susceptible to every pitfall at every step.

Sanitation is of primary importance since the overwhelming majority of wine difficulties are microbiological problems (excluding human error such as pumping from the wrong tank in blending, using the wrong fining agent or the wrong amount of the right fining agent, or mis-guaging the volume of a given tank). For this reason sanitation as a subsection of quality control has received most attention by the industry as a whole. The Wine Institute publishes a sanitation guide with periodic revisions and additions (5). This is intended for use by wineries in setting sanitation goals and guidelines within which to operate, and it should form the base around which any quality control effort is built.

Basic Quality Control Plan

Winery Phase. To develop a meaningful system, it is helpful to classify the winemaking operations. A convenient outline to use might be based on stages of processing, with each listed chronologically. One workable system is to divide the total winery quality control program into two major phases. A third phase is unique and it will be discussed later. Phase I, as defined here, will include the quality control effort from the vineyard harvest up to, but not including, the bottling room. Phase II begins at the bottling room and ends as the case goods are loaded onto rail cars or trucks for distribution.

Vineyard Phase. Quality wine production differs from other food production in that the state of the raw material is usually more critical for wines. It is generally true that the quality level of most preserved foods changes more or less directly with the original quality level of the corresponding raw foodstuff. This is true in wine for most of the same reasons that exist in all foods. However, additional factors in the case of fine wine processing occur which make the raw material quality level unusually and surprisingly critical to eventual wine product quality. It is not correct to say that two wines ought to be comparable because the grapes in each case were harvested at, for example, Brix levels of 22.7° and total acid levels of 0.75% (assuming identical processing in the winery).

Very marked differences occur in the quality of wine of the same variety grown on similar soil but under varying climatic conditions. Winkler (6) pointed out that even in the renowned wine-producing areas of Europe, with their varied soils, "heat summation [in the vineyard] must be accepted as the principal factor in the control of [wine] quality." Studies of this in the late 1930's led Amerine and Winkler to use heat summation to divide grape-producing areas into climatic regions. In retrospect, this was clearly one of the most important advances in enology/viticulture of this century. At the same time, the division into only five regions is not sufficient for a complete understanding and control of wine quality. It is increasingly apparent that wine quality is even more complicated in its relation to vineyard factors. It is now common to see references made to micro climates within the general climatic regions. Length of the growing season causes significant differences also. Grapes referred to in the paragraph above might reach that level of maturity on October 1st in Napa County and on November 1st in Monterey County. The resutling wine types will usually be different, although quality levels might be considered similar. This has great significance to the winery quality control effort but is outside the scope of this paper.

By any objective standard, the vineyard factors of wine quality control deserve at least as much weight as either of the winery phases described here. A given grape variety, grown in the central San Joaquin Valley of California, may yield greater tonnage but produce poorer wine than the same variety grown in one of the coastal areas. The price paid to the grower will usually reflect these factors. Cynical perhaps, but nevertheless a valid function of winery quality control is knowing the exact source of every grape used for every wine in the winery.

Matching the Quality Control Effort to the Size of the Winery. The quality control plan can be as deep or as shallow as the individual winery manager desires to make it. It is essential that he determine what his needs are from a quality control program. Sizing the quality control effort to fit the winery and fulfilling this effort is his major goal. Many wineries today turn the Phase I quality control effort over to departments that do more than quality control, partially because of small company size.

Examples of three different approaches to the problem are shown by these hypothetical wineries:

Winery A is very large and has a well staffed and departmented laboratory. It has a separate quality control department, but it is located in the main laboratory and performs its whole function within the confines of the bottling room, warehouse, and laboratory. In this example, the laboratory quality control department has Phase II as its total area of responsibility. A system of mechanical and routine inspections against a check list works perfectly.

Any functions which control quality of the product prior to bottling (Phase I) are done by someone else. These are usually the responsibility of the production manager who acts through individual winemakers, an operations manager, or individual plant managers where the company operates several winery facilities. Theoretically, it is always better to place the quality control function outside the production operation. Currently, however, this is rarely done. One of the reasons may be the difficulty most winemakers have encountered in trying to reduce their winemaking procedures to recipes. It usually ends up being more practical to retain control where the ultimate responsibility lies and hope that volume considerations don't overrule quality ones.

Winery B, by contrast, is much smaller and has no formal quality control department. The job of controlling product and package quality falls on the shoulders of the winemaster, who is also the production manager, plant manager, vineyard superintendent, and purchasing agent. Quality control effort in this type of organization, whether Phase I or II, is done by this man as part of his production job.

A third example, Winery C, is a relatively small company also, but somewhat larger than company B. Winery C is large enough to afford separate laboratory personnel, but only two or three. The primary functions of this laboratory are analysis, blend control, and inventory watching, a quality control function which can best be described as baby sitting the wine while it is in process in the winery. It is this baby sitting function that best defines the total quality control duty during Phase I winemaking, in all three of these example wineries.

Phase II quality control is often shallow and highly subjective in the smaller wineries, but it can be quite sophisticated in the larger ones. This is true in Europe as well as California and may be true in general. It is likely that the bottling room quality control in a large and sophisticated winery will differ little from bottling room quality control in any other large plant such as a brewery, soft drink, or fruit juice bottling plant. The application of statistical methods and criteria to the control of such things as glass, caps, and label manufacture and use might be copied directly with few modifications from any competent and well established bottling plant. Control charts, frequency distribution graphs, and other graphic methods for presenting data in more readable forms are commonly used in presenting the picture of existing product quality levels to management. Similarly, the various mathematical terms commonly used for interpretation of raw quality control data in other manufacturing industries will be found in the large-volume technically-sophisticated winery bottling room quality control department.

The reason should be obvious. By the time the wine product reaches the bottling operation, one is dealing with a manufactured product, no more and no less. Thousands of these manufactured products sail down the bottling line every hour, and they are just as subject to quality variations as any other manufactured product. Capacities and fill points of glass, the opening torque, liner and application uniformity of pilfer-proof screw caps, appearance defects, code dating, general appearance, and printing on labels and cases all are subject to the variations in quality normally found in manufactured articles. In all these, quality is a variable and the change in its magnitude is a frequency distribution. Probably the two most used criteria in recording or reporting a given value of quality in wine packages are fraction defective and range. Arithmetic mean and standard deviation are used, but to a lesser extent since they are less meaningful in most situations.

At least two large California wineries actually own and operate their own glass plants where much of their wine bottle requirement is manufactured. These are typical glassware manufacturing operations and both have typical glassware quality control programs. Likewise, some larger wineries manufacture their own aluminum pilfer-proof caps, and their quality control effort is included in Phase II as used here.

Most of the statistical quality control methods used in Phase II are impossible during Phase I. Winemaking is essentially a batch process and, since the product is a liquid, it can be made completely homogeneous by simply mixing the whole lot of a given wine in one tank. After this is done, the contents of the tank can be analyzed and tasted for bottling approval. When accepted for bottling, the entire contents of a tank can be transferred *via* pipelines to the filler bowl in the bottling room without fear of finding variations from one bottle to the next. Phase I quality control, therefore, usually insures that nothing adverse happens to any of the lots of bulk wine while they are being aged or otherwise processed in the winery. Thus, the term baby sitting for the quality control effort during the first phase of wine production.

Check Lists. Tables I and II list groups of quality control check points which should be monitored for each wine lot during the corresponding phases in the winery. The quality control manager should develop a list of check points to fit his individual situation. The lists shown here are examples and are by no means complete. In some situations, a quality control function might be the distribution of further check lists among workers for the operation of certain machinery, giving step by step procedures to follow. This would control quality by tending to eliminate human error in those exact areas where a particular error is most likely.

Also, some of these check points may be by-passed by some wineries. For example, where the winery manager is also operating the vineyard there will be no possible error in grape identification at the winery. Only one variety per day will usually be received. Conversely, very large wineries can expect to monitor grape receipt very closely for identification and source of each incoming load.

Some check points or groupings of check points are more critical during certain seasons than at other times. All checks in the bottling room are probably more critical during autumn even though the bottling area may be sealed off from the rest of the winery very well. The atmosphere in and around most wineries is literally full of migrant yeast cells during the fermenting season. Of course, some wines are completely dry when bottled while others, usually the standard or non-premium, large volume products, contain reducing sugar at bottling time. Obviously, quality assurance against microbial spoilage is a more difficult problem when the air contains excess yeast and the wine contains excess sugar. Less obvious, but definite is the effect on microbial spoilage of lower *vs.* higher alcohol and SO_2 content in the wine (7, 8). Even carbonation seems to inhibit subsequent spoilage in wine, but the mechanism is obscure. Similarly, the added flavors in certain special natural wines are known to inhibit spoilage (2).

Some winery requirements, such as corks and lead capsules, are not manufactured in the United States. These are purchased, along with a considerable percentage of the processing equipment used, from European sources, either directly, or through American sales representatives. Working through at least one middle man and two languages can complicate the quality control problem. Probably because the large economic loss absorbed by a European supplier shipping defective merchandise or by the American buyer who failed to order the exact thing he wanted, tends to guarantee that both seller and buyer fully understand and approve the specifications prior to shipment. Often actual product samples are submitted to the winery for approval prior to shipping the whole order. Also, more American enologists are visiting their European suppliers now than in the past, and close personal contact has minimized these potential problems. It is an excellent practice to meet the managers and see the manufacturing facilities of each current major supplier in

Europe. It is usually welcomed by the suppliers as much as the users, for the benefits accrue to both.

Check Points—Table I

For more detailed discussions of many of the individual winemaking points discussed here, *see* Amerine and Joslyn (7).

Grape Receipt. MATURITY OF GRAPES. Proper maturity of the grapes depends on what wine they will be used for. For most wines, an optimum sugar and acid content can be established based on the winemaker's experience for the particular vineyard and wine type. Sugar content should usually be estimated by refractometer in the field at weekly intervals beginning about three weeks prior to the assumed picking date, and the acidity checks need be started only when the sugar levels rise to within about 2° of the desired maturity level. Daily checks may have to be made throughout the week just prior to picking.

CONDITION OF GRAPES AS RECEIVED. *Mildew, Rot.* Check for mildew and rot by visual inspection of each gondolla or truck load.

Temperature. The must may have to be chilled immediately after crushing to remove excess field heat. Grapes should be delivered to the crusher as soon as possible after picking and at as cool a temperature as possible. Early morning picking is almost always best.

Foreign Matter. Leaves, excess stems, dirt clods, and clusters of unripe second crop grapes cannot be tolerated because they can downgrade wine quality significantly, even in small amounts. Rocks, tramp metal, and lost picking knives can damage equipment.

Juicing. The amount of juice in the bottom of the gondolla is related to berry firmness, temperature, and distance hauled. The color of the juice at delivery is critical.

Color of Grapes. Does the actual color match that expected for the variety and ripeness? Watch for sunburn and raisin formation as indexes of overripe condition and a reason to channel that lot of grapes into a lesser quality wine.

Insects. Watch for insects, especially fruit flies, bees, spiders, and moths.

CONFIRMATION OF VARIETY. Experience is required, but most varieties have sufficiently unique characteristics so that the variety can be recognized easily with experience. Check taste, berry size, shape, color, juiciness, acidity, type of leaf, and surface sheen.

SPRAY RESIDUES. As a general rule, little or no toxic material is used on vineyards after early summer. Finding a spray residue on grapes at the winery is highly unlikely, although many analyses are still made to detect this. Residues are usually traceable to wind drift from an adjacent field with another crop. Nevertheless, the prudent winery monitors this

Table I.   Quality Control Check Points—Phase I

Grape Receipt
Maturity of grapes
Condition of grapes as received
  Mildew, rot
  Temperature
  Foreign matter
  Juicing
  Grape color
  Insects
Confirmation of variety
Spray residues
Vineyard practices

Crushing and Fermentation
Daily equipment cleaning and sanitization
Sterilants
Proper, timely addition of SO_2
Temperature control during fermentation
Pure yeast cultures
Proper circulation during fermentation
Prompt removal of stems from crusher and pomace from presses
Insect control
Malo-lactic starter culture
Prompt removal of new wine from lees
Smoothness of fermentation curves

Processing and Aging
Constant supervision of wine tanks
Prompt and current chemical analysis
Organoleptic analysis before and after all wine transfers
Sanitation of tanks, hoses, equipment, and pipelines on a rigid schedule
Use of N_2 and CO_2 for oxygen control in bulk wine, in pipelines and in
  tanks
Oxygen control by timely topping of wine tanks
Control of temperature during storage
Quality control audit of incoming process materials such as filter powder,
  organic acids, cleaning agents, filter pads, fining agents, etc.
Proper use of equipment by personnel
Continuous monitoring of clarity and color during filtration
Inventory control by data processing (large wineries)
Barrel sanitation
Equipment maintenance
Final check before bottling

carefully. The first step is to survey all the source vineyards to record their exact schedule of pesticide use. According to a report on pesticide illness for 1971 issued by the California Department of Agriculture and Department of Public Health (9), there were 145 pesticide illness cases reported in the state during the year, none of which involved vineyard workers. Of the 145, 83 were occupational in origin with 68 among workers involved directly in pesticide application.

VINEYARD PRACTICES. A history should be developed during the growing season by periodic inspections of each vineyard for use and abuse of irrigation, fertilizers, sulfur, etc. Proper pruning to avoid over-cropping, avoiding heavy traffic on clay soils between vineyard rows, and physical damage to vines (tractor blight) should be monitored carefully.

Crushing and Fermentation. DAILY EQUIPMENT CLEANING AND SANITIZATION. All equipment used in the crushing operation must be cleaned and sanitized daily. This is especially important in keeping the local population of fruit flies to a minimum and in avoiding early build up of volatile acidity from *Acetobacter* infections. Winery waste tends to putrify rapidly if allowed to stand.

STERILANTS. Monitor the use of sterilants to ensure that the established procedures are being followed, especially during the fermenting season. Two chemicals universally used in wineries during the fermenting season, but which are violently explosive when inadvertently mixed (a reducing agent with an oxidizing agent) are potassium metabisulfite and calcium hypochlorite. These must be stored in areas remote from each other and left there. Only the small amount of each that is needed for the immediate purpose should be allowed out of the main storage areas.

PROPER, TIMELY ADDITION OF SO_2. SO_2 addition, as with other chemical usage, must be monitored often by spot checks. Analyze the tank contents for total SO_2 after each addition and mixing.

TEMPERATURE CONTROL DURING FERMENTATION. Temperature control during the fermentation process is one of the most important controls in winemaking. Smooth, controlled fermentation rates without off flavors being developed or volatilization of desirable flavor components can best be attained by precise temperature control. Also, the fermenting mass expands considerably during fermentation, and keeping it under control will keep it inside the tank where it belongs.

PURE YEAST CULTURE. Recheck population periodically for purity. If a wild yeast population builds up significantly, discard the culture and begin again from a slant. Some wineries are using mass pitching techniques with dry or frozen yeast very successfully. This eliminates the need for any monitoring of culture population. Wild yeasts often cause unreliable and erratic fermentation rates, and this is often accompanied by off odors and off flavors in the wine. Erratic fermentations also often invite bacterial contamination.

PROPER CIRCULATION DURING FERMENTATION. Check to see that the established procedure is being followed. Circulation is important in fermenting red wines for achieving proper extraction of color, skin tannin, and flavor components. However, often overlooked is the fact that timely tank circulation in both red and white fermentations facilitates temperature control.

REMOVING STEMS FROM CRUSHER AND POMACE FROM PRESSES. *See* Equipment Cleaning. Prompt removal of all waste from the fermentation area minimizes fruit fly buildup.

INSECT CONTROL. Keep screens in place over open tanks and in doorways and windows. Lights left on overnight above open fermentors can attract and trap moths inside wine tanks.

MALO-LACTIC STARTER CULTURE. This must be monitored with great care and precision. Most California wineries don't use *Leuconostoc* starter cultures for malo-lactic fermentations because the organisms are generally so unpredictable and difficult to control. Where used, in the cooler coastal areas, the results are worth the effort, but it must be stressed that this is not for amateurs. Precise control is absolutely essential. *See* Pilone and Kunkee (*10*), Tchelistcheff *et al.* (*11*), and earlier papers by Kunkee.

REMOVAL OF NEW WINE FROM LEES. Prompt separation of new wine from lees is basic to winemaking. Lees-like off flavors, H_2S, subsequent browning, instabilities, and several other wine disorders can be traced to abnormally long contact time between the new wine and its settled fermentation lees.

FERMENTATION CURVES. Fermentation curves are an index of the general well being of the fermentation processes taking place in the fermentation room. Abnormal yeast populations or improper temperature control often can be seen first in the plotted data of sugar content *vs.* time during fermentation.

Processing and Aging. Oxygen and heat are the two basic causes of most spoilage in the winery. The only answer is a constant check of every tank. Volatile acidity and SO_2 analyses are a good index of oxygen uptake and microbial spoilage. Be especially alert for film yeasts floating on the surface of the wine in tanks. With oxygen in the headspace and low SO_2, irreversible acetone-like off flavors can develop rapidly.

CHEMICAL ANALYSIS. It is suggested that a set procedure be established for running specific analyses according to a specific timetable. For example, a visual and taste inspection of all wines every two weeks (two months for wines in barrels), malic and lactic acid analyses at weekly intervals during the malo-lactic secondary fermentation, and SO_2 and color analyses of white wines at least every two weeks until the wine is bottled. Basic analyses to monitor are: V.A., SO_2, alcohol, pH, and T.A.

ORGANOLEPTIC ANALYSIS. Laboratory checks must be made to test all blends and fining prior to the actual operation in the plant. Rechecks of the finished tank by organoleptic and chemical analyses after the operation is completed in the cellar are the only good way to ensure that the specified operation was actually carried out and that the results were as predicted.

CLEANING. Do not wait until hoses look dirty to clean them. All hoses and pipelines, whether in use since the last cleaning or not, should be re-cleaned and sanitized at two week intervals.

OXYGEN CONTROL. CO_2, being heavier than air, is an excellent blanketing agent when properly poured onto the surface of a wine in a tank. N_2, on the other hand, being less soluble, is the preferred agent for sparging through a wine *via* a sintered surface to remove dissolved oxygen physically before it is able to react and downgrade the wine. Check these to see that they aren't being used interchangeably.

FULL TANKS. A sound tank that is kept adsolutely full cannot have any significant amount of spoilage from film yeast, aerobic bacteria, or the direct effects of oxygen reaction.

TEMPERATURE. Check the use of ventilating systems to see that they are open only when the outside air is cooler than the inside air. Be sure that lights near the ventilator openings are not left on since this can attract insects during the night. Walk through every room in the winery every day.

MATERIALS CONTROL. This is a standard check to insure that the right product has been received.

PROPER USE OF EQUIPMENT. Are personnel schooled in the proper techniques of equipment use? Do they use check lists for operation of complicated machinery? Will they report their own mistakes quickly so that the situation can be corrected before more damage is done? Silent covering of human mistakes has ruined many gallons of wine. Build trust in the workers for your quality control effort and attempt to remove fear of being caught in a mistake.

CONTINUOUS MONITORING. Prompt completion of any operation once it has begun in the cellar is important. Minimize the number of physical operations that any wine must submit to. When it can be seen very early that a given filtration is not providing the desired product clarity, report it at once. By changing the defective set-up promptly, you are eliminating the need to filter the wine twice.

INVENTORY CONTROL. It is often important to carry out a given operation promptly. Many situations tend to change certain wines rapidly, and whatever action is needed to correct it may be necessary within a short time. Being unaware that a needed fining agent or additive is out of stock can be damaging, and, in the case of large wineries, it is worthwhile to use data processing for inventory control. An added plus is the ability to order in economic lot sizes.

BARREL SANITATION. This is perhaps the most important single item in all of Phase I quality control for the winery which ages wines in barrels.

The consequences of contamination are largely irreversible, contagious, and calamitous. Unchecked bacterial spoilage in barrels can rapidly change the flavor of the whole wine inventory. Usually, because of the insulating nature of wood fiber, re-sterilization of badly infected barrels is impossible. The barrels should be burned and replaced without further delay. It is important to detect barrel spoilage early in the cycle.

EQUIPMENT MAINTENANCE. Any item which is to be used in contact with wine is suspect.

FINAL CHECK BEFORE BOTTLING. Winemaster approval of every lot of every wine prior to bottling, preferably after several tastings by experienced tasters. Once the wine is bottled, it must be salable or it will be an economic loss. This is the last chance to catch an error while it might still be corrected.

Table II.   Quality Control Check Points—Phase II

Bottling
  Quality control audit of incoming bottling room supplies, such as filtration media (membranes), glassware, corks, screw caps in non-premium wineries, labels, foil capsules or sheets, wire hoods, and plastic or natural corks in sparkling wine operations
  Sanitation practices
  Insect and dust exclusion from bottling room
  Sanitary design in equipment and floors
  Bottling room personnel cleanliness and work habits
  Container capacities
  Fill point variation
  Filler bowl wine temperature fluctuations
  Continuous (statistical) monitoring of sterile filtration units
  Bottle cleaner and CO_2 injector, if used
  Start up quality checks

Warehousing and Shipping
  Murphy's law errors
  Temperature and humidity inside warehouse
  Cork finish bottles stored with the cork pointed down
  Work habits of fork lift operators
  Stock rotation by chronology of bottling date
  Bottled goods library
  Inspection of railcars, loading, and packing procedures
  Follow-up information to distributors warehouse personnel

Check Points—Table II

Bottling. QUALITY CONTROL AUDIT. This ensures that the right packaging materials are in inventory.

SANITATION PRACTICES. All bottling room equipment should be sterilized each day before beginning operation of the line. Bottle rinsers, if used, must be monitored continuously during operation since they can be a primary source of infection. New wine bottles are normally sterile as received at the winery. Rinsers remove dust and carton fiber, but they can contaminate the product if care is not taken to ensure sterility of the rinse water.

INSECTS AND DUST. Check the room air filters and insect control devices daily. Air curtains or pressurized bottling room systems should be checked for proper operation before bottling room start-up daily. Some wineries enclose only the filling and cooking areas, but the quality control checks are the same as if the whole bottling room were isolated.

SANITARY DESIGN. Pipelines and equipment should be easy to clean and sterilize. Clean-in-place systems are recommended wherever possible. Use signs to constantly remind personnel about personal cleanliness and work habits.

CONTAINER CAPACITIES, FILL POINT VARIATION, STERILE FILTER MONITORING, START-UP CHECKS. These all can be checked easily on a statistical basis. The frequency of each check should be specified so that these become routine functions of quality control. Filler-bowl wine temperature fluctuations are extremely important since they may be a major cause of fill-point fluctuation in bottles. This fluctuation cannot be overemphasized because of legal ramifications. Specific standards of fill and headspace maxima are listed by the Bureau of Alcohol, Tobacco and Firearms in 27 CFR, part 4.

STERILE FILTRATION. In and out samples from sterile filtration units as well as finished product from the bottling line should be plated at periodic intervals, depending on wine type and sugar, alcohol, and SO_2 content. Hold all packaged goods until the incubator results of each corresponding plating are known.

START-UP CHECKS. This subgroup of checks forms the last line of defense against errors in the production operation. If an error or skew situation is found prior to bottling, the situation is usually correctable; however, product errors found after the product has been bottled nearly always result in dumping and total loss of product, package, and labor.

The idea behind this last line of defense is to detect errors which may have been made during the final filtrations, sterilization of bottling room equipment, startup, or movement of finished wine through recently emptied tanks, lines, hoses or machinery. Quality control personnel must be present at the start-up of each bottling run. They remove the first few bottles from the line and immediately run a quick but imprecise alcohol analysis to detect any dilution which would have occurred if any lines to the filler bowl were incompletely drained after washing and sterilization. Alternatively, sugar, SO_2, or acidity could be analyzed to give the same check. The same first bottle samples are tasted against a control at this time also as a final check on wine type and continuity.

These control samples were taken off the bottling line as finished goods and held in cool storage since the previous bottling of the same wine type. At the same time, the finished package is checked visually to ensure that the correct label and closure are applied properly and that the code dating systems on both labels and cases is working properly. In this way, quality control can restrict any loss from packaging an incorrect product to, perhaps, two or three minutes of bottling time. Clearly, the quality control effort that prevents bottling two or three days worth of defective wine has paid for itself.

Warehousing and Shipping. Murphy's Law Errors. Good warehousing and package marking technique avoids confusion between several different wine types packed in the same size, shape, and color cartons. Differing wine types should be easy for warehouse and shipping personnel to distinguish.

Cork Finish Bottles. It is believed widely that corks in bottles tend to crack and leak air more easily when bottles are stored upright than when stored upside down or on their sides so that the cork remains in contact with wine. Although no definitive study of this has been reported, it is recommended that the practice be continued.

Work Habits. Neatness in pallet rows facilitates a smooth operation as well as minimizing physical damage (fork lift blight) to cartons.

Stock Rotation and Bottled Goods Library. Chronological samples from each bottling should be stored under simulated store shelf conditions and an examination for microbial growth, color, or stability change after 1 week, 2 months, 6 months, and 2 years should be made. Tasting confirms or denies any apparent problem.

Feed-back from customers on per cent of breakage and geographical location of breakage in cars will help to minimize breakage on future shipments if the information is heeded. A continuing education process is necessary, to attain competence throughout the chain of product handlers.

Analysis

Management needs to know, but often does not know the accuracies of wine analyses as they are reported. Misunderstanding can be avoided, especially where identical analyses carried out on different dates indicate different results, by making clear what is meant when a subsequent analysis does not agree precisely with an earlier one. It might be that a wine is undergoing spontaneous malo-lactic fermentation, and the first evidence is a lower total acidity than previously found. Whether the difference is real or attributable to sampling or analytical inaccuracies should be reported any time a difference is found in a subsequent wine analysis. Using the uniform methods of analysis for wines as currently recognized by the American Society of Enologists (*12*), the following accuracies can be expected providing the work is done competently:

Alcohol by ebulliometer	± 0.25%
Alcohol by hydrometer	± 0.15%
Alcohol by pycnometer	± 0.05%
Brix by hydrometer	± 0.2°
Brix by refractometer	± 0.2° (juice only)
Total acidity by titration	± 0.01 grams/100 ml
Volatile acidity by titration	± 0.005 grams/100 ml

pH determination by meter depends on the meter used. SO_2 (free and total) analyses by iodine titration, though not the most accurate available, are in general use by a majority of the wine industry. No accuracy is reported because of the instability of the standard iodine solution, but certainly it is not better than ±5 ppm as actually practiced.

Quality Control During the Aging Process. This paper has alluded to organoleptic analysis and the use of subjective tasting for objective analysis. Although this is not necessarily an ideal situation, it ranks far above other alternatives. Many wine producers simply don't taste their own products often enough, nor do they taste their competitors' wine often enough. As yet, there is no valid qualitative analytical procedure to replace the winemakers' own experienced taste. This brings us to Phase III which is perhaps unique to wine. The winemaker cares very much about the eventual consumers' enjoyment of the wine product. He wants the product to be at its peak when the consumer tastes it. But wine has an unusual attribute which requires an unusual approach:

On the one hand, wine is a typical preserved food because its shelf life is affected by time and temperature during storage. Conversely, many wines are considerably improved by moderate storage after packaging. Much has been written about the development of bottle bouquet during the aging of wine in glass (*13, 14, 15, 16, 17*).

The same storage conditions which can turn a fine wine into a great one during a period of, say, 10 years can make that same wine poor in only an additional two or three years. We have no reliable means of measuring whether a given wine, just bottled, will reach its peak of perfection in two years or 10 years. In premium wineries the winemaker estimates (based on past experience, his own taste, and chemical analysis —in that unfortunate order) when the wine will enter and leave its period of drinkability. Figure 1 shows a curve which might represent the change in drinkability with time for a given white wine. It is a stylized picture of the bottle aging curve for this wine. Drinkability level A is the minimum acceptable level, and B is the maximum attainable for this particular wine under its own unique set of conditions. The winemaker must release his wine at the best time to get the greatest number of bottles to the greatest number of consumers when the wine is nearest to its peak of drinkability.

Consider the problem facing a winemaker who has bottled a white wine, for example, which he believes will develop a good bouquet and be organoleptically acceptable six months later and will continue to improve for two additional years; after that the fruitiness and bouquet will probably diminish rapidly so that within the third year, the wine will be overaged. The winemaker wants the wine to be drunk at or near its peak, in this case 18-30 months after bottling. Since that precision is impractical commercially, he determines from the sales department that

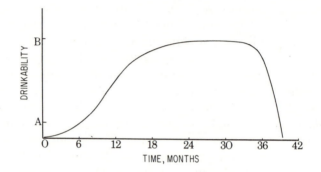

Figure 1.   Drinkability as a function of time after bottling for a white wine

his winery's distribution, sales, and consumption patterns for that particular wine type are estimated to be:

Storage in the distributor's warehouse: 0 to 9 months (average 3 months).

Storage on retail store shelves: 0 to 15 months (average 6 months).

Storage in the consumers cellar (here, the consumer can be an individual or a restaurant, club, etc.): indeterminate, but assumed to be 0 to 18 months (average 6 months).

Figure 2 shows these storage times plotted, beginning with 0 as the bottling date. Note the large difference in age at time of consumption between a bottle which undergoes average distribution speed and one which undergoes slow distribution. Figure 3 overlaps the two previous graphs to allow simultaneous study of the wine location and its drinkability. Several conclusions are apparent from this example:

1. If the wine is shipped immediately after bottling, some of it will get to the consumer immediately and he will judge the wine as poor, it being below a minimum acceptable drinkability level. This happens when a distributor ships it immediately on to the retailer, who sells it promptly to a consumer who opens it without further aging.

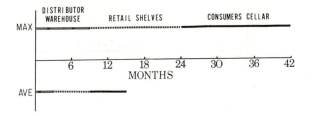

*Figure 2. Maximum and average times spent in
distribution channels for a white wine*

2. Some will not get to the retail store shelves until 9 months later
and then not to the consumer until an additional 15 months have passed—
but then will not actually be consumed for yet another 18 months. In
this case, the winery released the wine too soon, yet the customer actually
tasted it too late!

3. Between these two extremes, however, the "average" bottle was
released from the distributor's warehouse after 3 months of aging, removed
from the retail shelf at 9 months, and was opened and consumed at 15
months. At this time, it was considerably better than it had been immediately after bottling, but had not yet reached its peak of drinkability.

4. If the wine were simply held at the winery for 6 months prior to
shipping, no consumer would receive the wine in the "too young"
condition and the "average" bottle would be consumed near its peak
(Figure 4). However, a small percentage would not be consumed before
it had declined to below the acceptable level of quality through overage.

In this example, a six-month hold at the winery undoubtedly would
be preferred and would be used if other conditions permitted. The
situation described above is obviously an oversimplification of what really

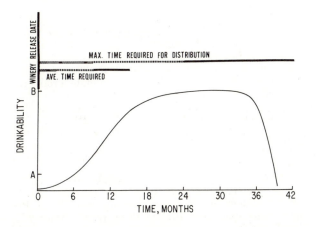

*Figure 3. Drinkability related to time in distribution
channels when the wine is shipped from the winery
immediately after bottling*

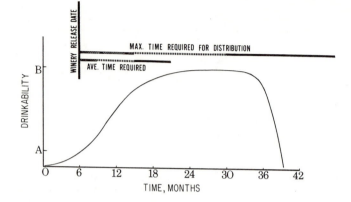

Figure 4. Drinkability related to time in distribution channels when the wine is bottle aged at the winery for 6 months

happens. It assumes, for example, that the storage conditions remain constant throughout all the various warehouses and in the consumers cellar. In fact, some wine always seems to get frozen in storage, or worse, it might be stored near a furnace or on a top shelf where the temperature destroys it before its time. But the principle is valid, even though this type of quality control consideration is not often found in the wine industry today. However, if the pursuit of excellence in wines is to reach its ultimate goal, then these are the types of details which must be considered.

Literature Cited

1. Amerine, M. A., "Quality Control in the California Wine Industry," *J. Milk Food Technol.* (1972) **35**, 373–377.
2. Peterson, R. G., unpublished data.
3. Acree, T. E., Sonoff, E. P., Splittstoesser, D. F., "Effect of Yeast Strain and Type of Sulfur Compound on Hydrogen Sulfide Production," *Amer. J. Enol. Viticult.* (1972) **23**, 6–9.
4. Peterson, R. G., Joslyn, M. A., Durbin, P. W., "Mechanism of Copper *Casse* Formation in White Table Wine," *Food Res.* (1958) **23**, 518–524.
5. Wine Institute, San Francisco, "Sanitation Guide for Wineries," revised, March 16, 1971.
6. Winkler, A. J., "General Viticulture," University of California Press, Berkeley, 1965.
7. Amerine, M. A., Joslyn, M. A., "Table Wines, the Technology of their Production," University of California Press, Berkeley, 1970.
8. Rankin, B. C., Pilone, D. A., "*Saccharomyces bailii*, a Resistant Yeast Causing Serious Spoilage of Bottled Table Wine," *Amer. J. Enol. Viticult.* (1973) **24**, 55–58.
9. California Department of Agriculture and California Department of Public Health, Berkeley, "Pesticide Illness Reported under Health and Safety Code," 1971.

10. Pilone, G. J., Kunkee, R. E., "Characterization and Energetics of *Leuconostoc Oenos* ML-34," *Amer. J. Enol. Viticult.* (1972) **23**, 61–70.
11. Tchelistcheff, A., Peterson, R. G., Van Gelderen, M., "Control of Malolactic Fermentation in Wine," *Amer. J. Enol. Viticult.* (1971) **22**, 1–5.
12. American Society of Enologists, "Uniform Methods of Analyses for Wines and Spirits," Davis, 1972.
13. Amerine, M. A., "The Response of Wine to Aging. I. Physical Factors Influencing Aging. II. Biological and Chemical Factors Influencing Aging. III. Bottle Aging. IV. The Influence of Variety," *Wines Vines* (1950) **31** (3), 19–22.
14. *Ibid.*, (1950) **31** (4), 71–74.
15. *Ibid.*, (1950) **31** (5), 28–31.
16. Singleton, V. L., "Aging of Wines and Other Spiritous Products, Acceleration by Physical Treatments," *Hilgardia* (1962) **32** (7).
17. Singleton, V. L., Ough, C. S., Amerine, M. A., "Chemical and Sensory Effects of Heating Wines under Different Gases," *Amer. J. Enol. Viticult.* (1964) **15**, 134–145.

RECEIVED July 25, 1973.

11

Chemical Aspects of Distilling Wines into Brandy

JAMES F. GUYMON

University of California, Davis, Calif. 95616

Wines for brandy distillation should be made from white grapes by fermentation of juice separated from pomace without adding SO₂ (to minimize aldehyde accumulation) at as low temperature as is practicable (less than 75°F, to minimize fusel oil formation) and should be distilled as soon as fermentation is complete. In California, wines are distilled into brandy in continuous plate columns arranged for separating a low-boiling heads fraction to eliminate aldehydes. Higher alcohols (fusel oil) are the most abundant group of congeners, and ethyl esters of C_8 to C_{12} fatty acids are next most abundant provided yeast cells are present in the wine distilled. Fatty acid esters and fusel alcohols accumulate in column tray liquids at 100° to 135° proof and distill most rapidly in the earliest fractions of simple pot distillation.

Brandy is a distillate of wine so its initial chemical composition depends on the compounds present in wine being sufficiently volatile to distill. Although ethyl alcohol and water are the two major components of any distilled spirit, aroma and flavor character depend on a multitude of minor compounds usually referred to as congeners or congenerics. Some of the congeners are derived directly from the grapes or other fruits that may be used for brandy. The most abundant congeners of brandy are minor products of alcoholic fermentation derived primarily from sugars but in part from components of the fruit. Distillation conditions affect the relative quantities of minor compounds recovered in the distillate, and some arise from chemical reactions during distillation such as heat-induced degradative changes.

Grape brandies and some fruit brandies are traditionally stored in oak casks for varying periods of time for aging or maturation. During aging in oak many substances are extracted from the wood, some of which react with compounds initially present, especially ethyl alcohol. Thus, the aging process produces a complexity of minor constituents which generally improves the palatability and enhances the aroma and flavor.

The overall chemical composition of brandy is derived from four general sources: the fruit, alcoholic fermentation, distillation, and aging in wood. The scope of this discussion is limited to chemical aspects of the components of wine and their behavior during distillation.

The term brandy as used here refers to distillates of grape wine before aging. Federal regulators require that distillates derived from other fruits must show the name of the fruit, for example, pear brandy.

Wines for Distillation

The production of grape wines destined for distillation into brandy is essentially the same as that for dry white table wines. However there are some desired characteristics of distilling wines which are relatively more significant than those of table wines produced for consumption. White grape varieties are preferred to red or black varieties even though white wines can be prepared from red grapes by separating the juice prior to fermentation. Red varieties tend to impart a measure of coarseness to the distillates and, compared with white varieties, generally lead to greater formation of higher alcohols during fermentation. This seems to occur whether the fruit is processed as white wines (fermentation of separated juice) or as red wines (fermentation of the entire must containing juice, skins, seeds, and pulp). Thus, the preferred procedure for producing distilling wines is to use non-pigmented grapes and to separate the juice from the marc or pomace prior to fermentation.

Sulfur Dioxide and Aldehydes. Sulfur dioxide is commonly added both before and after fermentation in preparing white table wines. It is an effective antioxidant as well as a selective inhibitor of unwanted microorganisms. However, sulfur dioxide, as the bisulfite ion in solution, combines with aldehydes, especially acetaldehyde, during fermentation giving an accumulation of aldehydes in the bound form of aldehyde–sulfurous acid.

$$CH_3CHO + HSO_3^- \rightleftharpoons CH_3CHOHSO_3^-$$

The accumulation of aldehydes as well as the presence of sulfur dioxide itself is undesirable in distilling wines for reasons given later. We have experienced no difficulty in producing sound wines (that is without bacterial contamination or significant oxidation) if good quality fruit is used. However, fruit that is significantly damaged by rain or other conditions whereby rot or molds have developed and increased polyphenol oxidase yield wines which oxidize and brown badly and are vulnerable to *Acetobacter* and lactic acid bacteria. In this case it is advisable to add 50–75 ppm of sulfur dioxide prior to fermentation. Amerine and Ough (*1*) showed that moderate levels of sulfur dioxide (50 ppm) increased aldehyde accumulation only slightly over wines fermented without sulfur dioxide.

Acetaldehyde content reaches a maximum at the time of fermentation that shows the most rapid rate of carbohydrate dissimulation, it falls to a low level at the end of fermentation, and then slowly increases (*2, 3*). Also aldehydes are formed by oxidation as evidenced by browning of wine when exposed to air. For these and other reasons, distillation should be carried out as soon as possible after fermentation is complete.

Higher Alcohols. The most abundant, volatile minor products of alcoholic fermentation are the higher alcohols or fusel alcohols. The most important are isoamyl (3-methyl-1-butanol), *d*-active amyl (2-methyl-1-butanol), isobutyl (2-methyl-1-propanol), and *n*-propyl (1-propanol) alcohols. It is now recognized (*4, 5, 6, 7*) that these higher alcohols are formed by decarboxylation of particular α-keto acids to yield the corresponding aldehydes, and these in turn are reduced to the alcohols in a manner analogous to the formation of ethyl alcohol from pyruvic acid.

$$RCOCOOH \xrightarrow{-CO_2} RCHO \xrightarrow{+2H} RCH_2OH$$

Higher alcohol formation by yeasts has long been attributed to the transamination of exogeneous amino acids to yield the analogous keto acid. For example:

$$\text{leucine} + \alpha\text{-ketoglutaric acid} \xrightarrow{\text{(transaminase)}} \text{glutaric acid} + \alpha\text{-ketoisocaproic acid} \quad (1)$$

$$\alpha\text{-ketoisocaproic acid} \xrightarrow{\text{(decarboxylase)}} \text{isovaleraldehyde} + CO_2 \quad (2)$$

$$\text{isovaleraldehyde} + NADH + H^+ \xrightarrow{\text{(alcohol dehydrogenase)}} \text{isoamyl alcohol} + NAD^+ \quad (3)$$

However, now it is well established that the appropriate α-keto acids giving rise to a particular higher alcohol arise mostly from carbohydrate sources through the synthetic pathways by which yeast synthesizes its amino acid requirements.

Peynaud and Guimberteau (8) estimated that no more than one-sixth of the leucine and valine in grape musts assimilated during fermentation gave rise to isoamyl and isobutyl alcohols, and since these amino acids are low in grape musts, the formation of higher alcohols by amino acid degradation is negligible. Instead, nearly all of the higher alcohols are derived from carbohydrate. However, Reazin *et al.* (9), in ^{14}C tracer studies during fermentation of grain mashes, concluded that the major proportions of higher alcohols were derived from amino acids, and lesser proportions were derived from carbohydrates.

Since the higher alcohols, especially the two amyl alcohols, constitute the most abundant minor components or congeners of brandy and other distilled spirits, they have a very significant effect on the sensory and taste character. Considerable variation in the concentrations of fusel oil alcohols in wines (10) and brandies (11) occurs in practice. Factors influencing the levels of higher alcohols produced during fermentation include composition of raw material, yeast strain (12), temperature of fermentation, suspended matter, and aeration (13, 14). In practice, it is desirable to minimize the formation of higher alcohols, especially the amyls. Although precise control is not fully feasible, highest levels are formed at medium fermentation temperatures of about 75°F (7, 12, 15, 16). The presence of suspended matter and aeration stimulate fusel oil formation (14). So the practice of draining the relatively clear free-run juice several hours after crushing is preferred to using the juice immediately after pressing, which yields more turbid liquid musts.

Several workers have reported the amounts of higher alcohols found in wines (8, 12, 17). While differences as much as fivefold occur considering diverse genera and strains of yeast, the variation is generally less for strains of *Saccharomyces cerevisiae*. Guymon and Heitz (10) found a mean fusel oil content of 25 grams/100 liters (250 ppm) for 120 samples of California white table wines and 29 grams/100 liters for 130 samples of red table wines. In our pilot plant scale production of experimental distilling wines, we obtain fusel oil levels ranging from about 16–30 grams/100 liters for white varieties with an average of 20. With red varieties, the levels may attain 40 or more grams/100 liters.

The appropriate fusel oil content of brandy distillate, though commercial producers are not in full agreement, ranges from 65 to 100 grams/100 liters at 100° proof (50 vol %). If 100 grams/100 liters is desired as the upper limit, a distilling wine containing 12 vol % ethyl alcohol should thus contain no more than 24 grams of higher alcohols per 100 liters, assuming all of the higher alcohols are recovered in the distillation process.

Fatty Acids and Their Ethyl Esters. The normal aliphatic acids and their ethyl esters are often the second most abundant group of congeners found in distilled spirits. After ethyl acetate, generally caprylate (C_8), caprate (C_{10}), and laurate (C_{12}) esters predominate (*18*), although those with both fewer and more carbon atoms are found. The amount of suspended yeast cells in the wine at the time of distillation significantly affects the amount of fatty acids and esters in the distillate. The higher molecular weight acids and esters are closely bound to the cells of yeast whereas those with lower molecular weight tend to be secreted into solution (*19*).

Thus, the concentrations of fatty acids and esters found in brandy distillates are greatly affected by the nature of the wine at the time of distillation, particularly the time interval between fermentation and distillation since most of the yeast cells settle out fairly quickly after fermentation has ceased. Obviously the degree of resuspension of the settled lees into the wine when distilled affects the amount of fatty acids and esters recovered in the distillates. The method and techniques of distillation are also very important since this class of congeners, having relatively high boiling points and weak solubility in water, exhibit wide ranges of volatility as affected by the alcohol content of the liquid volatilized.

The desirable properties of distilling wines include wines: a) made from white varieties by fermentation of separated juice clarified as much as practicable, b) fermented without addition of sulfur dioxide, c) fermented with strains of yeast which form comparatively low amounts of fusel alcohols at temperatures below 75°F, and d) distilled as soon as possible after fermentation.

Distillation

Many processes are used throughout the world for distilling wine into brandy. Continuous distillation in plate columns, discontinuous or batch distillation using pot stills (with and without adjunct rectifying columns), and even distillation apparatuses involving both continuous and discontinuous aspects are used.

In France, the famous cognac brandy is distilled in simple direct-fired pot stills. The *Méthode Charentaise* (*20*) includes two successive distillations. In the first, wine is distilled into low wines (*brouillis*) containing 26–30% alcohol or approximately a threefold concentration. The low wines are redistilled, and three fractions are collected, *i.e.*, a small heads cut amounting to about 1 vol % of the charge; the brandy *coeur* as the alcohol content decreases from about 80 to 60 vol % average 70 vol %; and the tails (*seconde*) to recover the remaining alcohol in the charge. The heads and tails fractions are recycled.

In the Department of Gers, the Armagnac brandies are distilled in an apparatus known as the *Verdier système*. Wine, preheated by con-

densing vapor, flows continuously to the top of a short plate column
and is partly stripped of alcohol as it transverses the column and flows
into the top section of a two-tiered, direct-fired boiler. The vapor generated in the lower section by the furnace is caused to bubble through the
liquid layer in the top section and on up the plate column. Vapor from
the top (feed) tray is condensed by the preheater and condenser into
brandy containing 50–54 vol % alcohol. Intermittently most of the spent
wine in the lower level of the boiler is drawn off to waste, and then the
same volume of liquid from the upper level is drawn into the lower part.
Thus the process is both continuous and discontinuous.

In California, distillation is usually carried out in a continuous single
or split column unit as shown in Figure 1. The brandy distillate, at a
maximum of 170° proof as required by federal regulations for standards
of identity, is drawn as a liquid sidestream from a tray in the middle or
upper region of the concentrating section. A small portion of the over-
head distillate from the vent condenser is withdrawn as a heads cut at a
rate of about 5–15% of the brandy rate to remove most of the low-boiling
aldehydes, sulfites, esters, etc. Some producers also separate another
small fraction (*ca.* 10–20%) of the product rate as a low oils or high-
boiling stream from a level one or two trays below the brandy tray to

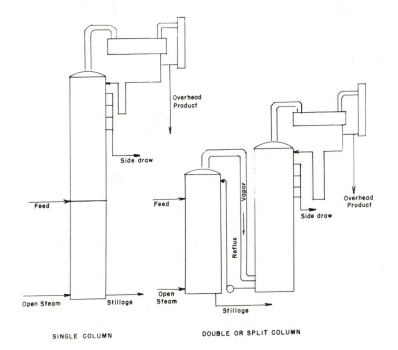

Figure 1.   Single and split columns for brandy distillation

reduce the level of fusel oil in the product. The concentration of higher alcohols in the tray liquids of a concentrating section producing brandy at 168°–170° proof is usually at its maximum at two trays below the product level (21). This maximum concentration of higher alcohols is about 2 to 2½ times that of the product and occurs where the tray liquid indicates approximately 120°–135° proof. Thus the separation of a low oils fraction at a 20% rate should reduce the fusel oil content of the brandy by about 30%. In practice it is usually less. The disadvantage of this technique is that the amount of brandy, produced from prime quality wine, is reduced by the amount of low oils fraction removed which is usually redistilled to recover the alcohol as neutral spirit.

Most distillation units in California have been installed with the capability of producing high proof wine spirit (189°–192° proof). Accordingly the concentrating sections may contain 20–40 bubble cap trays whereas only three or four ideal trays are required to concentrate alcohol from 10 to 85 vol % (the strength of beverage brandy distillate). In beverage brandy production practice, perhaps two-thirds of the trays lie between the feed and product trays, and the remaining one-third of the trays at the top are used for concentrating the low-boiling impurities into the heads cut. Recent installations generally provide for at least 12-inch tray spacings.

The stripping or wine section typically contains 18–22 sieve or perforated trays with 15–18 in. tray spacing. Direct or open steam is the usual heat source although one major brandy producer uses reboilers. Column diameters are commonly 5–6 feet though some larger and smaller are in use. Feed rates to the stripping section are typically in the range 3000–8000 gallons/hr.

Some recent developments in design of distilling units include the use of valve and sieve trays in lieu of bubble caps in the concentrating sections, air-cooled instead of water-cooled condensers, stainless steel in lieu of copper construction, energy economizers which incorporate a thermocompressor to transfer heat energy from the bottoms or stillage to the entering steam, and location of distilling units entirely outside of buildings. However, most of these developments pertain to the more elaborate recovery of neutral wine spirit from residue materials.

Some three- and a few four-column units are used. The third column of multicolumn units is generally an aldehyde concentrating column in which the heads cut from the brandy concentrating column is concentrated (Figure 2) as much as 20-fold; the alcohol, stripped of aldehydes and other low-boiling components, is recycled to the main unit. Another three-column arrangement provides for the brandy to be distilled as a top product without a heads cut; the entire brandy is then fed into an aldehyde stripping and concentrating column. The low-boiling substances

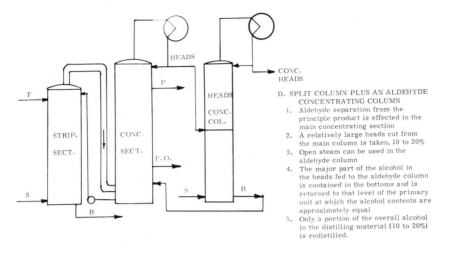

Figure 2.   *Split column with a heads concentrating column*

D. SPLIT COLUMN PLUS AN ALDEHYDE
   CONCENTRATING COLUMN
1. Aldehyde separation from the
   principle product is effected in the
   main concentrating section
2. A relatively large heads cut from
   the main column is taken, 10 to 20%
3. Open steam can be used in the
   aldehyde column
4. The major part of the alcohol in
   the heads fed to the aldehyde column
   is contained in the bottoms and is
   returned to that level of the primary
   unit at which the alcohol contents are
   approximately equal
5. Only a portion of the overall alcohol
   in the distilling material (10 to 20%)
   is redistilled.

are thus removed as the top product of the aldehyde column and the brandy is removed as the bottom stream. A reboiler, to avoid dilution, is therefore required to supply vapor (Figure 3).

Some ingenious designs have been developed for producing beverage brandy with a substantial removal of the amyl alcohols without exceeding 170° proof in the primary distillate. In one such case a low oils cut, about 20% of the brandy rate, is fed into a fusel oil concentrating column to concentrate the alcohol in the low oils from approximately 130° to 190° proof. A fusel oil cut from this column is thus sufficiently concentrated to separate into oil and aqueous layers in a conventional fusel oil decanter (22). The 190° proof alcohol (thus very low in higher alcohols) is recycled to the main column. Another arrangement effects a partial separation of higher alcohols, especially the amyls, by concentrating the brandy well above 170° proof in the primary concentrating column which permits some separation of fusel oil in its decanter; the brandy stream is subsequently reduced to 170° proof or less by blending it with a low proof stream drawn from an appropriate plate low in the heads concentrating column which, of necessity, is heated with open steam.

Only a small amount of pot-still brandy is produced in California. It is used primarily for blending, in small percentages, with continuous still brandy after aging. Equipment used for this purpose consists of pot-rectifiers. In at least one case, the fractionating column is so arranged that it can serve either as the concentrating section for the pot or as a part of a continuous still unit.

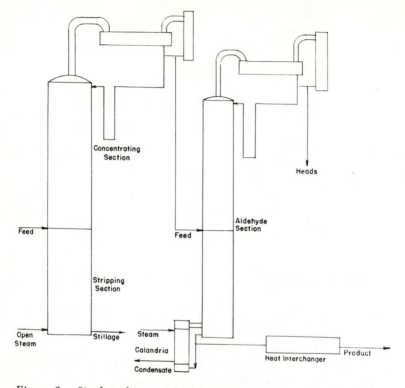

Figure 3. *Single column with aldehyde separating column heated by reboiler—brandy is bottom product*

Aldehydes and Other Low-Boiling Components. As mentioned, a low-boiling fraction, called heads, is normally taken from the vent condenser during the distillation of wine into brandy. The principal impurities removed are acetaldehyde, diethyl acetal, ethyl acetate, and acetaldehyde–sulfurous acid.

The heads fraction may be highly acidic and corrosive if sulfur dioxide or bisulfite is used in the production or preservation of the wine. The reversible reactions involved are:

$$H_2O + SO_2 \leftrightarrows HSO_3^- + H^+$$
$$HSO_3^- + CH_3CHO \leftrightarrows CH_3CHOHSO_3^-$$

The acetaldehyde–sulfurous acid compound has the properties of a sulfonic acid with C–S linkage rather than an ester structure as once assumed. It is properly 1-hydroxyethane sulfonic acid (*23, 24*) and is highly acidic (*25*). Samples of concentrated heads from commercial aldehyde columns having aldehyde contents of 5–13% gave pH values of 0.7–0.9 and contained high levels of copper (*25*).

Whitmore (*26*) and Deibner and Bénard (*27*) claim that sulfonic acids are capable of distilling intact. However, it would seem more likely that at the temperature of the distilling process, 1-hydroxyethane sulfonic acid in the wine would dissociate into the very volatile free acetaldehyde and therefore rapidly distill along with free sulfur dioxide and recombine at the top level of the fractionating column.

Another reaction of interest is the production of acetal by the reaction of 2 moles of ethanol and 1 mole of acetaldehyde. This reaction is strongly catalyzed by hydrogen ion and during distillation depends

$$CH_3CHO + 2C_2H_5OH \overset{[H^+]}{\underset{\rightarrow}{\leftrightarrows}} CH_3CH(OC_2H_5)_2 + H_2O$$

on the presence of bisulfites or the 1-hydroxyethane sulfonic acid to establish the necessary acidity. Equilibrium is rapidly established at low pHs and the degree of acetalization obviously depends on the concentrations of the two major components, ethanol and water. However, attempts to determine an equilibrium constant are unsuccessful presumably because the reaction proceeds in two steps and the rate-determining step is the formation of hemiacetal from one mole each of ethanol and acetaldehyde (*24*).

Ethyl acetate is a product of yeasts and a normal component of wine. Its level can be increased by *Acetobacter* contamination, although most wines showing excess volatile (acetic) acid do not necessarily contain excess ethyl ester initially. It is quite possible to obtain brandy of normal composition and quality by continuous distillation of newly fermented wine containing excess acetic acid, e.g., 0.1%. On the other hand, ethyl acetate can be formed in continuous columns, particularly if the distillation conditions provide for a relatively high ethanol concentration on the feed tray or immediately below. Since acetic acid is weakly volatile in all mixtures of ethanol and water, it does not appreciably distill upward. Therefore there is no opportunity for acetic acid to combine wtih ethanol in tray liquids normally of high ethanol concentration.

Ethyl acetate is the major low-boiling impurity of heads fractions from continuous columns if bisulfites are absent in the distilling material. A heads fraction from a typical brandy column usually contains less than 1% of these volatile impurities, although the concentrated heads from an aldehyde-concentrating column may contain as much as 10–15% aldehydes. In either case, ethyl alcohol is the major component of the heads cut, and its recovery in usable form has been a troublesome processing problem.

A simple method, based on the strongly reducing conditions during alcoholic fermentation, was developed in our laboratory (*25, 29, 30*). The heads liquid is recycled into an actively fermenting dilute grape juice or sweet pomace wash, initially containing sugar equivalent to about 10° Brix. The titratable aldehydes, including acetal, are very effectively reduced to alcohols, and apparently about half of the ethyl acetate present is metabolized. Guymon and Nakagiri (3) found that acetaldehyde or its equivalent concentration of acetal added before fermentation in excess of about 0.1% caused important delays in fermentation. But up to 0.35% acetaldehyde or 0.9% acetal could be added to a vigorous fermentation without significant effect on fermentation rates.

Ethyl acetate, added prior to fermentation in amounts up to 3%, caused delays in the onset of fermentation proportional to the amount added, but once begun, the fermentation rate was unaffected. When the same levels of ethyl acetate were added to an active fermentation, the fermentation period was sluggish.

This simple method of heads disposal by refermentation eliminated the earlier deterrent to taking an adequate cut to control volatile aldehydes to the desired level, now considered to be approximately 10 ppm or less in new brandy distillates.

Other aliphatic aldehydes have been identified and quantitatively measured in various brandies (31, 32). These include small amounts of formaldehyde, propionaldehyde, isobutyraldehyde, isovaleraldehyde, furfural, etc. Furfural is a normal component of pot-still distillates, but its presence in continuous still brandies is negligible before aging in wood.

Higher Alcohols and Fatty Acid Esters. Because of the remarkable utility of gas chromatography for analysis of minor volatile components, an extensive literature on the composition, mechanism of formation, and variables affecting concentration of higher alcohols for various kinds of distilled spirits has appeared during the past 15 years. Reviews include Stevens (33), Suomalainen and Ronkainen (34), Lawrence (35), Webb and Ingraham (5), Guymon (6), Suomalainen et al. (32), and Äyräpää (7). Data generally limited to the influence of distillation will be reported here.

Guymon (21) reported the composition of tray liquids for brandy distilled in continuous column, respectively, at 130°, 170°, and 181° proof. The maximum level of fusel oil occurred on the tray nearest in proof to about 130°. This is the second tray below the product tray for the customary 170° proof of distillation of the product.

Later we measured the concentrations of n-propyl, isobutyl, and the combined isoamyl and active amyl alcohols (3-methyl-1-butyl alcohol and 2-methyl-1-butyl alcohol) in distillation tray liquids using a gas chromatographic method with n-butyl alcohol as the internal standard. The distribution of the higher alcohols in the 14-tray concentrating section of a 12-inch pilot column during a run in which the product from tray 7 was 169° proof is shown in the upper portion of Figure 4. The

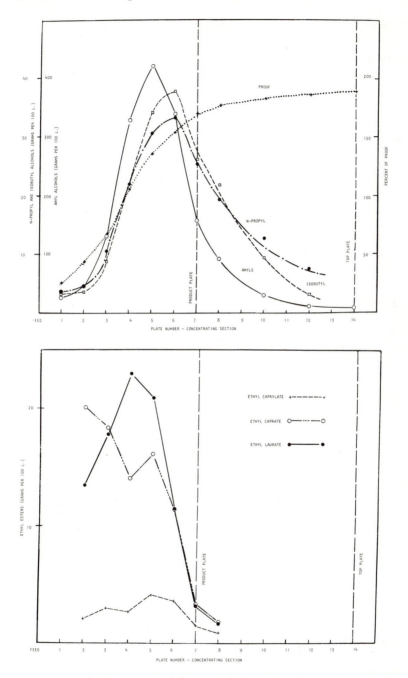

Figure 4. Distribution of higher alcohols (above) and ethyl esters of three fatty acids (below) in tray liquids of concentrating section producing brandy at 169° proof from tray No. 7

highest level of n-propyl and isobutyl alcohols was found on tray number 6 (152° proof) or one tray below the product tray. The more abundant amyls (plotted on a scale one-tenth the size) were most concentrated on tray number 5 (at 135° proof), two trays below the product tray.

Guymon and Crowell (18) developed a quantitative procedure for extracting (with methylene chloride) the ethyl esters of the straight chain fatty acids from brandy or distillates of wine, and concentrating and measuring by programmed temperature gas chromatography the three most abundant ethyl esters: caprylic, capric, and lauric acids. Ethyl pelargonate was used as the internal standard.

The lower portion of Figure 4 shows the distribution of three ethyl esters in the same set of tray liquid samples. These high boiling esters tend to concentrate at slightly lower proof and tray levels than the higher alcohols, but they all overlap. Consequently, a fusel oil or low oils layer drawn from a column will include both higher alcohols and these fatty acid esters.

The higher alcohols and ethyl esters for a set of plate samples from a run during which the product at 173° proof was withdrawn as the top product, i.e., from the condenser, are given in Table I. In this case, the higher alcohols and ethyl esters were most concentrated on the two uppermost trays.

Table I. Distillation of Beverage Brandy Drawn from Condenser

Tray No.	°Proof	Higher Alcohols[a]			Ethyl Esters[a]		
		n-Propyl	Iso-butyl	Amyls	Capry-late	Caprate	Laurate
Product	173.1	26.4	30.2	134	1.6	2.3	1.1
14	147.5	32.2	39.6	319	4.4	25.6	10.4
13	107.5	24.8	27.5	305	2.8	14.2	13.3
12	74.2	13.6	13.1	150	4.3	7.7	5.5
11	44.2	6.5	5.9	46.6	3.6	8.9	3.5
10	28.8	Trace	Trace	11.4	1.4	3.2	1.8

[a] Grams per 100 liters at existing proof.

The analyses for higher alcohols in a set of tray samples in which the primary product was drawn as a sidestream at 186° proof, three trays below the top, are given in Table II. Under the operating conditions for this run, all of the higher alcohols were most concentrated in the lower four trays of the section with an increasing percentage of amyl alcohols on the lower trays.

A set of sequential samples taken during the simple batch or pot distillation of a charge of low wines into brandy were analyzed for higher alcohols and fatty acid esters as shown in Table III. These represent the

Table II. Distillation of Wine Spirits as Sidestream Product from Tray 11

Higher Alcohols, grams/100 liters[a]

Tray No.	°Proof	n-Propyl	Iso-butyl	Amyls	n-Pro-pyl, %	Iso-butyl, %	Amyls, %
Heads	189.8	—	2.3	—	—	100	—
14	189.2	—	7.1	2.4	—	74.7	25.3
13	188.6	11.8	11.8	5.9	40.0	40.0	20.0
12	187.5	19.9	16.4	14.0	39.6	32.6	27.8
Product	186.0	25.4	25.4	25.4	33.3	33.3	33.3
11	186.6	25.7	22.2	23.4	36.0	31.1	32.9
10	184.0	42.5	42.5	64.0	28.5	28.5	43.0
9	181.9	62.3	65.6	158	21.8	23.0	55.3
8	179.2	65.0	85.0	207.5	18.2	23.8	58.0
7	176.8	88.0	110.0	354	15.9	19.9	64.2
6	171.9	107.5	150.5	688	11.4	15.9	72.7
5	166.4	120.5	177	1000	9.3	13.7	77.0
4	160.2	135	200	1500	7.4	10.9	81.7
3	148.5	116	190	2150	4.7	7.7	87.6
2	128.5	128	208	3210	3.6	5.9	90.5
1	94.4	53	71	1178	4.1	5.4	90.5

[a] Grams per 100 liters at existing proof.

Table III. Pot Distillation of Beverage Brandy—Composition of Sequential Samples

Sample	No.	°Proof	Higher Alcohols[a]			Ethyl Esters[a]		
			n-Propyl	Iso-butyl	Amyls	Cap-rylate	Cap-rate	Lau-rate
Heads		158.6	24.1	49.7	116.5	6.1	11.1	8.2
Brandy fractions	1	156.4	25.1	49.4	117.2	3.5	7.8	6.3
	2	153.2	23.7	44.5	115.5	1.2	2.9	2.5
	3	151.4	23.1	41.4	110.8	0.3	0.6	0.6
	4	149.0	23.0	38.2	108.9	0.1	0.1	0.1
	5	144.6	19.8	34.4	102.3			
	6	139.8	19.8	28.2	91.3			
	7	121.8	16.7	19.6	57.3			
	8	112.0	14.4	10.4	25.2			
Tails fractions	9	84.8	9.0	5.0	10.5			
	10	55.2	3.9	2.0	3.6			
Low wines (charge)		57.0	10.7	9.1	22.1	0.3	0.5	0.4
Aggregate product		142.4	23.8	32.8	100.8	0.7	1.7	1.5

[a] Grams per 100 liters at existing proof.

conditions during the second distillation step of the Charente method used for cognac brandies and for the all-malt highland whiskies of Scotland. The unit used was a 350-gallon pot of copper metal heated by steam condensing in a heating coil. Both the higher alcohols and fatty acid esters are distilled over at the highest rate at the start of the run. The ethyl esters of the C_8, C_{10}, and C_{12} fatty acids, all having boiling points greater than the higher alcohols of fusel oil, distilled over almost completely in the early part of the run so none were detectable in the later product fractions nor in the tails. n-Propyl alcohol distilled over more uniformly than any other component measured.

Though not reported in Table III, a peak for phenethyl alcohol emerged in the gas chromatograms of the ester concentrates of the later brandy fractions and increased in size into the tails fractions. Phenethyl alcohol is barely detectable, if at all, in corresponding concentrates of continuous still brandies.

The behavior of higher alcohols and fatty acid esters during distillation is clarified by Figures 5 and 6, prepared by Unger (36) from vapor–

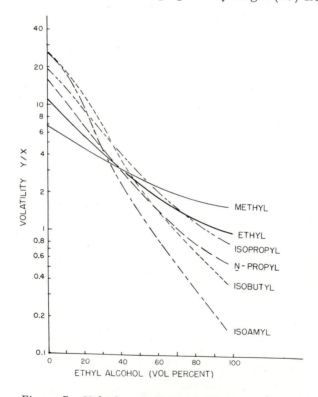

Figure 5. Volatility of aliphatic alcohols at various ethyl alcohol–water concentrations

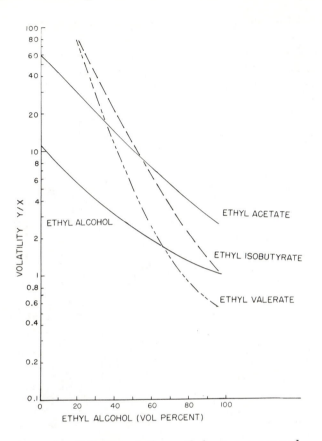

Figure 6. Volatility of three ethyl esters compared with ethyl alcohol at various ethyl alcohol–water concentrations

liquid equilibrium data obtained by Williams (37). The volatility, or concentration of a component in the vapor phase, divided by that in the liquid phase at equilibrium for the several alcohols or esters added in small amounts (less than 1%) to various strengths of ethyl alcohol–water solutions are shown as a function of the proof strengths of the equilibrium liquid. A higher molecular weight and high boiling alcohol such as isoamyl alcohol is markedly less volatile than ethyl alcohol at high proof strengths since isoamyl alcohol is soluble in ethyl alcohol in all proportions. Normal binary solution behavior for isoamyl–ethyl alcohol mixtures is exhibited. But with water, in which isoamyl alcohol is only slightly soluble, the volatility of the isoamyl alcohol is greatly enhanced, characteristic of a partially miscible binary system which exhibits a minimum boiling point azeotrope.

The proof strength at which the volatility curve of a particular component intersects that of ethyl alcohol indicates that proof at which the minor constituent will be concentrated in a fractionating column at the limiting condition of total reflux. For practical conditions, the maximum concentration of a particular congener or minor component occurs at a proof in the column at which its' volatility is approximately equal to the internal reflux ratio (L/V where L is the molal liquid or overflow rate and V the molal vapor rate). This can be established by the technique

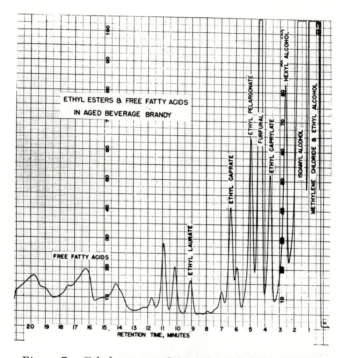

Figure 7. Ethyl esters and free fatty acids in methylene chloride extract of brandy. FFAP column, 6 ft × ⅛ in., programmed linearly from 100–215°C at 7.5°C/min.; 3 μl; internal standard, ethyl pelargonate.

of writing material balance or operating line equations with respect to the particular congener, as in the well-known McCabe-Thiele method in which molal liquid and vapor rates are assumed constant. The internal reflux ratio of the column when the data for Table II were taken was approximately 0.70. Reading from the curves of Figure 5 at 0.7 volatility, the maximum concentrations should have been obtained as follows: amyl alcohols, 128° proof; isobutyl alcohol, 158° proof; n-propyl alcohol, 170° proof. The values agree fairly well with Table II.

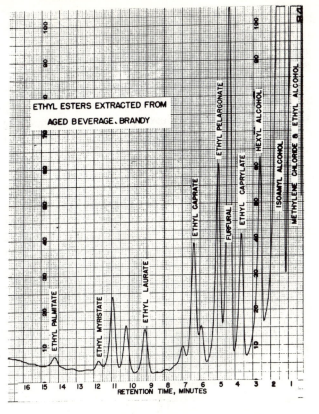

Figure 8.   Ethyl esters extracted from brandy after removal of free fatty acids. Same conditions as for Figure 7.

The ethyl esters of caprylic, capric, and lauric acids are not included in Figure 6. However, they should all fall to the left of the curve for ethyl valerate and exhibit a somewhat steeper volatility gradient.

Nordstrom (*19*) demonstrated that esters are formed primarily by a direct biosynthetic process during fermentation in which acyl–CoA compounds containing the particular fatty acid moiety combine with alcohols of the medium, which explains the predominance of ethyl esters. Ester formation during fermentation does not appear to be direct esterification between alcohols and free fattty acids. However, some direct esterification may occur on the plates of a distilling column where acids and alcohols are most concentrated.

Free Fatty Acids. In our gas chromatographic analyses of brandy extracts prepared for fatty acid ester measurements, free fatty acids also emerged as broad tailing peaks at the end of the programmed temperature run. Free fatty acids have been identified as components of various fermented and distilled beverages by several workers. Nykänen *et al.* (38) gave quantitative levels of acetic acid and relative proportions of 21 other volatile fatty acids in a variety of brandy, rum, and whiskey samples. They found that acetic, caproic, caprylic, capric, and lauric acids were generally most abundant. They measured acids by gas chromatography after conversion to methyl esters.

Crowell and Guymon (39) developed a gas chromatographic procedure for quantitative measurement of the predominant free fatty acids by successive extractions of brandy with methylene chloride and then with dilute alkali to separate the free acids from esters. The acids were converted to their *n*-butyl esters for programmed temperature gas chromatography. Figure 7 shows the chromatogram of ethyl esters extracted from a brandy without removing the free acids. Figure 8 shows the ethyl ester chromatogram with the free fatty acids removed and converted to *n*-butyl esters and chromatographed as shown in Figure 9. Ethyl pelar-

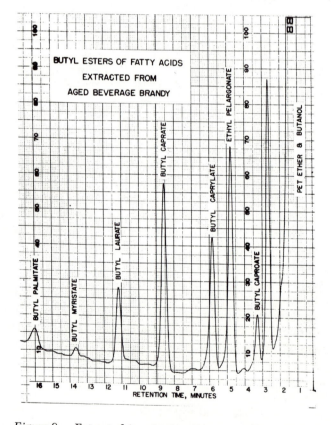

Figure 9. Extracted free fatty acids converted to n-*butyl esters. Same material and conditions as Figures 7 and 8.*

gonate is used as the internal standard for the naturally occurring ethyl esters or for the free acids after conversion to their *n*-butyl esters.

We found approximately 2–4 grams of caprylic, 2–5 grams of capric, and 1–2 grams of lauric acid/100 liters in our experimental continuous still beverage brandies but usually less than 1 gram of each per 100 liters in unaged commercial brandy distillates (*11*). New commercial brandy distillates made from fortified wine contain less than those distilled from straight dry wine. Distillates from a continuous vacuum column and a pilot pot still also contain less.

During aging of brandy and other distilled spirits in wood, very significant changes in composition occur, most of which are well known (*40, 41, 42, 43*). During aging in oak, acids (primarily acetic), esters (primarily ethyl acetate), color, tannin, extract, and other components increase to a degree dependent on the kind and activity of the wood, length of aging, environmental conditions, etc.

Literature Cited

1. Amerine, M. A., Ough, C. S., "Studies with Controlled Fermentation VIII. Factors Affecting Aldehyde Accumulation," *Amer. J. Enol. Viticult.* (1964) **15**, 23–33.
2. Ribéreau-Gayon, J., Peynaud, E., Lafon, M., "Recherches sur la genese des produits secondaires de la fermentation alcoolique," *Bull. Soc. Chim. Biol.* (1955) **37**, 457–473.
3. Guymon, J. F., Nakagiri, J. A., "Effect of Acetaldehyde, Acetal, and Ethyl Acetate upon Alcoholic Fermentation," *Amer. J. Enol.* (1957) **8** (1), 1–10.
4. SentheShanmuganathan, S., "The Mechanism of the Formation of Higher Alcohols from Amino Acids by *Saccharomyces cerevisiae*," *Biochem. J.* (1960) **74**, 568–576.
5. Webb, A. D., Ingraham, J. L., "Fusel oil," *Advan. Appl. Microbiol.* (1963) **5**, 317–353.
6. Guymon, J. F., "Mutant Strains of *Saccharomyces cerevisiae* Applied to Studies of Higher Alcohol Formation during Fermentation," *Devel. Ind. Microbiol.* (1966) **7**, 88–96.
7. Äyräpää, T., "On the Formation of Higher Alcohols by Yeasts and its Dependence on Nitrogenous Nutrients," *Sv. Kem. Tidsk.* (1971) **83** (10), 1–12.
8. Peynaud, E., Guimberteau, G., "Sur la formation des alcools supérieurs par les levures de vinification," *Ann. Technol. Agr.* (1962) **11**, 85–105.
9. Reazin, G., Scales, H., Andreasen, A., "Mechanisms of Major Congener Formation in Alcoholic Grain Fermentations," *J. Agr. Food Chem.* (1970) **18**, 585–589.
10. Guymon, J. F., Heitz, J. E., "The Fusel Oil Content of California Wines," *Food Technol.* (1952) **6**, 359-362.
11. Guymon, J. F., "Composition of California Commercial Brandy Distillates," *Amer. J. Enol. Viticult.* (1970) **21**, 61–69.
12. Rankine, B. C., "Formation of Higher Alcohols by Wine Yeasts and Relationship to Taste Thresholds," *J. Sci. Food Agr.* (1967) **18**, 583–589.
13. Guymon, J. F., Ingraham, J. L., Crowell, E. A., "Influence of Aeration upon the Formation of Higher Alcohols by Yeasts," *Amer. J. Enol. Viticult.* (1961) **12**, 60–66.

14. Crowell, E. A., Guymon, J. F., "Influence of Aeration and Suspended Material on Higher Alcohols, Acetoin and Diacetyl during Fermentation," *Amer. J. Enol. Viticult.* (1963) **14**, 214–222.
15. Ough, C. S., Guymon, J. F., Crowell, E. A., "Formation of Higher Alcohols during Grape Juice Fermentations at Various Temperatures," *J. Food Sci.* (1966) **31**, 620–625.
16. Hough, J. S., Stevens, R., "Beer Flavor IV. Factors Effecting the Production of Fusel Oil," *J. Inst. Brew. London* (1961) **67**, 488–494.
17. Radler, F., "Uber den Gehalt von Isoamylalkohol und Isobutylalkohol in Weinen von Kulturrebensorten und Neuzüchtungen," *Vitis* (1960) **2**, 208–221.
18. Guymon, J. F., Crowell, E. A., "Gas Chromatographic Determination of Ethyl Esters of Fatty Acids in Brandy or Wine Distillates," *Amer. J. Enol. Viticult.* (1969) **20**, 76–85.
19. Nordström, K., "Studies on the Formation of Volatile Esters in Fermentation with Brewer's Yeast," *Sv. Kem. Tidsk.* (1964) **76**, 1–34.
20. Lafon, R., Lafon, J., Couillaud, P., "Le Cognac-sa Distillation," 4th ed., J. B. Baillière, Paris, 1964.
21. Guymon, J. F., "The Composition of Plate Samples from Distilling Columns with Particular Reference to the Distribution of Higher Alcohols," *Amer. J. Enol. Viticult.* (1964) **11**, 105–112.
22. Guymon, J. F., "Principles of Fusel Oil Separation and Decantation," *Amer. J. Enol.* (1958) **9** (1), 64–73.
23. Joslyn, M., Braverman, J., "The Chemistry and Technology of the Pretreatment and Preservation of Fruit and Vegetable Products with Sulfur Dioxide and Sulfites," *Advan. Food Res.* (1954) **5**, 97–160.
24. Guymon, J. F., Nakagiri, N. A., "The Determination of Combined Aldehydes in Distilled Spirits," *J. Ass. Off. Anal. Chem.* (1957) **40**, 561–575.
25. Lichev, V. I., Panaitov, I. M., "Identification of Aldehydes in the Spirit of Cognac," *Vinodel. Vinograd. S.S.S.R.* (1958) **18** (4), 10–15. *Chem. Abstr.* (1958) **52**, 1901.
26. Whitmore, F. C., "Organic Chemistry," pp. 151, 191, 2nd ed., D. Van Nostrand, New York, 1951.
27. Diebner, L., Bénard, P., "Separation of Aldehyde–Sulfurous Acid Compound Contained in Alcoholic Beverages," *Ann. Fals. Fraudes* (1954) **47**, 9.
28. Peynaud, E., Maurie, E., "L'acetal dans les eaux-de-vie, son dosage volumetric-reactions d'acetalisation," *Bull. Int. Vin* (1938) **11** (18), 42–46.
29. Guymon, J. F., Jaber, M. S., "Reduction of Aldehydes during Alcoholic Fermentation. Application to Processing of Heads," *Agr. Food Chem.* (1959) **7**, 576–578.
30. Guymon, J. F., Pool, A., "Some Results of Processing Heads by Fermentation," *Amer. J. Enol.* (1957) **8** (2), 68–73.
31. Guymon, J. F., Nakagiri, J. A., "Utilization of Heads by Addition to Alcoholic Fermentations," *Amer. J. Enol.* (1955) **6** (4), 12–25.
32. Suomalainin, H., Kauppila, O., Nykänen, L., Peltonen, R. J., "Branntweine. Handbuch der Lebensmittelchemie," vol. 7, pp. 496–653, Springer-Verlag, Berlin-Heidelberg, 1968.
33. Stevens, R., "Beer Flavour I. Volatile Products of Fermentation. A. Review," *J. Inst. Brew. London* (1960) **66**, 453–471.
34. Suomalainen, H., Ronkainen, P., "Aroma Components and Their Formation in Beer," *Tech. Quart. Master Brew. Ass. Amer.* (1968) **5**, 119–127.
35. Lawrence, W. C., "Volatile Compounds Affecting Beer Flavor," *Wallerstein Lab. Commun. Sci. Pract. Brew.* (1964) **27**, 123–152.
36. Unger, E., private communication, 1967.
37. Williams, G. C., "Vapor–Liquid Equilibria of Organic Homologues in Ethanol–Water Solutions," *Amer. J. Enol. Viticult.* (1962) **13**, 169–180.

38. Nykänen, L., Puputti, E., Suomalainen, H., "Volatile Fatty Acids in Some Brandy of Whiskey, Cognac and Rum," *J. Food Sci.* (1968) **33**, 88–92.

39. Crowell, E. A., Guymon, J. F., "Studies of Caprylic, Capric, Lauric and Other Free Fatty Acids in Brandies by Gas Chromatography," *Amer. J. Enol. Viticult.* (1969) **20**, 155–169.

40. Boruff, C. S., Rittschof, L. A., "Effects of Barreling Proof on the Aging of American Whiskeys," *Agr. Food Chem.* (1959) **7**, 630–633.

41. Liebmann, A. J., Scherl, B., "Changes in Whiskey while Maturing," *Ind. Eng. Chem.* (1949) **41**, 534–543.

42. Valear, P., "Brandy," *Ind. Eng. Chem.* (1939) **31**, 339–353.

43. Valear, P., Frazier, W. H., "Changes in Whiskey Stored for Four Years," *Ind. Eng. Chem.* (1936) **28**, 92–105.

RECEIVED July 25, 1973.

12

Some Aspects of the Wooden Container as a Factor in Wine Maturation

VERNON L. SINGLETON

Department of Viticulture and Enology, University of California, Davis, Calif. 95616

The relationship of modern wine styles to aging wines in wooden barrels is described. The physical and chemical characteristics of different woods lead to the conclusion that white oak is best for wine cooperage. European cooperage oak samples had 161% of the extractable solids of North American samples and 154% of the phenol per unit of extractable solids, but American oak contributed more oak flavor to wine per unit of extract. Data are presented on the density and extractable phenol content of summer and spring oak wood, the characteristics of rapidly vs. slowly grown oak, the surface per unit volume effects of container size and shape, the variability among trees in flavor and extract content, the analysis of wood extract in wine, and the depth of penetration of wine into staves.

For a long time wooden containers were probably the most common type of storage vessel for bulk wine. Under proper conditions, they still may be very desirable wine cooperage for a number of reasons. However, wooden containers are relatively permeable and are subject to leakage and contamination unless very carefully made and maintained. For these reasons, with the development of stainless steel and other relatively inert and impervious tanks or tank linings, some wineries, at least in the United States, boast that they use no wooden containers for any of their wines.

This shift from wood to stainless steel cooperage has been general and a factor in changing production practices around the world, particularly for red wines. A wine well protected from oxidation in an impervious tank will not change rapidly in tannin content. If it was vinified so as to be too tannic for acceptance by the consumer, it is likely to remain so for a long time unless otherwise treated. Conversely, a wine subjected to considerable oxidation, as is generally the case during long storage in small barrels, must have a fairly high initial phenol content to survive the treatment in good condition (*1*). Thus, the widespread use of stainless steel or similar cooperage has shortened the average time that red wines are vinified in the presence of pomace. The red wines which result from short fermentations on the grape skins are saleable very young and tend to be fresh and fruity. This style of wine is attractive for pink and many white table wines. For such wines, maturation in wooden barrels is not necessary and is often considered detrimental.

Many of the world's red wines, many dessert or appetizer wines, and a few white table wines owe part of their character, complexity, and quality to maturation in wood. If all wineries limited their production to fresh, fruity styles which can be produced without benefit of wood aging, we would lose an important part of the world's wines, often the most prestigious and valued. Recognizing this, many modern wineries have retained the use of wooden cooperage for certain wines. In the United States, the consumer's willingness to pay premium prices for high quality wines of diverse styles and types has led several wineries to purchase additional wooden cooperage and expand their program of wood maturation of certain wines. Because of the high costs of high quality new cooperage and the extra costs of increased labor and evaporative losses inherent in aging in wooden barrels, there is renewed interest in optimizing the conditions for maturation of wine in wood. Furthermore, there is continued interest in achieving maturation of wine similar to the effects of several years in barrels by methods adaptable to less expensive wines, methods which do not tie up capital as long and incur all the costs of prolonged storage in barrels.

Different Woods for Cooperage, the Primacy of Oak

Many woods have been used for tight cooperage. To make a barrel, the wood needs a certain strength, resilience, and workability. Generally, fairly large diameter trees are needed for economical conversion, and the grain must be straight for the length required by the staves for the particular size of the containers to be built. The wood must be free of defects leading to leaks, and it should not contribute undesirable flavors to the contents. Within these limitations alone many woods would suffice.

In the United States, woods which have been used for tight cooperage include white oak, red oak, chestnut oak, red or sweet gum, sugar maple, yellow or sweet birch, white ash, Douglas fir, beech, black cherry, sycamore, redwood, spruce, bald cypress, elm, and basswood (*2, 3, 4*). In Europe, cooperage for wine or brandy has been made from oak, chestnut, fir, spruce, pine, larch, ash, mulberry, and a number of additional species imported from Africa, South America, and Australia (5, 6, 7). Australian woods which have been used for wine cooperage include acacia, karri (*Eucalyptus diversicolor*), jarrah (*E. marginata*), stringybark (*E. obliqua* and *E. gigantea*), and she oak (*Casuarina fraseriana*) (8).

Still other woods are used for tight cooperage in the orient such as *Albizzia odoratissima* in Ceylon and Burma (8) and *Cryptomeria japonica* in Japan (9), but what their effect would be on European-type wines is not known.

This extensive list of woods rapidly dwindles, however, if only the woods considered most satisfactory for alcoholic beverage containers are taken. A number of the woods can contribute flavors that, when detectable, are objectionable. For such timber to be used it must either be in large tank sizes with a low surface-to-volume ratio, be so well leached by treatment or use that very little extractives reach the wine, or be used for so short a time or for a wine of such low standard that the wine remains acceptable. Redwood (*Sequoia sempervirens*) is a conifer with an unusually high extractable phenol content, 10–15% of the virgin heart wood is hot-water extractable (10). It is not bendable to make barrels, but it occurs in California as large, straight grained, easily worked timber. It makes fine straight-stave tanks 3000 liters and larger that have been used for wine for many years. However, a concentrated aqueous alcoholic extract of redwood is not very pleasant in flavor, and long storage of wine in new or small redwood containers is not practiced. Douglas fir, spruce, pine, and other conifers may contribute resinous flavors, and they are used as temporary receptacles if at all (6). Eucalypts also are either used as large casks for temporary storage or coated with paraffin or other substance to minimize pickup of flavor by wine from the wood. Eucalypts and acacia are considered unsuitable for brandy (7) because brandy extracts more flavor from wood than does wine, but it is clear that the flavor is considered undesirable at least when recognizably strong.

Some woods are too porous although they may be used if coated inside with parafin or other suitable substances. Red oak, ash, and gum should be coated (2). Chestnut, perhaps the second most used wood, is excessively porous, high in tannin, and low in durability compared with good oaks (5, 6, 7). Acacia imparts a yellow pigment to wine (11) and is generally reserved for ordinary wines or shipping containers (7). Still other woods are no longer readily available. Bald cypress was once a preferred wood for large tanks, but because of cost and scarcity it has been largely displaced by redwood in the United States (4).

After one eliminates all those woods that are less suitable, not readily available, poorly tested, or suitable only in larger sizes, the white oaks remain. Hankerson (2) states that white oak combines in about ideal proportions the characteristics desired in tight cooperage. Brunet (5), Huber and Kehrbeck (6), Veiga (7), and others consider oak best for wine cooperage although the relative merits of different oaks are subject to more discussion.

What Makes Oak the Best Tight Cooperage Wood?

Both their structure and chemical composition make selected oaks highly suitable for barrels for beverage storage. Oak wood is the standard for strength against which other woods are often compared. It is resilient and bendable. Although sometimes hard, it is generally suitably workable. Some oaks are only shrubs, and others are excessively gnarled or branched, but those suitable for cooperage grow large, and in forests they produce straight-grained stave bolts. The structural features that make certain oaks outstanding for tight cooperage are their rays and tyloses.

The rays of wood lie along a radius of the trunk cylinder from the pith to the bark and form the diffusion channel in the living tree for horizontal translocation (12, 13). When a tree trunk is cut into sections of the length desired for staves, it is then split lengthwise through its center into quarters. The staves are cut from the faces of each quartered bolt (quartersawed) so that the rays are parallel to the width of the stave and thus perpendicularly opposed to the diffusion of liquid through the sides of the barrel. The diffusion of water through most wood across the gain is about 25% greater in the radial direction along the rays than across them in the tangential direction (14). The difference in oak is probably even greater because of its large rays.

Quartersawing also minimizes the dimensional change of the width of the stave as it shrinks on drying or swells with wetting. This is probably the reason early barrel makers learned to quartersaw staves. Oak, in drying from green to air-dry, shrinks only about 0.15% in length but about 13% in volume, with the radial dimension shrinking about 4% and the tangential 7–8.5% (13, 15).

Rays one cell wide (uniseriate rays) as in conifers are too small to be readily seen, but those two or more cells wide (multiseriate) become a significant part of the pattern of lumber such as oak, especially when quartersawed. Oaks, like most dicotyledons, have both types of rays, but their multiseriate rays are so unusually wide compared with other hardwoods, that they have been considered a separate type: compound rays (12, 13, 16). These large rays are higher along the grain than the uniseriate rays and taper to a point at top and bottom (12). Their height tends to be greater as the tree gets older and larger and ranges from about 0.5 to greater than 5 cm.

In *Quercus alba* the rays represent 28% of the wood volume and in other oak species 19–32%; in most other hardwoods the rays occupy about 15% but only 8% of the wood volume in the conifer *Sequoia sempervirens* (*13*). The large rays of the oak are so spaced and numerous that a molecule of water diffusing through the side of a barrel must cross five or more large rays if it exits on a straight path or follow a much extended path if it is to go around the rays interposed. These large rays no doubt contribute to the strength and bendability of oak as well as to its dimensional stability and relative impermeability.

Hardwoods are porous compared with softwoods because they contain conductive vessels or tubes (*13*). Cooperage oak is ring porous. The wood formed in the spring includes one to three rows of open vessels as large as 0.3 mm in diameter while the summer wood has much smaller vessels ("pores" in cross section) not visible to the naked eye. If these pores were not plugged by the growth of tyloses, the staves would readily leak from the ends. Preferred oaks for cooperage are naturally rich in tylosis development. Tyloses are formed by the ballooning of the cell wall of adjacent living cells through pits into the channel of the vessel (*17*). The tyloses enlarge and form in sufficient numbers in cooperage oak to completely fill the vessel and effectively block the penetration of liquid along these vessels. The complete plugging of the vessels with tyloses ordinarily coincides with the conversion of sapwood to heartwood (*12, 13*) and is one reason only heartwood is used for staves. Not all species of oak normally produce tyloses, and such oaks (red oak for example) are therefore too porous for unlined barrels.

The composition of oaks also seems especially well suited to barrel use. Hardwoods average about 45% cellulose, 25% hemicelluloses (predominantly xylans), 23% lignin, 7% acetyl groups, and 3–5% extractives including tannins (*18*). Heartwood of American white oak analyzed about 50% cellulose, 22% hemicellulose, 32% lignin, 2.8% acetyl groups, and 5–10% hot-water extractables, and European oak is similar (*19, 20, 21*). The high lignin and tannic extract content seem significant to barrel quality. Other woods used for cooperage, notably chestnut, redwood, and some eucalypts, often contain even more extractable phenols and tannins (*10, 20*). This high tannin content, among other effects, tends to make the container less subject to attack and decay. The tannin of oak wood is readily extracted into wine or brandy and along with other extractives is an important contributor to flavor development during the maturation of wine in wood.

Excessive levels of flavor derived from oak leads to oaky or tannic wines which are lesser in quality. Subtle levels, however, are probably the most essential contribution of maturation in oak cooperage to wine quality. As timber, oak is classified as unscented and tasteless since these features have no identification value in contrast to camphor, sassafras, incense cedar, *etc.* (*13*). Rather, the odor of oak is much like any good carpenter shop aroma, an odor pleasing to most people and one well suited to increasing wine quality by subtly increasing its complexity (*22*). Oak odor as such has been studied little, although recently the woody odor of whiskey has been attributed to nine-carbon γ-lactones that were also found in the woods of white oak and two Asian *Quercus* species (*23, 24*).

Vanillin, syringaldehyde, and related phenolic products derived from lignin appear to be important in the desirable odor contributed to beverages by oak barrels (*25, 26, 27*). These substances are present in the wood, but more appears in the beverage than simple extraction seems to allow. Also the odor of vanillin seems to increase with time and mellowing of the beverage compared with a fresh extract of new wood (*25*).

Ether extraction of wood removes resins and other less polar substances like these aldehydes. The ether extract of heartwood is concentrated in the ray tissues and thus would help make them less water permeable (*20*). In white oak heartwood the ether extract is about 0.62–0.71% of the weight of dry wood (*19, 20*), but after alkaline hydrolysis it was 1.40% for *Q. alba* (*28*). From the ether extract after alkaline hydrolysis of 100 grams dry wood, 18 mg vanillin, 33 mg syringaldehyde, 36.5 mg vanillic acid, 59 mg syringic acid, and 53 mg ferulic acid were obtained. *Q. lyrata* was similar except about half as much syringaldehyde was found. Many other hardwoods showed the same products usually in similar proportions (*28*). Hydrolysis and oxidation of lignin are believed to account for production of these compounds during aging of alcoholic beverages in barrels (*25, 29, 30*). Lignin of small molecular size extracted from oak barrels is also an important component of brandy and, presumably, of wine.

Many other odorous or volatile compounds, particularly furfural derivatives, nitrogenous compounds, and other phenols, have been identified in distilled beverages after barrel aging and not before (*23*), but their role is not yet clear. So far it appears that the components contributed by oak to wine and brandy probably are common to most mildly pleasant but not distinctively odorous hardwoods. They appear to be agreeably combined with a moderate tannin level and generally unaccompanied by any unpleasant tastes and odors in cooperage oak.

Which Oak is Best?

This question cannot be answered very well because there are so many uncontrolled variables, conflicting opinions, and little good experimental data. Some clarification appears possible, however. A variable one would expect to be significant would be the botanical species of oak. Others may be the conditions of growth of the trees, individual variations from tree to tree and within a tree, and variations in the seasoning practices of different coopers.

Variations within a tree may be large. In one example from *Q. robur*, the tannin of the bark was 10.9%, sapwood 1.0%, and heartwood 7.7% at the periphery, 6.6% intermediate, and 6.6% at the center, all at the base of the tree which was 50 cm in diameter and 90 years old (*31*). At the top of the same trunk (34-cm diameter, 71 years) the same respective tissues had tannin contents of 5.7, 0.8, 5.3, 4.6, and 4.7%. This pattern of higher extractable tannin in the heartwood nearest the sapwood and lower near the pith, higher as the tree gets older, and higher in the base of the tree than the top is general (*21*).

Large as these differences are, the practices of barrel making should tend to eliminate them and also much of the tree-to-tree variation. Staves should not include sapwood and, being radially cut, they represent a heartwood cross section. The individual staves and pieces of heading pass through several operations with many opportunities for the wood to mix from different tree sections (*2, 5, 6, 7, 31, 32*). The typical American barrel of about 190 liters (50 gal) is made of about 31 pieces of wood (*2*). Each piece is not unlikely to represent a different tree, and the staves almost certainly would be from trees different from the heading. Therefore, each barrel should approach a random sample of the general source of trees being drawn on by the cooper. There are still barrel-to-barrel differences in effects on wine or brandy, but ordinarily these are not pronounced in a group of barrels made by the same cooper at the same time—*i.e.*, from the same stock of timber.

Commercial white oak furnished 90% of all tight cooperage in the United States as early as 1908 (*3*). Tight cooperage reached its maximum production in the United States in 1929 at nearly 107 million m³ of wood consumed, dropped until 1952, and then rose slightly to about 26 million m³ in 1962 (*33*). Present production is believed similar. These figures illustrate the decline in the use of wooden cooperage, but they also indicate that, with less demand for barrels and their use primarily for premium products, the oak used can be restricted to the best. The situation in Europe appears similar.

The production or use of barrels has not been confined to oaks indigenous to the vicinity of the cooperage or the winery. With consolidation and mechanization of cooperages, relatively few producers supply a large part of barrel production, and they may ship barrels considerable distances. Staves, heading, or timber for them are also often shipped long distances before use. Theron reported in 1947 that South

Africa imported about $500,000 worth of stave wood annually, the bulk from the United States and the rest from Europe (*34*). A mill near St. Louis, Mo. organized exclusively for export sent staves to Spain, Portugal, Italy, France, England, Ireland, and Scotland. They processed timber grown in Missouri, Iowa, Illinois, Kansas, and Nebraska. In Portugal, imported oak is preferred to domestic, and oak from Poland, France, Brazil, and North America is used (*7*). Hardman reported in 1878 that American (believed North American) oak was exclusively used for sherry and Marsala (*35*). Costa Rican oaks are now being imported to Spain where they are reportedly esteemed for sherry barrels (*36*). European barrels or staves are being imported into California by several wineries for maturing white table wines from Chardonnay and red table wines, particularly from Cabernet Sauvignon and Pinot noir grapes. American barrels are widely used also. With so much transfer of oak and the potential to influence wines differently by using different oak sources, it behooves the winemaker to consider foreign as well as domestic oaks.

In the United States most tight cooperage is used for bourbon whisky, and barrel production is geared to that market. An American wine barrel is ordinarily a bourbon barrel which has been left plain and not charred inside. Bourbon barrels are exclusively heartwood of select commercial North American white oak. Nearly 60 species of oak in the United States are trees and about 23 of these are commercially important (*37*). The red oak group (*Erythrobalanus*) usually lacks tyloses and is not suitable for tight cooperage (*13*). This group is not found in the Old World (*31*). The tylosis-bearing white oak group (*Leucobalanus* or *Lepidobalanus*) includes about 10 American species, but not all of these are used for cooperage. Wood from *Q. virginiana*, an evergreen oak, is recognizably different, being diffuse porous (*29*). Chestnut oak, *Q. montana*, lacks tyloses in many pores and is not suitable (*13, 31, 34*). The Pacific coast *Q. garryana* has been but is not now used for tight cooperage (*38*), and its particular flavor contribution is not known.

Oak species which may be included in select commercial white oak for American cooperage are *Q. alba*, *Q. prinus*, *Q. bicolor*, *Q. muehlenbergii*, *Q. stellata*, *Q. macrocarpa*, *Q. lyrata*, and *Q. durandii* (*13, 31, 37*). *Q. durandii* is geographically limited to south Atlantic and Gulf of Mexico coastal areas removed from the centers of barrel production in Kentucky, Missouri, and Arkansas. The other species are listed in decreasing order of estimated comparability or slight inferiority to *Q. alba*. *Q. alba* is not only considered the best of the group, it is by far the most significant. It is estimated to constitute 45% of the standing timber of the group (*37*), and it grows over the widest area from southern Canada nearly to the Gulf of Mexico and from the Atlantic Coast to Iowa,

Kansas, Oklahoma, and Texas (*31, 37*). Considering the respective geographical distribution of these oak species (*37*), staves from Kentucky, Missouri, and north would be expected to include *Q. alba, Q. bicolor* and *Q. macrocarpa,* but not much *Q. prinus* or *Q. lyrata.* Staves from Arkansas and south could have *Q. prinus, Q. lyrata,* and probably more *Q. stellata* but less *Q. macrocarpa,* and little or no *Q. bicolor.* The preferred habitat of these species varies from dry hills to marshy areas and also affects which oaks are obtainable from what locality (*31, 37*).

It is not possible to distinguish between these species by examining the staves (*13*). The trees are often difficult to separate into species even by trained botanists. Furthermore the species not only are rather polymorphic and variable related to different growing conditions and regions, but they also often interbreed (*31, 37*). American oak barrels as a group impart fairly constant character to wine. It appears safe to conclude either that these oaks have very similar physical and chemical composition or that such a large proportion of the average barrel is *Q. alba* and the other species are so small and so mixed that they do not cause major differences in stored beverages. Nevertheless study of botanically identified trees in this regard would be desirable.

The European oak species preferred for cooperage are *Q. robur* and *Q. sessilis* (*31*). Where these species are plentiful, others do not appear to be used for casks. *Q. sessilis* has been considered a variety of *Q. robur* because they are so similar (*31*); it has also been called *Q. petraea* and *Q. sessiliflora,* and common names include rouvre, liverneau, durmast, dry oak, hill oak, Traubeneiche, Steineiche, roble, and winter oak. *Q. robur* has been called *Q. pedunculata,* and common names include pedunculate oak, rouvre, valley oak, feminine oak, French oak, English oak, summer oak, Stieleiche, and roble albar. The two species differ in that *Q. robur* acorns are on a long peduncle and in *Q. sessilis* they are attached directly to the twig. According to Camas (*31*), two files of fine vessels in the summer wood often join together in late summer to give a bifurcated appearance in a cross section of *Q. robur* wood while they remain in separate files in *Q. sessilis.* If this feature is diagnostic, a very high proportion of the European barrel staves examined came from *Q. robur.* Otherwise the wood is very similar and, like all cooperage oaks, not readily distinguished by appearance from other oak species including American.

The geographical range of *Q. robur* includes essentially all of that of *Q. sessilis* but extends further north in Scandinavia, much further east into Russia (to the Urals), and further south in Turkey, Soviet Georgia, and Portugal (*31, 39*). They are found together over the rest of Europe from the British Isles to beyond Warsaw and nearly to Odessa, and from Scandinavia to the Balkans, Italy, and northern Spain (*31, 39*). Since they often grow side-by-side and hybridize, it is no wonder that their wood is not usually distinguished in commerce. When they share the same habitat the wood is essentially equal in properties (*31*). However, *Q. robur* is more demanding and tends to predominate in richer, deeper, heavier, and more moist soils. *Q. sessilis* is better able to grow on shallower, dryer hill soils. For this reason *Q. robur* wood tends to be a little more dense and harder than *Q. sessilis*. However, *Q. sessilis* is usually preferred in replanting and forest management since it tends to have a more compact crown, fewer low branches, and a taller, more regular trunk than *Q. robur*. *Q. sessilis* is more resistant to certain insects and *Oidium* infection as well as more tolerant of poor soil and less sunlight (*31*). *Q. robur* stands flooding better and is suited to stream valleys. It is less vulnerable to spring frosts and a little more resistant to winter cold. Neither species likes dry spring weather or great heat which limits their growth in the Mediterranean area. The oaks of Stellenbosch, South Africa are *Q. robur*, according to Camas (*31*).

Many other species of oak have been used to make staves or to flavor wine in Europe, North Africa, and the Near East, particularly when the two important European species were lacking or where they grow in the same forests and are not commercially distinguished. They include *Q. cerris, Q. afares, Q. macranthera, Q. longipes, Q. imeretina, Q. iberica, Q. pedunculiflora, Q. lanuginosa* (*Q. pubescens*), *Q. farnetto* (*Q. conferta*), and *Q. mirbeckii* (*7, 31*). In no instance are these species reported as superior to or preferred to *Q. robur* or *Q. sessilis*. Specific comments on their flavor contribution to wine or brandy were not found. In Asia, *Q. mongolica* has been used for staves, but it is not esteemed (*31*). "Amazon oak" (*Quercus* is not native to the Amazon) is reported to contribute high tannin and bitterness unless pretreated (*7*). The Costa Rican oak exported to Spain for sherry barrels is reported to be *Q. costaricensis* and *Q. eugeniaefolia* (*36*), but these are *Erythrobalanus* species, and *Q. copeyensis* may be involved. A sample received through Guymon of our department gave less (but not objectionable) flavor than either American or European oak.

From the considerations outlined, it appears one can usefully divide oak species having a reputation for high quality wine or brandy barrels into only two groups: North American oak derived from *Q. alba* and related species and European oak derived from *Q. sessilis* and *Q. robur*. Woods derived from other oak species seem either to be inferior or to have been insufficiently tested to demonstrate their qualities. These two groups, North American and European, have sufficiently different extractives that it is often possible for an experienced taster to state which was used for wine or brandy storage. Within each group, however, the flavor effects are similar although they vary considerably in intensity, depending on length of storage, history of the barrel, etc.

Within the groups, American or European, coopers and winemakers have had preferences for oak from certain sources. These sources are sometimes named for the particular forest but often only for the port or railway center from which the timber was shipped. A number of interesting opinions and comparisons have been published (5, 7, 31, 33, 35, 40). Taken together these observations seem to indicate that within the two groups the major factor in the quality of oak for wine barrels is the size and regularity of the trees and perhaps how rapidly they grew. These factors depend on the time since the forest was last harvested and the terrain and climate which are variable between localities. Therefore, it does not seem useful to seek the best geographical source of oak since it is likely to change with the particular trees being harvested. Rather, choosing a dependable supplier and specifying the desired qualities in the staves and heading would seem more productive. It should also be kept in mind that the factors which make a stave better from a cooper's viewpoint (soft, easily worked, etc.) may not have much direct significance to the barrel user who wants good flavor contribution, low permeability, etc.

Fast vs. Slowly Grown Oak

Winemakers generally believe that slowly grown oaks make better barrels than those grown rapidly, but they do not seem to have a clear reason for this opinion. Since the growth rate can be judged from the annual rings in the staves, this is a feature which can be checked by the winemaker whereas most other information on the wood must depend on the supplier. A high proportion of spring wood and a low proportion of summer wood is characteristic of slow oak growth, and the reverse characterizes rapid growth. Since the spring growth contains the large vessels, slowly grown oak should be less dense, softer, more easily worked, but probably more permeable than rapidly grown oak. These properties should be desirable to the cooper but not to the barrel user. Also, slow growth is often characteristic of large, mature trees desirable to the cooper. In this case there might also be an advantage to the user since older trees have taller large rays and thus should be less permeable.

Experiments were conducted to test some features of spring vs. summer wood and the relative growth rate of stave wood. Sections of barrel staves of American and French origin were tested in an Instron universal testing machine for the force necessary to cause the wood to yield under pressure of a blunt point 0.5 mm in diameter. The force necessary in the summer wood in the same radial direction as fluid would pass in the barrel averaged 3.61 kg for the European oak and 4.78 kg for American. In paired, adjacent spring wood the respective averages were 1.86 and 1.72 kg. This clearly shows the greater penetrability and lesser density of spring wood compared with summer wood. Although there was considerable variation from stave to stave, the highest of the spring wood values was lower than the lowest of the summer wood values. Wood technologists have already shown that penetration force on a larger section is proportional to density, and increased wood density is strongly correlated to strength, hardness, and other properties of the wood.

Other data were obtained by sawing parallel to the annual rings so that a saw cut included only summer wood or spring wood, collecting all the sawdust and using the cut volume and wood weight to determine density and extractive content (Table I).

Table I. Composition of Summer *vs*. Spring Wood in Oak Staves

Stave	Growth, cm/10 yrs	Density, Air-Dry, gram/cm³		Extractable Phenols, grams gallic acid equiv./ 100 grams (Oven-Dry)	
		Spring Wood	Summer Wood	Spring Wood	Summer Wood
American Oak					
1	4.45	0.448	0.675	1.04	0.72
2	3.74	0.591	0.855	2.98	2.14
3	2.93	0.762	0.930	2.15	1.94
4	2.52	0.607	0.907	1.40	1.33
Average	3.32	0.602	0.842	1.90	1.54
European Oak					
1	3.80	0.347	0.487	6.87	5.25
2	1.62	0.624	0.967	7.64	4.33
Average	2.71	0.486	0.727	7.26	4.79

Several points seem apparent even though the number of samples is not large. The density of the summer wood was invariably greater than (averaging 144%) that of the spring wood in the same piece of wood. However, the faster the tree grew the less dense both forms of wood became. Apparently the tree which is growing fast deposits less wood substance per unit volume of wood. The same tendency appears, but is less clear, in the extractable phenols of the wood. The fastest growing trees, both European and American, had less extractable phenols than the slower growing ones. The trend was not uniform, but a confounding factor here is the fact that growth rate varies from time to time in the same tree, and oak heartwood extractives are often deposited about 10–15 years after the growth rings are formed (*31*). Invariably, however, the phenolic extractives were greater in the spring wood (average 132%) than in the adjacent summer wood of the same stave. This shows that, other things being equal, the higher proportion of spring wood (slow growth) would lead to a greater extractable phenol content in the wood. The tendency of European oak to have higher extractable phenols than American oak is also illustrated in Table I. Slowly grown oak stave wood may be better than fast-grown because it would ordinarily contribute more extractives to wine or brandy.

Cooperage Size and Wine Maturation

There are three reactions which distinguish maturation of wine in wooden cooperage from storage in impermeable containers: evaporation of alcohol and water through the sides of the container, admission of oxygen into the wine as a result of the permeable sides, and extraction of substances from the wood into the wine. All are surface related and would be more intense with larger surface per unit contents of the container. The fact that small cooperage affects wine more rapidly than does large because of its larger surface-to-volume ratio is well recognized, but it does not appear to have been considered sufficiently .

Calculations have been made of the surface of several shapes of container per liter of contents over a series of volumes. The shapes chosen were a sphere (the least possible surface per unit volume), a cube (counting all six faces), a cylinder with a height twice its diameter, and a barrel with typical inside measurements (2, 5) of the bilge (maximum) diameter 1.25 times the head diameter and the distance between the heads 1.3 times the bilge diameter. The barrel was calculated as two frustums base to base and also as a truncated ellipsoid of rotation.

Table II. Surface per Unit Volume for Different Container Shapes and Capacities

Container Volume	Sphere, cm²/liter	Barrel Ellipsoid, cm²/liter	Barrel Frustum, cm²/liter	Cylinder, cm²/liter	Cube, cm²/liter
5 liter	283	307	313	340	351
10 liter	224	244	248	270	278
20 liter	178	193	197	214	221
50 liter	131	149	145	158	163
2 hl[a]	83	90	91	99	103
20 hl	38	42	42	46	48
50 hl	28	31	31	34	35
100 hl	22	24	25	27	28
1000 hl	10	11	12	12	13
10000 hl	4.8	5.2	5.3	5.8	6.0

[a] Hectoliter.

The data (Table II) demonstrate the great decrease in container surface per unit contents as the volume of the container becomes larger, regardless of container shape. Figure 1 makes clear the fact that a plot of the logarithm of the container surface per liter *vs.* the logarithm of the container volume gives a straight line for containers of a given shape and constant proportions. The fact that these lines are parallel demonstrates that the surface per unit volume of containers of different shape is a certain multiple of that of a sphere of the same volume. In this series the ellopsoidal barrel had 1.085, the double frustrum barrel 1.106, the cylinder 1.201, and the cube 1.240 times the surface per unit contents of the sphere with the same volume. It therefore becomes easy to compare the surface per volume and the expected aging effects of containers of different shapes and dimensions. Most container shapes used in the wine industry fall between the sphere and the cube of similar contents with regard to surface per unit volume.

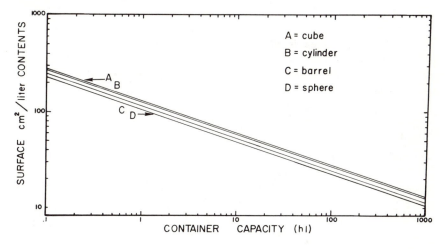

Figure 1.   Surface per unit of contents for containers of different shapes and sizes

Further examination of Table II gives more insight into wine maturation. Since we believe that surface effects control aging in wooden barrels, we can consider that extraction from a given surface area of wood, evaporation through it, etc., are constant and independent of gradient and other effects at least in the earlier stages of aging in a given type of container and assuming stave thickness, temperature, humidity, etc., remain constant for containers of different sizes. Note (Table II) that increasing the container volume 10-fold gives a little less than half as much surface/liter volume, and containers 1000 times as large in volume have exactly 1/10 as much surface/liter volume.

Suppose it has been decided that three years in 200-liter barrels is optimum for a certain wine. The winemaker decides he cannot afford to store all this wine in 200-liter barrels and wishes to store in 2000-hl oak casks instead. If extending that surface/liter is the only consideration, we can tell him that it should take 30 years in 2000-hl casks to equal 3 years in 200-liter barrels. Clearly the change would be impractical or the aging could not be equivalent. The winemaker then asks, suppose I age some of the wine in smaller barrels and blend to reach the same effect? If half the wine were put in 20-liter barrels and the other half in a large impervious tank, predictions from surface area alone would be that blending after a little over three years should give the same effect as aging all the wine in 200-liter barrels. This procedure would save no time and be more costly because to put half the wine in barrels 1/10 as large would require five times as many barrels. Thus it appears that the fact that the most commonly used barrels for aging of red wine are about 190–230-liter capacity all over the world is partly the result of a search for the optimum size for aging effect and not just representative of what two men can lift or one man can roll, as commonly assumed. If the barrels were smaller, about 10 times as many would be needed to cut the time of aging in half, and if they were larger the time of desired maturation rapidly becomes impractically long.

An appealing idea is to use pieces of oak within impervious containers to supply oak extractives equivalent to those which would be contributed to the wine at the optimum combination of barrel surface/ liter and time of contact during maturation in barrels. If oxygen is necessary it could be metered into the container, and much of the evaporative losses could be prevented. Treatment of wine with oak chips or extracts has generally been considered less satisfactory than barrel aging, but it can be better than no wood maturation at all. Both approaches give information helpful to understanding and improving the maturation of wines. I believe treatment with oak chips deserves to be more popular and will become so.

An Analysis of Wine for Tannin Extracted from Wood

Kramling and Singleton (41) developed an analysis that differentiated flavonoids from nonflavonoids in wine based on the ability of formaldehyde to precipitate the former but not the latter from solution under proper conditions. The total phenols (42) were determined colorimetrically before and after this precipitation. They (1, 41) also showed that the nonflavonoid content of all young wines was relatively low and constant.

Since the phenolic substance extracted from oak heartwood are predominantly gallo- or ellagitannins and lignin derivatives which are not

flavonoids (*43, 44*), it appeared possible to use the increase in nonflavonoid content as a measure of wood aging. Such a procedure would be
very useful to monitor aging in wooden cooperage and to study its effects.
This procedure has now been tested, and its utility is indicated (*45*). The
average of 17 American oak samples indicated 6.4% of the dry wood was
extractable with 55% aqueous alcohol. Each gram of extracted solid
contained 365 mg total phenols calculated as gallic acid equivalents
(GAE) of which 320 mg or 87.7% was nonflavonoid. The similar average
values from, respectively, European oak (7 samples), redwood (5 samples), and cork (1 composite sample) were: per cent extractable—10.4,
15.9, 2.4; total phenols, mg GAE/gram solid extracted—561, 568, 141;
and nonflavonoid—87.7, 97.2, 97.5%. These values show that the majority
of the phenolic substances extracted by brandy or wine from these woods
would add to the nonflavonoid content of the wine. The data also confirm that European oak contributes both more extractable solids (161%)
and more phenol per unit of extracted solids (154%) for a total of about
double (248%) the phenol contribution to brandy and probably to wine
(*45, 46*). At 1 gram dry wood per liter of beverage, typical American
oak should contribute about 20 mg and European oak about 50 mg
nonflavonoid.

When added to wine, the alcoholic extract of wood raised the nonflavonoid content, and the apparent per cent recovery of the added
nonflavonoid was 61% for a white wine and 86% for a red wine (*45*).
The addition of various levels of oak chips to a red wine gave a linear
relationship between chip dosage and nonflavonoid content (Figure 2).
That extraction from oak chips about 1 mm thick was essentially complete after one to eight days, depending on pretreatment and alcohol
content of the extracting solution, had been shown earlier (*47*) but was
verified at about four days in these studies (Figure 3) with wine at room
temperature (*45*).

The nonflavonoid analyses and standard deviation on this red table
wine were 219.2 ± 8.7 mg for a more experienced analyst and 224.8 ±
9.9 mg for a less experienced analyst. Since a statistically significant
analytical difference between single analyses should require four standard
deviations or 40 mg GAE nonflavonoid/liter, this would represent about
2 grams American or 1 gram European oak chips/liter wine. Preliminary
tests with American chips indicated that this minimum analytically
detectable difference was about the same as the chip dosage to give a
recognizable level of oak flavor (*45, 47*).

Several wines known to be free of wood contact were analyzed for
nonflavonoid content. The white wines averaged 157 mg GAE/liter
nonflavonoid and the reds, 271 mg. The highest value obtained for a
white wine was 211 and for a red wine, 448 mg GAE nonflavonoid/liter.

On the basis of these analyses it can be assumed that if a sample of the original wine is not available for direct comparison, white wines which exceed 200 mg/liter and red wines that exceed 450 mg GAE nonflavonoid/liter have contacted wood or another source of extra nonflavonoid phenols. A series of 26 wines with a known history of cask aging was analyzed, and of these, nine either were known to be free of wood contact or contained little nonflavonoid, and contact was not indicated by analysis. The remaining 17 wines all were correctly indicated by analysis to have

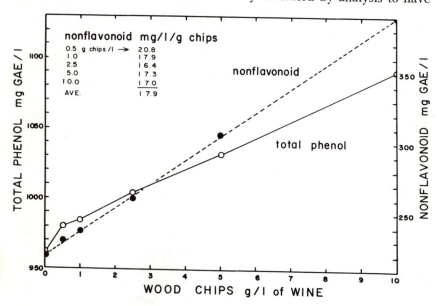

Figure 2.   Contribution of phenols to red table wine by American oak chips at various doses

been stored in wooden cooperage. An estimate of the amount of American oak chips it would have taken to produce the same analytical values ranged from 0.4 to 15.5 grams/liter wine. There was at least a rough correlation between the wood extract indicated by analysis and both the sensory oakiness of the wine and the expected oak extract content based on barrel size or previous use. Although further testing of the usefulness of this analytical method will be required, the fact that it can detect and monitor storage of wine in contact with wood under practical conditions is proved.

Oak Flavor Magnitude in Wine

A series of extracts of samples of American oak was prepared in 55% ethanol in 1961. These extracts were stored at about 25°C until

1971 in half-empty bottles with dry glass stoppers. They were analyzed in 1961 and again in 1971 for chemical and sensory properties. Three composites had nearly identical properties, but individual staves showed considerable variation (Table III). The 1971 analyses for extracted solids content averaged 102% of the 1961 values, indicating some evaporation during storage. A few samples dropped slightly in solids content because of precipitation. However, both the extracted solids and phenol content (Table III) of the samples remained rather constant.

The extracts were added to a port wine and compared in duo-trio blind testing, the same wine treated with the same amount of aqueous alcohol containing no oak. The same single taster and the same wine (stored in 20-liter glass bottles) were used in 1961 and 1971. Threshold values (the oak-treated sample was distinguished from the untreated sample in 50% above the number of trials expected from chance) were determined from a plot of the log of the concentration of added extract *vs.* the per cent above chance probit plot based on a minimum of 40 blind trials. The threshold values obtained in 1971 were very similar to the 1961 values. This demonstrates that the flavor intensity of the extracts did not appreciably change during nine years with opportunity for oxidation. There was a general impression that the quality of the odor was uniformly oaky with a slight citrus note in 1971 and more variable with some medicinal or creosote notes in 1961. These data seem to refute

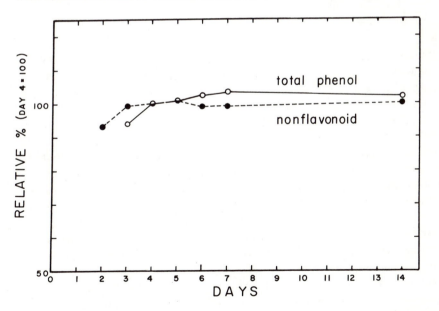

Figure 3.   Extraction with time of phenols from oak chips into red table wine (4-day value = 100%, five levels of chip treatment combined)

Table III.   Analysis and Flavoring Value of Extracts of American Oak

Sample		Solids Extd. by 55% EtOH, grams/100 grams dry wood	Phenols in Extract, grams/100 grams dry wood	
			1961	1971
Composite	1	5.3	2.2	2.4
	2	6.1	2.6	2.5
	3	6.0	2.3	2.3
Stave	1	8.7	3.9	3.9
	2	8.4	3.3	2.8
	3	8.4	5.3	3.9
	4	3.5	1.2	0.9
	5	5.5	1.1	2.1
	6	10.5	4.5	4.6
	7	7.7	2.1	2.7
	8	6.6	2.4	2.2
	9	4.6	1.5	1.3
	10	4.6	1.3	1.1
	11	5.8	2.0	1.9
	12	7.1	2.6	2.3
	13	8.3	2.4	3.1
	14	8.8	3.5	3.4
	15	8.6	3.6	3.6
Average		6.91	2.66	2.60

[a] Per liter of wine, 50% above chance.

statements from winemakers that oak flavor in chip-treated wines becomes stronger with time. Their observations may be explained on the basis of fatigue and familiarity when tasting during chip treatment compared with later tasting of the finished wine.

The amount of oak wood equivalent to a flavor level detectable in 50% of the tests over chance level was variable stave to stave, 100–1400 mg dry wood/liter wine. A good part of this variability was attributable to differences in extractable solids and phenols in the various oak samples. Correlation coefficients between the grams of oven-dry oak wood to give threshold flavor and the extractable solids and phenols in that wood were, respectively, −0.47 and −0.56 in 1961 and −0.42 and −0.61 in 1971. These values are significant at the 1% level for phenols and at lower levels for the solids. An average of 34% of the variation of flavoring contribution by different oak samples was explained by the relative content of extractable phenol. The more extractable phenol the wood contained the more flavor it contributed to wine, not only because of the phenols themselves but because the mixture of extractives tends to be of uniform composition.

Judged in the Same Port-Type Wine by the Same Taster in 1961 and 1971

Oak Dosage to Give
Threshold Flavor [a]

grams Dry Wood		mg Solids		mg Phenols	
1961	*1971*	*1961*	*1971*	*1961*	*1971*
0.43	0.40	23	22	9.7	9.4
0.57	0.59	35	35	14.9	14.9
0.46	0.38	27	21	10.6	8.7
0.26	0.32	22	31	10.1	12.7
0.39	0.32	33	27	12.7	8.9
0.33	0.34	28	30	17.5	13.2
0.61	0.59	22	17	7.6	5.6
0.56	0.59	31	36	6.5	12.4
0.10	0.16	11	18	4.5	7.4
0.59	0.96	45	106	12.3	26.2
1.01	0.72	67	46	24.7	15.7
0.48	0.53	22	18	7.0	6.8
1.14	1.43	53	56	14.8	15.3
0.75	1.33	43	74	14.8	25.6
1.00	0.52	71	35	26.3	11.9
0.72	0.23	60	23	17.0	7.0
0.60	0.38	52	31	21.2	12.8
0.52	0.43	45	39	18.5	15.1
0.58	0.57	38	37	13.9	12.8

The 55% ethanol extracted 6.91% of the weight of average oven-dry American oak (Table III). In flavoring 1 liter of California port wine for one taster, an average of 575 mg of this wood, 38 mg of its solid, or 13 mg of its phenols extractable by 55% alcohol produced a just detectable difference. A series of five samples of European oak in comparable analyses averaged 11.37% extractable solids, and 1 liter of the same port wine was just detectably flavored by 515 mg of oven-dry wood, 51 mg of extractable solid, or 30 mg of phenol. The fact that European oak contributes more extract and more tannin to wine and yet, per unit of extract or phenol, less flavor is clear from these data. This agrees with taster's opinions and is believed to be because American oak contributes considerably more oak odor per unit of tannin. The amount of flavor per unit of wood is about the same, but the European oak counteracts less flavor per unit of extract by its high extract content. In other tests, 12% alcohol removed 49% as much extract as did 55% alcohol from American oak and 71% as much from French. In earlier tests (47) about 63% as much extract was obtained from American oak by 12% alcohol as by 55% alcohol. Whether these extracts would have the same flavor value has not yet been studied.

Experiments of the same type were made on the same port wine and wood-free light red and white table wines to estimate the flavor contribution of an extract of a composite chip sample of American and another of European oak. A 10-member expert panel was used to get a better estimate of the population's sensitivity to oak flavor and to compare with the single taster's results. The American extract had 5.3 grams soluble solids and 2.8 grams GAE phenol/100 grams oven-dry wood and the European extract, 11.3 grams solids and 7.3 grams GAE phenols, comparing well with the average oak of each type. The amount of equivalent oven-dry wood to flavor detectably 1 liter of wine for the panel (50% above chance) for American oak was 352 mg for the white table wine, 315 mg for the red table wine, and 851 mg for the port. European oak gave values of white 213 mg, red 266 mg, and port 504 mg. For the port, the comparable values for the single taster were 399 mg American and 297 mg European which reflect greater practice and perhaps slightly greater sensitivity on the part of the single taster.

How Much Wood is Extracted in a Barrel?

A new 22-liter American barrel and a 29-liter European barrel were calculated to have total surfaces of 4107 and 4824 cm², respectively. When filled with 12% alcohol and 0.2% tartaric acid solution, the American barrel contributed 2.4 grams GAE phenols in 55 days and the European barrel, 13.6 grams GAE phenols. A sample of new wood rasped from the outside of the barrels representing all staves in proportion (but not heads) gave in the same solution 1.9 grams phenols/100 grams oven-dry wood for the American and 4.3 grams for the European barrel. Thus about 125 grams of wood were extracted in the American barrel and 320 grams in the European barrel. If we assume an average density for dry oak of 0.85 gram/cm³, this represents complete extraction to a depth of about 0.4 mm in the American and 0.8 mm in the European barrel.

Calculation with the barrel surface/liter (Table II) shows that for each millimeter that wine penetrates into the side of the barrel, it would extract about 16.4 grams wood/liter wine in a 20-liter, 12.7 in 50-liter, 9.6 in 100-liter, 7.6 in 200-liter, and 5.6 in a 500-liter barrel. These data mean that the typical oak barrel near 200 liters in size can contribute 3800 mg wood for extraction at a depth of penetration of 0.5 mm or in about two months for a new barrel. Based on the tasting results this is about three to 10 times the amount necessary to produce a tastable difference in wine. We have measured the depth to which wine has visibly penetrated into staves from used wine barrels and found that 6-mm penetration is not unusual. This could represent extraction of 45.6 grams wood/liter wine in a

200-liter barrel. This could give about 100 times the minimum detectable level of flavoring for most wines or, properly done, the barrel could be filled and emptied 100 times and detectably flavor all that wine. A 55-mm thick stave from a large old oak tank showed wine penetration to 15 mm. Extracts from the wood in different layers showed both the progressive depletion of nonflavonoid wood phenols and diffusion into the wood of wine flavonoids.

This high possible contribution of oak extractives and flavor to wine has not, I believe, been properly appreciated. American oak chip treatments for flavoring wine have been generally recommended in the range of about 1–10 grams/liter, depending on the wine and the intensity desired. These levels compare reasonably well with the minimum detectable difference of about 0.2–0.9 gram/liter indicated in the work just described. The much higher equivalent dosage from barrels no doubt explains all the conditioning treatments which have been used to get new barrels ready to receive wine. These treatments depend on prior leaching of a new barrel to lower the flavor level contributed to the wine. This was sensible in the days when nearly all wine was stored in wooden cooperage for its entire maturation period. With today's costs and the limited use of barrels it would seem preferable to remove the first wine from a new barrel in a few days, and by blending or residence time controlled by analysis, avoid wasting of valuable barrel extract. Extract diffusing from oak into wine will change somewhat as the surface is exhausted because larger molecules will take longer to diffuse (47). However, a molecule capable of diffusing at all through the semipermeable wood system should eventually be extracted, and it is perhaps not surprising that even after long continued use of a barrel or cask it still may contribute detectable oak flavor to wine stored in it for sufficient time.

Acknowledgment

This paper was originally presented at the Third International Oenological Symposium in Cape Town, South Africa, March 6–10, 1972. The Cape Wine and Spirit Institute, Stellenbosch, South Africa is thanked for permission to publish the paper in this volume. The California Wine Advisory Board is gratefully thanked for funds supporting the research. C. Kramer, D. Draper, and K. Singleton are thanked for assistance. E. B. Roessler is thanked for advice in calculation of barrel surfaces.

Literature Cited

1. Singleton, V. L., Esau, P., "Phenolic Substances in Grapes and Wine, and Their Significance," Academic, New York, 282 p., 1969.

2. Hankerson, F. P., "The Cooperage Handbook," Chemical Publishing, Brooklyn, New York, 182 p., 1947.
3. U.S. Dept. Agr. Forest Service, Advan. Sheets; Chem. Abstr. (1908) 2, 701.
4. Panshin, A. J., Harrar, E. S., Baker, W. J., Procter, P. B., "Forest Products, Their Sources, Production, and Utilization," McGraw-Hill, New York, 549 p., 1950.
5. Brunet, R., "Manuel de Tonnellerie," Librairie J.-B. Baillière et Fils, Paris, 284 p., 1948.
6. Huber, H., Kehrbeck, E., "Fachbuch für Küfer," Verlag C. F. Müller, Karlsruhe, 109 p., 1950.
7. Veiga, J. C. N., "Tanoaria e Vasilhame," Livraria sa da Costa, Lisbon, 259 p., 1954.
8. Howard, A. L., "A Manual of the Timbers of the World," 3rd Ed., Macmillan, London, 751 p., 1951.
9. Yamada, M., Bull. Agr. Chem. Soc. (Japan) (1928) 4, 18.
10. Luxford, R. F., J. Agr. Research (1931) 42, 801.
11. Prillinger, H., Mitt. (Klosterneuburg) (1965) 15A, 21.
12. Esau, K., "Plant Anatomy," 2nd Ed., Wiley, New York, 767 p., 1965.
13. Brown, H. P., Panshin, A. J., Forsaith, C. C., "Textbook of Wood Technology," McGraw-Hill, New York, Vol. 1, 562 p., 1949; Vol. 2, 783 p., 1952.
14. Stamm, A. J., Forest Prods. J. (1960) 10, 524.
15. Peck, E. C., Forest Prods. J. (1957) 7, 235.
16. Langdon, L. M., Botan. Gaz. (1918) 65, 313.
17. Sachs, I., Kuntz, J., Ward, J., Nair, G., Schultz, N., Wood Fiber (1970) 2, 259.
18. Locke, E. G., U.S. Dept. Agr. Forest Prods. Lab. Rept. (1960) 2179, 1.
19. Ritter, G. J., Fleck, L. C., Ind. Eng. Chem. (1923) 15, 1055.
20. Hillis, W. E., "Wood Extractives and Their Significance to the Pulp and Paper Industries," Academic, New York, 513 p., 1962.
21. Howes, F. N., "Vegetable Tanning Materials," Butterworths, London, 325 p., 1953.
22. Singleton, V. L., Ough, C. S., J. Food Sci. (1962) 27, 189.
23. Nishimura, K., Masuda, M., J. Food Sci. (1971) 36, 819.
24. Guymon, J. F., Crowell, E. A., Amer. J. Enol. (1972) 23, 114.
25. Baldwin, S., Black, R. A., Andreasen, A. A., Adams, S. L., J. Agr. Food Chem. (1967) 15, 381.
26. Guymon, J. F., Crowell, E. A., Qual. Plant. Mater. Veg. (1968) 16, 320.
27. Bricout, J., Mouaze, Y., Ann. Technol. Agr. (1971) 20, 217.
28. Pearl, I. A., Beyer, D. L., Johnson, B., Wilkinson, S., Tappi (1957) 40, 374.
29. Skurikhin, I. M., Efimov, B. N., Vinodel. Vinograd. SSSR (1968) 28(7), 8.
30. Skurikhin, I. M., Efimov, B. N., Tekhnol. Piishch. Prod. Tast. Proiskhozhd. (1966) 146; Chem. Abstr. 67, 107370f.
31. Camas, A., "Les chênes, monographie du genre Quercus," Paul Lechevalier, Paris; Vol. 1, Genre Quercus, sous-genre Cyclobalanopsis, sous-genre Euquercus (sections Cerris et Mesobalanus), text, 686 p. (1936-1938), Atlas, 93 p., 78 plates (1934); Vol. 2, Genre Quercus, sous-genre Euquercus (sections Lepidobalanus et Macrobalanus), text, 830 p. (1938-1939), Atlas, 177 p., plates 79-236 (1935-1936); Vol. 3, Genre Quercus, sous-genre Euquercus (sections Protobalanus et Erythrobalanus, Monographie du genre Lithocarpus et Addenda aux tomes 1, 2, 3), text, 1314 p. (1952-1954), Atlas, 163 p., plates 237-522 (1948).
32. Larue, P., Rev. Viticult. (1923) 58, 249.
33. Hair, D., Wooden Barrel (1964) 32(10), 6, 9, 15.
34. Theron, C. J., Wooden Barrel (1947) 16(2), 4, 5, 20.
35. Hardman, W., "The Wine-Growers and Wine-Coopers Manual," Wm. Tegg, London, 166 p., 1878.

36. Kukachka, B. F., *U.S. Dept. Agr. Forest Próds. Lab. Research Paper* (1970) **125**, 1.
37. Betts, H. S., *U.S. Dept. Agr. Forest Service* (1959.) 1.
38. Schoonover, S. E., "American Woods," Watling, Santa Monica, Calif., 250 p., 1951.
39. Krahl-Urban, J., "Die Eichen, Forstliche Monographie der Traubeneiche und der Stieleiche," Paul Parey Verlag, Hamburg & Berlin, 288 p., 1959.
40. Titmus, F. H., "A Concise Encyclopedia of World Timbers," The Technical Press, London, 2nd Ed., 264 p., 1959.
41. Kramling, T. E., Singleton, V. L., *Amer. J. Enol. Viticult.* (1969) **20**, 86.
42. Singleton, V. L., Rossi, J. A., Jr., *Amer. J. Enol. Viticult.* (1965) **16**, 144.
43. Hennig, K., Burkhardt, R., *Weinberg Keller* (1962) **9**, 223.
44. Seikel, M. K., Hostettler, F. D., Niemann, G. J., *Phytochem.* (1971) **10**, 2249.
45. Singleton, V. L., Sullivan, A. R., Kramer, C., *Amer. J. Enol. Viticult.* (1971) **22**, 161.
46. Guymon, J. F., Crowell, E. A., *Wines Vines* (1970) **51**(1), 23.
47. Singleton, V. L., Draper, D. E., *Amer. J. Enol. Viticult.* (1961) **12**, 152.

RECEIVED July 23, 1973.

13

The Chemistry of Home Winemaking

A. DINSMOOR WEBB

Department of Viticulture and Enology, University of California, Davis, Calif. 95616

Home winemaking chemistry and technology are considered for the production of dry white, dry red, sweet table, and flor sherry wines. Small size lots with the resulting higher surface-to-volume ratios and the home winemaker's lack of specialized equipment and knowledge are considered the main problems mitigating against the production of high quality wines in the home. Unsuitable fruit and yeast strains also are frequently problems. Basic procedures and equipment are discussed in a preliminary section followed by detailed descriptions of production of each of the four specialized wine types feasible for the home winemaker. Establishment and operation of a four-stage film yeast solera by the home winemaker are described in detail.

The chemistry involved in the production of wines in the home is exactly the same as that involved in making wines in the commercial winery—the biochemistry of the conversion of hexose sugars to alcohol and carbon dioxide. There are differences in the attention required and the results obtained that reflect primarily the difference in scale. For instance, the home winemaking operation is faced continually with the higher surface-to-volume ratio of small lots of wine and the concomitant greater tendency toward oxidation because of oxygen absorption. The home winemaker usually does not have equipment designed to perform the various required operations quickly and without aeration, nor, frequently, does he have the necessary experience or knowledge to perform the winemaking steps at the optimum times. The home winemaker seldom has cooling equipment adequate to cope with the heat produced by fermentation. Finally, with few exceptions, the home winemaker has little or no control of the kind and quality of fruit to be fermented. Certain varieties of grapes and other fruits are more suited to wine production than others. It is difficult to identify varieties after they have been harvested, so in the market, high-yielding sorts may be confused with the more desirable shy bearers. The quality of wine produced within one variety can vary widely with the level of cropping. The grower who sells on the open market naturally wants to have as large a crop as possible. In spite of these many difficulties, many home winemakers do produce acceptable wines and, on occasion, quite good ones.

Basic Aspects of Winemaking

Crushing–Maceration–Pressing Operations. Fundamental to making any type of wine is the conversion of sugar to ethyl alcohol by the enzymes present in living yeast cells according to the equation established by Gay-Lussac,

$$C_6H_{12}O_6 \rightarrow 2\ C_2H_5OH + 2\ CO_2$$

The many operations performed during wine production are principally concerned with manipulating the fruit or other raw material to make sugar readily available to the yeast while preserving and protecting the alcoholic solution resulting from the yeast's action. Thus, the first step following harvesting or procurement of the fruit is a crushing–maceration–pressing series to make a juice or mash from which the yeast cells can easily absorb the sugar.

Liberating the sugar from the intact fruit or berry is not the complete story. To obtain a wine it is necessary also to extract some of the color, tannin, and flavor compounds from the skin and pulp of the fruit. The exact nature of the crushing–maceration–pressing steps is modified according to the type of fruit used and with respect to the type of wine being made. Red-colored wines are produced by fermenting the skins, pulp, and juice together in the fermentation vessel so the pigments may be extracted from the skins into the partially fermented wine. White wines are made from fruits without red pigments or by carefully straining the pigment-containing skins from the juice before fermentation begins.

Grapes are unique among fruit because of a relatively large portion of stems which can give the wine a bitter or astringent flavor if they are not separated from the must before fermentation. In all cases it is important not to break or crush seeds before fermentation as the oils extracted from them by the new wine are likely to give the aged wine a rancid fat flavor. The length of time the new wine or juice stays in contact with the pulp determines the amount and balance of flavors extracted. The optimum time to separate the new wine from the pulp is determined by careful tasting as the fermentation continues. Experience is necessary in this tasting as it is difficult to estimate flavor intensity in the presence of the sugar and carbon dioxide of the partially fermented wine. Records of impressions on tasting preserved in a cellar book are very useful in guiding fermentations in subsequent years.

In producing wine from *vinifera* grapes, three pieces of equipment are convenient for the crushing–maceration–pressing operation. A machine to crush or break the grape berries is constructed with two corrugated rollers mounted so that there is about 3/16 to 1/4 in. clearance between them. The grape clusters are introduced from a hopper mounted above the rollers, and they are crushed as they pass between the rollers. The second piece of equipment is a relatively coarse screen to separate the crushed berries from the stems. Frequently this screen is constructed as an open-ended rotating cylinder mounted at a small angle. The crushed berries and stems are introduced from the crusher rollers at the higher end. As the perforated cylinder rotates, the berries drop through the perforations but the stems are carried to the lower open end of the cylinder and drop into a separate container. A finer-meshed screen to separate the juice from the skins is the third piece of equipment required. This finer-meshed screen is frequently combined with some type of press so that a larger proportion of the juice or wine may be separated from the skins than would be possible by simple straining. All equipment to come in contact with the juice and wine should be stainless steel, glass, plastic, or wood. Iron, copper, brass, bronze, tin, galvanized iron (zinc), and aluminum all tend to dissolve in fruit juices and wines and cause clouding or other problems. Some plastic vessels and tubes are formulated with a softener and coloring materials that are slightly soluble in wines.

For making wines from fruits other than the *vinifera* grape, it is convenient to have a fairly large food chopper or mill in addition to the above equipment. The object in all cases is to reduce the fruit to a mash or puree in which the sugar and other soluble materials are available to the yeast.

Addition of Sulfur Dioxide. Certain fruits and some of the white varieties of *vinifera* have a tendency to brown during crushing and other early processing operations because of oxidation. This oxidation may be promoted by enzymes in the fruit, or it may be a direct reaction between phenolic material of the fruit and oxygen from air. Sulfur dioxide (SO_2) is a strong enough reducing agent that it is oxidized in preference to the phenolics of the fruit juice. Sulfur dioxide may also function by denaturing the oxidizing enzymes. Therefore, to prevent browning, add 25–200 ppm SO_2 to the fruit immediately after crushing. The quantity of SO_2 is governed by the ease of browning of the particular juice being vinified. SO_2, in addition to preventing oxidative browning in juices, inhibits growth of bacteria and wild yeasts. Thus, it provides a more nearly sterile field for the action of the desirable

Table I.   Sulfur Dioxide to be Added to Juice[a]

	Fruit Condition	
Browning Tendency	Poor; Warm, Infected, Some Decay	Good; Cool, Fresh, Sound, Clean
High (white juices)	200–300	100–150
Low	75–125	0–25

[a] Mg per liter.

yeast starter added by the enologist. The quantity of SO_2 to be added to the juice is varied according to the condition of the fruit—clear, cool, sound fruit fresh from the vineyard requires very little while fruit that is in poor condition and warm needs more. The amounts of SO_2 to be added to a juice can be estimated from Table I.

SO_2 is a pungent and unpleasant smelling, dense gas at normal temperature and pressure. Under moderate pressure it condenses to a liquid which can be stored in steel cylinders. The large winery usually adds SO_2 to the crushed grapes by carefully metering a small stream of the liquid from a cylinder to to the inlet line of the pump that transfers the must from the crusher to the fermenting tanks; this ensures that SO_2 is uniformly mixed into the mass of crushed fruit. For the small winery and the home winegrower, however, the relatively small amounts of SO_2 required are difficult to measure and transfer as liquid, so either water saturated with SO_2 or a SO_2-liberating salt is used.

Water saturated with SO_2 gas at room temperature contains 5–6 wt % SO_2 depending on the temperature. While the SO_2-saturated water solution is still very pungent and unpleasant smelling, it does not present the handling and measurement problems of pure liquid SO_2.

The sodium and potassium salts of SO_2 are simpler and more pleasant to use as they do not have the odor of the pure liquid or the 5% water solution. They are rapidly soluble in must where they react with a small portion of the natural acid present to liberate SO_2. There are two sodium salts of SO_2 available, Na_2SO_3 (neutral sodium sulfite) and $NaHSO_3$ (sodium acid sulfite). The latter compound introduces less sodium into the wine and removes less acid from the wine for an equivalent amount of SO_2 liberated. Potassium acid sulfite and potassium pyrosulfite (potassium metabisulfite) are the two salts of potassium with SO_2 that are readily available, soluble in grape juice, and capable of yielding SO_2 upon reaction with the acid of the juice. Potassium salt is recommended when it is desired to keep the wine low in sodium ion content for diet reasons. The salts should be edible or food product grade, that is, free of heavy metals and other toxic impurities. They must be stored in tightly closed containers or they will react with the water vapor and carbon dioxide of the air to yield sodium or potassium carbonate and SO_2—thus losing their effectiveness as sources of SO_2 when added to the grape juice.

The required dose of SO_2 should be estimated conservatively and measured precisely because excessive amounts of SO_2 destroy the aroma and taste of the wine and can delay the onset of fermentation. Also SO_2 in excess interferes with the natural development of bouquet in red table wines and diminishes the intensity of the red color. One should always use only the minimum amount of SO_2 required to inhibit bacterial growth and counter oxidation—more definitely is not better.

Yeasts and Bacteria. One of the purposes of adding SO_2 is to inactivate bacteria and wild yeast so that the fermentation may be conducted with a chosen desirable strain of yeasts. Fortunately the wild yeast and the bacteria on grape berries (frequently confused in the older literature with the wax-like bloom which is naturally present on some berries) are susceptible to inactivation by relatively low doses of SO_2. A clear field is thus available to the large inoculum of SO_2-tolerant pure culture yeast added by the enologist.

It is true that wines were made for thousands of years before it was known that yeast was responsible for the fermentation. It is also true that in certain regions of the world wines are still made without SO_2 and pure yeast starters. These latter regions are generally those in which the yeast-containing sediments and press residues from the winery are returned to the vineyards and worked into the soil. Over many years it is likely that this procedure has resulted in the natural selection and stabilization of a mixed culture of yeasts which is carried from the vineyard to the winery and back and that the particular mixture contains enough of the desirable types to produce good wines in most years. It is also true that in years of cold summers and rainy harvest seasons many of the wineries normally relying on spontaneous fermentations use SO_2 and pure-culture starters. Today nearly all standard quality wine (*vin ordinaire*) and probably the majority of fine wines of the world are vinified using SO_2 and pure-culture yeast starters.

The bacteria which are found on sound grapes as they come from the vineyards are few in types and normally no problem in wine production as the acid, tannin, and alcohol of the wine stop their growth. The wild yeasts cannot be trusted to produce a good fermentation, however. In comparison with selected strains of SO_2-adapted yeasts, defects of wild yeasts are the inability to multiply rapidly in the relatively concentrated sugar solution of grape juice, a sensitivity to alcohol which prevents completion of the fermentation, a tendency to form excessive amounts of odoriferous esters or other non-alcohols, and the characteristic of remaining dispersed throughout the wine rather than aggregating and falling to the bottom of the container. The advantages to the home winegrower to be derived from the use of a selected yeast are obvious.

About 3 vol % of actively fermenting pure-culture yeast starter is required. A clean juice which has had a low dose of SO_2 will start and ferment satisfactorily with a lower inoculum, but the 3% level usually results in a quicker starting fermentation. For the home winegrower the simplest way to get the gallon or so of starter required is from a nearby winery. One has no choice of yeast strain and no guarantee of purity by this method, however. Winery supply agencies can usually furnish some strains of desirable wine yeasts such as Montrachet and Champagne in lyophyllized or freeze-dried form. These can be added directly to the SO_2-treated juice and probably represent the optimum solution to the starter problem for the home winemaker. If it is desired to use a yeast strain that is not readily available in either of the above-mentioned forms, a small pure culture of the desired strain will have to be obtained from a biological laboratory supply house or research laboratory maintaining a yeast collection. The small culture next must be multiplied until enough cells are present to inoculate the grape juice in the large fermenting tank. Sterile medium is required for the multiplication. Juice from a white grape variety of low flavor, such as Thompson Seedless, heated 30 min at 15 lbs per square inch pressure (2 atmospheres) in a pressure canner, serves very well. The small culture is transferred from the original tube to about one pint of the cooled, aerated, sterilized juice contained in a sterilized quart jar or bottle. Avoid contamination from the hands or the surroundings. The sterile jar should be covered or plugged so that air can penetrate but dust and cells of undesirable organisms cannot—a plug of sterile absorbent cotton works well. The jar should be placed in a room or cupboard at 70°–80°F, and it should be shaken gently at intervals. Within a day or two, growth and fermentation should be evident. The juice will foam and bubble, particularly when the jar is shaken. When the culture is actively fermenting, it is transferred into 1–2 gallons of sterile juice containing 100 ppm SO_2 which after a day or two will be actively fermenting and constitutes enough starter for 25–50 gallons of SO_2-treated juice. Successive fermentations can be inoculated from large batches that have fermented without difficulty although there is always the possibility of some contamination of the pure culture.

Yeasts, along with the algae, lichens, and other fungi, are known as thallophytes, a term which means they are undifferentiated plants or ones which do not have separate roots, stems, and leaves. Wine yeasts, along with most brewer's, distiller's, and baker's yeasts, are classed in the genus *Saccharomyces* or sugar fungus. The classification of yeasts is based on microscopic observation of their shape and forms, the way they divide during growth, and the way they respond when subjected to different test solutions of sugars or other chemicals. As scientists develop newer tools, such as the electron microscope, and as they study and classify more and more types of yeasts, it is desirable to develop further and to modify the older classification systems. Most of the wine yeasts are today put into the species *cerevisiae* with several strains being recognized by enologists. Many of these were formerly known as strains of S. *cerevisiae* var. *ellipsoideus*. It is quite likely that further study of the many species, varieties, and strains of wine yeasts will result in further refinements of the classifications.

Conversion of Sugar to Alcohol. Winemaking is basically concerned with the fermentation of the sugar in fruit juice solutions by yeasts. Some understanding of the chemistry involved in the conversion of sugar to alcohol and carbon dioxide is significant not only because it engenders an appreciation of the beauty of natural processes but because it also lets us understand and control certain factors affecting the quality of the wine.

The suspension of yeast cells will be added to the must a few hours after adding sulfur dioxide—a time long enough to permit most of the SO_2 to react with juice constituents or to volatilize. The low level of SO_2 and the aeration during the mixing in of the yeast starter permit the cells to start their action in an oxygenated environment, a condition which favors their conversion of some of the sugar to carbon dioxide and water with a high yield of energy for building many new yeast cells. The yeast population increases rapidly from the inoculation level of about one million cells per milliliter to about one hundred to two hundred million cells per milliliter, one to two days after inoculation. Then, nearly all of the oxygen will have been taken from the juice by the yeast cells, cell multiplication will slow dramatically, and conversion of sugar to carbon dioxide and ethanol becomes the main chemical reaction.

Fruit juices, depending on the type of fruit, contain one or more of the three sugars, sucrose, glucose, and fructose, in relatively high concentrations. Other sugars are present in trace to small amounts. Most yeasts have an invertase enzyme on the outer layer of their cell walls which rapidly converts the sucrose to glucose and fructose. These simpler sugars are carried rapidly through the cell wall by active transport. This is not understood fully, but it is known that glucose and fructose get into the cell interior faster than they should by simple diffusion.

Inside the yeast cell the hexoses are converted principally to ethanol, carbon dioxide, and adenosinetriphosphate (ATP) with the liberation of waste heat. The ATP is an energy source in cell metabolism; the ethanol and carbon dioxide diffuse across the cell wall to the exterior where the ethanol dissolves in the juice and the carbon dioxide bubbles to the surface. Excess heat must be removed to prevent the self-pasteurization of the wine, as most yeasts cease fermentation at 40°–45°C. Minor amounts of numerous other compounds are formed as by-products.

In addition to the carbon and nitrogen which are necessary to yeast for building enzymes, a few elements such as phosphorus, potassium, magnesium, manganese, and possibly traces of others, and a few vitamins are required for growth and fermentation. Normally, grape or other fruit juice will contain all substances necessary for yeast growth and fermentation. In preparing certain special flavored wines where the main component of the mixture for fermentation may be pure sucrose, it is necessary to add a yeast food—usually a mixture of ammonium acid phosphate with some autolyzed yeast—as a source of materials required for growth and fermentation.

The Course of Fermentation. The fermentation can be followed, in a rough way, by the bubbling in the fermentation tank since carbon dioxide is a product of the reaction. However, this doesn't indicate the extent or degree of completion of fermentation. Under some conditions, fermentation will stop before all the sugar is transformed, leaving the new wine subject to bacterial spoilage; therefore it is desirable to have a simple way to follow the loss of sugar. Water solutions of sugars are more dense than pure water while water solutions of alcohol are less dense than pure water. Density determinations performed daily thus provide one measure of fermentation.

Normally a stem or hydrometer is used to determine density. Hydrometers may be scaled in many different units. In the United States, grape juice and wine densities are usually measured in Brix or Balling degrees which are density units reflecting the weight per cent of sucrose in sucrose-water solutions.

As densities vary with temperature, and as hydrometers are calibrated to be accurate at different temperatures, the fermenting solution should be warmed or cooled to near the calibration temperature for the particular hydrometer used; for precise determinations, the actual temperature should be measured and the measured density should be corrected.

In theory the fermentation could be followed equally satisfactorily by measuring the alcohol content of the solution. In fact, however, alcohol determinations are much slower and more complicated than density determinations, so they are seldom, if ever, used. It is possible for the fermentation to stop—successive density determinations showing the same value—while there is some sugar left in the solution, although this is not normal behavior for fermentations. It is good practice to analyze for low levels of sugars in all wines when they have apparently completed their fermentations.

Equipment Cleaning. The home winegrower will find that much of the work involved in making wines is that required in cleaning the area and the equipment. To reduce this unpleasant task as much as possible, both the equipment and the processing–fermenting area should be designed to make the cleaning as quick and easy as possible. In general this involves securing equipment of stainless steel or plastic in which all parts and surfaces can be easily reached for washing and designing the winery work area so that the floors are sloped toward drains and will drain dry after washing. Hot and cold water should be available. Foaming detergents are to be avoided as they tend to get into the wines in small quantities and cause trouble in racking and bottling. Foam is appropriate for beer and champagne but out of place on table wines. Chlorine-releasing cleaners are valuable in removing red-grape stains from equipment and in thoroughly sanitizing the winery. There should be draining racks upon which hoses, buckets, funnels, and the like can drain dry after washing.

If possible, the storage area of the winery should be walled off from the crushing–fermenting work area as it will not need such frequent cleaning. There will also be less trouble with mold growth on the barrels and on bottles, corks, and labels if the humidity in this area can be kept at a lower level than is possible in an area needing frequent washing. The storage area should be, ideally, a room of constant, fairly low temperature—50°–65°F. Slow changes in temperature, as from summer to winter, are less troublesome than daily variations in temperature. This is because temperature changes cause the steel barrel hoops to expand and contract which in turn increases and decreases the volume of the barrel. As the temperature (and thus the volume) of the wine in the barrel lags behind that of the hoops there is a tendency to suck air into the barrel when the cellar temperature rises and a tendency to force wine out through small leaks in the barrel when the temperature falls. This problem can result equally from inappropriate design of air conditioning for a storage cellar—what is needed is good heat insulation for the cellar rather than a large capacity short-time-cycling cooler. There will be much less trouble with mold growth on the barrels and on the corks and labels of the bottles in storage if the relative humidity in the cellar is 30–60%, that is, fairly dry.

Barrels. Oak barrels or kegs are necessary for full development of traditional aroma and flavor in red table wines, flor sherries, and dessert wines such as ports. Caring for them, however, is difficult and they are relatively expensive. In general, the larger the size the longer the barrel is likely to last, but barrels larger than 40–50 U.S. gallons are heavy and difficult for the home winegrower to wash because of their size. Also, in general, the thicker the staves and heads the longer the barrel will last, but the cost will be greater. The main difficulty for the home wine-grower is that he is not likely to have wines to keep the barrels full all the time, and empty barrels are likely to become moldy on the inside, to dry out and develop leaks between the staves, or to have the staves crack. In the commercial winery empty cooperage is maintained in good condition by a strictly observed schedule of frequent washings to keep the wood from drying out and by displacing the air from the barrel or cask by filling it with SO_2 gas (frequently generated by burning some elemental sulfur in the tightly closed container) to keep bacteria and fungi from growing inside. The cask will be refilled with wine as soon as possible, however. For the home winegrower, then, it is advisable to buy only the number of barrels that can be kept full of wine; they should be of the best quality oak with thick, split (rather than sawed) staves, and they should be 20–50 U.S. gallons.

Barrels can be purchased from coopers who are usually located in the oak-forested central southern part of the United States. As these firms are principally concerned with making barrels for aging Bourbon whiskey, it is very important to the enologist that he specify an uncharred barrel. New charred barrels tend to give wines odd flavors reminiscent of Bourbon whiskey. Wines aged in new uncharred American white oak barrels acquire the oak wood aroma and flavor rapidly. The vanillin aroma of the wood is quite pleasant, but the bitter and astringent tannins, while agreeable tasting at low to moderate levels, can reach intensities that unbalance the taste of the wine. Thus, the first wine to be aged in a new barrel should be tasted critically at frequent intervals to determine when it has dissolved enough of the flavor substances from the oak. One year in a new American white oak container of moderate size (20–50 U.S. gallons) is usually the maximum. Wines stored in the same barrel in subsequent years will pick up the oak flavor at a more moderate rate and can stay in the barrel longer than 1 year.

Kegs and small barrels coopered of various European oaks are imported and sold by some winery equipment firms. In general, these are strictly first class wood and construction. The flavors produced in the wines seem to vary slightly with the area of growth of the oak. In general, the vanillin aroma is less apparent and the character of the tannin is less green and less bitter than with American-grown oak. Care must be taken to taste the wines frequently during the first year of aging in new European oak, nevertheless, because it is quite possible to extract wood flavor into a light-bodied wine in excessive amounts.

Opinion varies as to the best treatment of new oak before wine is stored in it. Some persons favor leaching with hot sodium carbonate solutions while others recommend hot citric acid–sulfur dioxide solutions. Hot basic or acidic solutions will extract a good portion of the most readily soluble lignin and other flavor materials from the wood, thus decreasing the amount available to the first wine put into the barrel and thereby reducing the danger of getting an over-woody wine during the first year of use. On the other hand, recognizing that it is the extractable flavor constituents in the wood that justify the greater cost of the oak barrel in contrast with a plastic aging container, it seems foolish to discard them by barrel pretreatment. A single washing with warm water and a few hours draining before filling with wine would seem to be the best practice for the home winegrower who will taste the wine frequently during the first year's use of the new barrel and thus avoid the problem of production of an over-oaky wine.

Federal Regulations. The making of wine at home is controlled by the Bureau of Alcohol, Tobacco, and Firearms of the Internal Revenue Service of the Treasury Department, the federal agency which regulates commercial brewing, winemaking, and distilling in the United States. Form 1541, which declares a householder's intent to make wine, must be filed with the nearest office of the Internal Revenue Service each year before the wine is made. The law authorizes making up to 200 gallons of wine per year by the head of a family for his family use, and it specifically prohibits selling or giving away the wine. There is no fee for filing Form 1541. It is to be noted that the law does not permit distillation of the wine. The home winegrower is to record the kind and quantity of each wine made.

Types of Wine. Many kinds of wine can be made at home, and the choice will be determined by the type of fruit available and the ambition and skill of the home grape grower–enologist. Perhaps the simplest wines to make are the dry (all sugar fermented) white and red wines from *vinifera* grapes grown in regions where they contain about 22% sugar and 0.8% acids, expressed as tartaric acid. Nearly as simple are similar wines vinified from grapes or fruits in which it is necessary to adjust the sugar and acid levels slightly before the fermentation. Somewhat more difficult to ferment and finish are wines based principally on solutions of sugar in which some flower, herb, or vegetable contributes the flavor and a mixture of buffering salts and yeast food is required to ensure proper yeast growth.

Sweetened, low-alcohol wines are a difficult challenge to the home winegrower. It is simple to add the sugar but difficult to treat the wine so that the added sugar will remain unattacked by yeast or bacteria without having the treatment damage the flavor of the wine. Cloudy sediments, gas, and odd off-flavors characterize too many of these wines produced by home winegrowers.

Flor sherry can be made by the skillful and patient home winegrower although its production probably should not be attempted until one has had experience with making dry white wines. Further, the requirements of cooperage and temperature control of the storage cellar are factors to be carefully considered before commencing a flor sherry *solera*.

White Dry Wines

Dry white table wines differ from red table wines in characteristics other than color. They contain lower concentrations of the puckery and astringent tannins. They usually have a higher acid content which makes them somewhat lighter in body and fresher on the palate. Dry white wines also are usually simpler in their aroma in contrast to the complicated aroma and bouquet that can develop in a red table wine. While it is possible to make white table wines from red-skinned varieties, for the home winemaker this task is very difficult because it is almost impossible to remove all traces of the pigment-containing skins from the juice quickly enough to prevent some reddish or amber color from developing.

To ensure that all the sugar ferments, it is recommended to start with a must of 20–22% sugar. If white grapes are being used, this sugar level is found in grapes of about 21.5–23.5° Brix or Balling (weight per cent of dissolved solids). For other fruits, the natural sugar should be supplemented by adding pure sucrose so the total will be 20–22 wt %. Most grapes will have a suitable level of natural acidity to produce a balanced tasting dry wine. It may be desirable to add tartaric or malic acids if the fruit is below 0.5 gram per 100 ml of acid expressed as tartaric acid. With most other fruits it will be necessary to dilute the naturally present acid with water to reduce it to a palatable level. In general, one adds water and dry sugar until the acid level is diluted to 1.0 gram per 100 ml expressed as tartaric in the juice before fermentation.

Techniques for extracting juice from the fruit vary with the type of fruit. A complete breakdown of the cellular structure is desired so the sugar is available to the yeasts. With most fruits, however, it is not desired to conduct the entire fermentation in the presence of the cellular material as this permits the extraction of excessive amounts of tannins. One needs equipment that will separate the maximum quantity of relatively clear juice from the fruit before or shortly after the start of fermentation.

White wines are normally fermented in containers that can be partially closed so that the surface of the fermenting wine is protected from the air by a blanket of carbon dioxide. For the home winemaker, 5-gallon carboys serve admirably. An hour or so after adding SO_2, one adds approximately 3% of an actively fermenting pure-culture starter yeast.

The carboy mouth is covered by a double layer of cheesecloth which permits the exit of carbon dioxide but serves to keep out dust and vinegar flies.

Temperature of Fermentation. Heat is a by-product of fermentation. Temperature in the fermenting wine approaching 40°C can damage the yeast cells and stop the fermentation, thus spoiling the wine. Even if the yeasts are not damaged by the heat, temperatures above 15°C detract from the quality of white table wines. The fruity fresh flavors of these wines become less and less perceptible with higher fermentation temperature. For the maximum fruity character and lightest color, the wine should not be permitted to rise above 15°C during fermentation. Much lower temperatures, however, will result in very slow fermentations and the accumulation of yeasty flavors. With small fermenting vessels (5–25 gal) the simplest way to control fermentation temperature is to place the containers in an air-conditioned room. Lacking this, one can immerse the bottles in larger vessels of cooled water. In emergency situations runaway fermentations can be cooled by transferring the wine to larger-mouthed containers in which plastic-wrapped blocks of ice can be immersed in the wine, or small pieces of solid carbon dioxide can be added to the overheated wine.

Racking. When the fermentation nears completion, determined by the cessation of violent bubbling and the tendency for the solids to drop to the bottom of the container, the wine should be transferred to a clean bottle of size such that it is nearly completely filled. The bottle should be lightly stoppered or fitted with a fermentation bung—a bubble trap filled with SO_2 solution through which the gas may pass. There must be an opening for CO_2 to escape, but it is equally important that oxygen of the air not have access to the new wine. No bottle should be tightly stoppered until chemical tests have shown that all of the sugars have fermented and that you have not inadvertently produced a glass bomb.

At the end of the fermentation the wine will be turbid and muddy from the suspended yeast cells and the debris from the fruit. Most of this material will settle quickly, forming a more or less thick layer in the bottom of the bottle. The new wine should be syphoned away from the deposited sediment promptly to avoid off flavors from autolysis of the yeast in the sediment. Also at this time wine acquires the defect of H_2S which is produced by reduction of the elemental sulfur dust applied to the grapes as fungicide by the decomposing yeast cells in the thick layer of sediment. The first transfer of the new wine from the sediment should occur very soon after active fermentation, the second about two weeks later, and the third and possibly a fourth two to six months later. These rackings may be conducted under a blanket of nitrogen or carbon dioxide if the particular wine oxidizes easily.

Storage. The new wine should be nearly brilliantly clear after six months, and either it can be put into smaller bottles for easier storage in the cellar or, in certain cases, it may be put into oak barrels for a short period of wood age. The greater proportion of white wines are fruitier and better balanced without adding oak flavor, but varieties such as Chardonnay, White Riesling, Semillon, and Sauvignon Blanc sometimes benefit from a short period of aging in oak. Most of the fruit wines, in contrast, have their greatest intensity of aroma immediately after fermentation and are most tasty within the first year. It is doubtful that the complication of oak aroma is desirable in these cases. The situation with respect to the direct-producing hybrids still requires experimentation to determine which will benefit from the addition of oak aroma and some bottle age.

Grape juice is a nearly saturated solution of potassium acid tartrate, a compound which is less soluble in alcohol solutions than in water or sugar solutions. Therefore, most grape wines immediately after fermentation are supersaturated solutions. Normally, most of the excess potassium acid tartrate will precipitate fairly quickly, and for stabilization of the new wine against later tartrate precipitation the wine is simply siphoned away from the precipitate as it forms for several weeks. To be certain that the wine will not precipitate more crystals of potassium acid tartrate after it has been put into bottles for long-time storage, two to three weeks of very low temperature ($-7°$ to $-10°C$) storage followed by siphoning the wine away from the crystals at the low temperature is required.

After the white wine is bottled it should be stored with the bottles lying on their sides so that the corks stay damp and tight enough to prevent the ingress of air to the wine. The wine will be preserved best in a cool, quiet, dark environment. White wines are particularly sensitive to heat and sunlight.

Red Dry Table Wines

Color and Tannins. Red- or purple-colored wines contain natural pigments, anthocyanins, which are responsible for the attractive colors. In nearly all red wine grapes these pigments are located in the skin—the pulp is essentially colorless. During fermentation the pigments are dissolved from the skins and liberated into the wine. It is primarily the alcohol produced by the early stages of fermentation that is responsible for dissolving the pigments.

The skins also contain tannins and aroma substances. These are soluble in the dilute alcohol of partially fermented wines. Certain tannin materials—those which give rise to harsh or hard tastes—tend to be dissolved in the later stages of fermentation when the alcohol content of the wine has reached its maximum.

The secret of making good red table wine, if there is one, is knowing when to separate the partially fermented wine from the skins. Frequently, one must compromise. The quantities of pigment, tannin, and aroma extracted at a given time may not produce a harmonious taste. One may have to accept a wine with less color than desired to avoid a bitter tannin, or conversely, one may have to accept the rough tannin to have the deep color desired. One criterion of the suitability of a red table-wine grape variety to a particular region is how nearly it comes to having optimum balance of color, tannin, and aroma extraction at the time of withdrawal from the skins. A certain amount of harsh tannin can be removed from the wine by later fining treatments. For this reason it is preferable to err on the side of extra tannin extraction rather than on the side of insufficient color.

For the home winemaker, the color intensity may be estimated by eye. It is essential to strain or filter the bulk of the yeast and pulp particles before examining the wine because these suspended materials make the wine appear much lighter in color than it actually is. Color intensity is estimated by looking down through the wines (sample and a control wine) contained in equivalent glass test tubes which are illuminated from the bottom. The depths of the wines are adjusted until the two wines appear to have the same brightness, even though the hues may differ considerably. The amount of color in the test wine varies inversely with its depth in the tube. Thus, if the depth of the test wine is twice that of the standard control wine then it contains only one-half as much color.

Tannin and aroma levels are estimated by tasting the partially fermented wine. The presence of yeast, sugar, and carbon dioxide in the fermenting wine complicates the tasting considerably. While the sediment can be removed by filtering a small sample for tasting, one must beware of introducing strange tastes from the filter papers. The carbon dioxide can be removed by shaking the sample in a partially filled bottle or by pouring from one glass to another several times. One must simply learn, however, to estimate tannin by taste in the presence of the sugar for there is no quick simple way in which the sugar can be removed. In general, sugar tends to counteract the palate impression of tannins so that a partially fermented wine will taste less astringent and bitter than it will when all of the sugar has fermented.

Until recently, red table wines were usually fermented dry on the skins. That is, the partially fermented wine was not separated from the skins until all of the sugar had been converted to alcohol and carbon dioxide. The very rough and astringent young wines produced by this technique required years of aging and, frequently, drastic fining in the cellar to make them palatable. The more recent method of making red table wines is to avoid the problems of excessive astringency and bitterness by separating the skins from the partially fermented wine when the fermentation is approximately one-half completed, using visual observation of color and tasting for tannin and aroma to decide exactly when the separation is to be made.

For the home winemaker, the separation is facilitated if a cylindrical stainless steel screen closed at the bottom and long enough to extend from the top level of the cap to the bottom of the fermentation vessel is obtained. The screen is pushed down through the layer of skins until the screen bottom is in contact with the bottom of the fermenter. The wine is then siphoned from the inside of the screen into the containers in which the fermentation will finish. When no more free run wine can be obtained, the residual skins may be pressed to augment the yield considerably. Excessive pressure is to be avoided in this step lest the press wine be overly astringent and bitter. The free run wine should be kept separate from the press wine until the fermentations are finished. Tasting at this time will dictate whether the two can be blended without impairment of quality or whether the press wine is to be kept as a second quality wine.

Procedure. The grape berries are separated from the stems and gently crushed. The juice and skins are placed in a fermentation vessel, to no more than two-thirds full. The foam and floating cap formed during fermentation require the remaining volume. The fermentation vessel should have a large opening at the top so the skins may be removed easily. Glazed crocks of 10–20 gallons make fine red wine fermenters for home use. Small wooden barrels or kegs from which one head has been removed serve well also. Stainless steel and plastic containers of 20–30 gallons are lighter and therefore easier to move about, and they are not breakable. They are usually more expensive than crocks, however.

The calculated amount of SO_2 is added and mixed thoroughly with the grapes, and the opening of the vessel is covered with a cloth to prevent the entry of insects. After 2–4 hrs the pure-yeast starter culture is added—again with stirring so that there is thorough mixing. The cloth cover is replaced. Active fermentation will be apparent within a day or so.

For the color, tannin, and aroma compounds to be transferred from the skins to the fermenting wine it is necessary to mix the floating skins with the wine. This punching down the cap should be done about every 6–8 hrs with a wooden or stainless steel plunger. The less color in the grapes, the more frequently the juice and skins should be punched so there be maximum color extraction.

Punching down the cap has the secondary effect of aerating the fermenting mass. This aeration is most desirable during the period immediately after the inoculation because it is then that the multiplying yeasts have the greatest demand for oxygen. After fermentation is obviously going vigorously, the aeration should be minimized and the punching should be confined to immersing the cap in the wine to as great an extent as possible. There is no benefit from growing more yeast cells than are required for fermentation.

Fermentation. Perhaps the simplest way to follow the fermentation is by successively measuring the density of the juice with a hydrometer. A graph of the density of the grape juice as a function of time during fermentation gives an S-shaped curve. There is little change in density at the start during the time of rapid yeast cell growth. When fermentation becomes rapid the density decreases at a faster rate, and finally, as the sugar nears exhaustion, the rate of density decrease slows once more. The density of the final wine will be less than the density of water because of the alcohol it contains. With a hydrometer stem calibrated in terms of per cent sugar, the reading when wine is measured will indicate less than no sugar. This seeming absurdity is explained by the effect of the low density of alcohol on the hydrometer.

The partially fermented wine which is separated from the skins should be placed in clean wooden kegs, 5-gallon carboys, or other vessels which can (later) be tightly stoppered. The container should be only loosely stoppered while fermentation continues, however, because the pressure generated by the fermentation carbon dioxide is sufficient to break the vessel if it is gas tight. In general, only two to five days are required to finish fermentation. This stage of development is recognized by the cessation of carbon dioxide production and, usually, by the fact that the yeast begins to drop to the bottom, leaving the wine clear. Within a few days (two to five) of the clearing, the new wine should be siphoned away from the layer of heavy yeast sediment in the bottom of the container. At this time it is necessary to put the wine into bottles or kegs that can be completely filled and tightly stoppered. The oxygen of the air will change the wine to vinegar—the alcohol being oxidized to acetic acid—if the wine is not kept away from air.

Malo–lactic Bacteria. Promptly separating the new wine from the thick layer of yeast and pulp particles is most important. Many wines are spoiled by procrastination. The yeast cells, if left in a thick layer, will begin to digest themselves and produce bad-smelling materials. Some breakdown products from this action have odors reminiscent of rotten eggs—the odor of hydrogen sulfide. Once the wine has acquired a rotten egg odor it is very difficult to remove it.

The malo–lactic fermentation occurs in some red table wines near the end of the sugar-to-alcohol fermentation. This conversion changes one molecule of malic acid to a molecule of lactic acid and one of carbon dioxide. Many types of bacteria are capable of producing the malic-to-lactic conversion, but they vary greatly in the other changes they produce in the wine. Some change the aroma only slightly, others add desirable odor characteristics, while still others produce unpleasant odors reminiscent of spoiled sauerkraut. The substitution of a molecule of lactic acid for one of malic acid reduces the tartness of the wine. For excessively acid wines, this is a desirable change, but for wines which are low in natural acidity the loss may be disastrous to the flavor balance.

The growth of malo–lactic bacteria in wines is favored by moderate temperatures, low acidity, very low levels of SO_2, and the presence of small amounts of sugar undergoing fermentation by yeast. It is frequently possible to inoculate a wine with a pure culture of a desirable strain of bacteria and obtain the malo–lactic fermentation under controlled conditions. The pure-culture multiplication of the selected strain of bacteria is difficult, however. It is also difficult to control the time of the malo–lactic fermentation—sometimes it occurs when not wanted, and at other times will not go when very much desired. For the home winemaker it is probably most satisfactory to accept the malo–lactic fermentation if it occurs immediately following the alcoholic fermentation. The wines should then be siphoned away from deposits, stored in completely filled containers at cool temperatures, and have added to them about 50 ppm SO_2. If the malo–lactic fermentation does not take place spontaneously and the wine is reasonably tart, the above described regime of preservation will likely prevent its occurrence. When the malo–lactic transformation takes place in wines in bottles, the results are nearly always bad. The wine becomes slightly carbonated, and the spoiled sauerkraut flavors are emphasized.

The home winemaker who is interested in determining whether his wine has undergone the malo–lactic conversion can use paper chromatography to analyze for the presence of malic acid. The equipment required is simple and the procedure is easy. The solvent for developing the chromatogram is a mixture of four volumes butyl alcohol, one volume formic acid, and water added until a second layer starts to form. Very small spots of the wine to be tested and of dilute solutions of known malic, tartaric, and lactic acids are placed along a line near the bottom edge of the paper and dried. The chromatogram is developed ascending and the spots are revealed by spraying the paper with a dilute alcohol solution of bromothymol blue which has been made alkaline by adding a few drops of base. Where the acids are present on the paper the the indicator will show a yellow spot against the blue background. The

various spots in the wine sample will correspond to tartaric acid, succinic acid, and either or both malic and lactic acids.

Aging. The agreeable odors and tastes of a fine red table wine reflect
not only the grape variety and a careful fermentation but both aging
in oak and bottle aging. For the home winemaker 20–50-gallon oak kegs
are probably most satisfactory as they are large enough for satisfactory
aging of the wine but not so large and heavy that they are difficult to
move and wash. After the initial racking (siphoning the clear wine
away from the sediment) the wine may be placed in the oak kegs to
start the aging. Depending on the amount of sediment deposited, the
new wine should be racked two or three times more during the first year.
At the time of each racking the wine should be critically tasted with
particular attention to the amount of oak flavor acquired. When enough
oak flavor has been obtained, the wine is racked into glass containers to
continue the aging process. Before the wine is put into small bottles for
long-time storage, it should be subjected to cold treatment to stabilize
it against the precipitation of potassium acid tartrate in the bottle. This
procedure is the same for red table wines as was described for white
table wines. Also, corked bottles of red wine should be stored on their
sides in a cool, constant-temperature, dark cellar.

Sweet Table Wines

The production of sweet table wines at home presents many more
problems than does the production of dry wines. Low alcohol (10–13
vol %) sweet wines are susceptible to further fermentation, and the
sugar is also capable of supporting the growth of several undesirable
types of bacteria. When these further transformations occur after the
wine has been put into bottles, the result is the production of unsightly
clouds, the trapping of the generated CO_2, and the production of unpleasant flavors.

Enological techniques have developed over the past 10–20 years so
that it is possible now for the commercial-scale winery to ensure the
biological stability in the bottle of young sweet table wines. This has
made possible the production and marketing of inexpensive sweet white
table wines and the introduction of balanced, pleasant-tasting, low-sugar
reds as well. It has led further to the introduction of the very sweet
Concord or berry-flavored red table wines and to the development of a
host of sweet wines of various colors specially flavored with mixtures of
extracts of fruits, berries, and herbs.

Commercial wineries employ two basically different techniques in
making sweet, low-alcohol wines. The traditional procedure involves

starting with very sweet grapes and conducting the fermentation and racking or filtration cellar operations slowly so that by the time there is 12–14% alcohol in the wine it will be depleted of one or more substances essential to the growth of yeast. It will thus be stable against further fermentation. The cellar steps employed usually involve fermentation at very low temperatures (25°–35°F), treatment with fining agents which tend to remove micronutrients, and filtrations through very small-pore filters which reduce the concentrations of yeast cells present. Because of the time and cost involved in the traditional procedure, it is normally used only with premium quality grape varieties and only in years when they reach an exceptionally high sugar level.

The quick or competitive procedure for making sweet table wines depends on sweetening a dry table wine with a very sweet blending wine or with sugar or grape concentrate and some procedure of sterilization which ensures that no viable yeast cells are present in the closed bottle. The simplest is pasteurization of the blend after it has been bottled. This technique, however, is not favored as it tends to give the wine a cooked aroma and taste.

Sorbic acid, 2,4-hexadienoic acid, is a compound which inhibits yeast growth at fairly low concentrations and which can be used to preserve sugar in low alcohol sweet wines. Sorbic acid is not effective against growth of bacteria, however, and it has a flavor that can be detected by some consumers even at the low levels effective in preventing fermentation. Another method, involving a bit more risk, depends on filtering the sweetened wine through very fine-pored filters that remove all of the yeast and bacteria and sterilizing the bottles, corks, and bottling machinery with steam and SO_2. In both the traditional and the competitive techniques, SO_2 is added to the blend just before bottling to help avoid oxidation of the wine during the bottling operation. The amount added is usually too small to have much inhibiting effect on yeast, but it does counter the possible growth of bacteria as well as functioning as an oxygen scavenger.

The home winegrower is probably wise to try first adding sugar to a dry wine for sweetening and pasteurizing the bottle with a pressure canner for stabilization. This is so because of lack of specialized winery equipment in the usual home winery and because of the great difficulty in achieving true sterility under even the best circumstances.

Diethylpyrocarbonate (DEPC), an ester of ethyl alcohol and carbon dioxide, kills yeast and bacteria. This ester also decomposes within a few hours after addition to wine. Thus, if a small amount of diethylpyrocarbonate were added to the sweet wine blend just before bottling, the ester would kill all organisms inside the bottle and then decompose to ethyl alcohol, carbon dioxide, and the simpler ester, ethyl carbonate.

However, diethylpyrocarbonate also will react with the traces of ammonia in wines under high pH conditions to produce a few parts per billion urethane. Therefore, the compound is no longer used in any foods produced commercially, and its use by the home winemaker cannot be recommended.

The more experienced and venturesome home winemaker may wish to attempt making a sweet table wine from sweetened must or juice. The first step in this project is sweetening the juice that has been pressed or drained from freshly crushed grapes. Pure sucrose should be added slowly and with thorough mixing until the Brix or Balling degree equals 28 wt %. As an alternative, the chemical sugar can be determined in the juice, and the amount of sugar can be calculated to reach a concentration of 26 wt %. The 2% difference represents the non-sugar soluble solids of the juice. SO_2 is added and mixed into the juice—the quantity depending on the ease of oxidation of the particular juice and on the amount of contaminating yeasts, molds, and bacteria present. Three to four hours after SO_2 is added, 1–3 vol % pure yeast culture is thoroughly mixed into the juice. There may be some advantage to starting the fermentation with a yeast strain such as *Sauternes* which is not tolerant of high alcohol levels and which ferments slowly. The difficulty in trying to maintain a pure culture fermentation over a long time in the home winery is such, however, that it is nearly inevitable that the wine will actually be fermented by a mixture of yeast strains.

The inoculated juice should be kept at 60°–70°F for about 12 hrs and then cooled to 30°–40°F. The yeast population will increase rapidly at the higher temperature while the lower temperature will slow the rate of fermentation and permit stopping it with some sugar left. If the fermentation is conducted in a small refrigerated room, one must take some precautions against the danger of asphyxiation with trapped carbon dioxide. With a small slow fermentation, opening the door to the room once a day should provide sufficient ventilation to prevent a high level of carbon dioxide accumulation. Usually the fermentation will slow to a rate low enough that a layer of yeast cells and grape pulp particles will accumulate in the bottom of the fermentation vessel. The clearer fermenting wine should be siphoned away from the layer of sediment at bi-weekly intervals at the start and less frequently during the latter stages of the fermentation. The racking should be done with the minimum amount of aeration possible; the delivery end of the syphon tube should discharge below the surface of the accumulating wine, and stirring or other agitation should be avoided.

The fermentation should be followed by determining the density or degree Balling daily or semi-weekly. When the degree Balling drops to a value near +2°, which should happen from two to six months after the start of fermentation (depending on temperature), the wine is stabilized by a final racking and addition of 100 ppm SO_2. Storage for several more months at near 20°–25°F in completely filled but loosely stoppered jugs is recommended as there may be slow evolution of small amounts more of carbon dioxide. The wine should be stored in the larger jugs at the low temperature until it is brilliantly clear, and all evolution of carbon dioxide has ceased. It should be racked at intervals of six months or so and may be bottled at two years of age with fair confidence that it will not recommence fermentation in the bottle.

Flor or Film Wines of Sherry Type

The wine yeast, *Saccharomyces fermentati,* is able to form a film or veil on the surface of dry white wines of about 15–16% alcohol. This yeast produces agreeable smelling and tasting substances which dissolve in the wine and give it the aroma and flavor characteristic of Spanish fino sherries. To provide itself with energy for growth while in the film form on the surface of the wine, the yeast utilizes some of the oxygen from the atmosphere above the wine in the partially filled butt or barrel to oxidize some of the ethyl alcohol from the wine. The ethyl alcohol of the wine is not completely metabolized to carbon dioxide and water, however, but is oxidized to acetaldehyde—probably the principal compound in the complex mixture responsible for the aroma of this type of appetizer wine.

Growth of film-forming yeasts is inhibited by ethyl alcohol concentrations higher than about 16 vol %. Fortunately, most of the bacteria and undesirable yeasts which will spoil wines when oxygen is available and which could be a great problem in the production of film wines are inhibited or killed by alcohol levels above 14%. The flor sherry producer thus has a range of about 1½–2% ethyl alcohol content in his wines within which it is reasonable to expect film sherry rather than vinegar to result.

The principal grape variety used in producing film sherries in Spain, and also in most of the other regions of the world where they have been made satisfactorily, is the Palomino, a variety of *Vitis vinifera.* This variety is characterized by a neutral aroma and flavor, a moderate sugar and acid concentration at maturity, and a tough, pulpy flesh covered by a thick skin. These latter characteristics, together with the loose, straggly cluster, help the Palomino withstand disease and insect attacks, particularly during the periods of rainy or cloudy, cool weather. The tendency to reach only moderate sugar concentrations is a handicap where it is desired to make wines of 15–16% alcohol, and this characteristic of Palomino has led to the picturesque custom in Spain of partially drying the harvested clusters on grass mats for a day or two to raise the sugar concentration by evaporating some of the water from the berries. The home winemaker can much more easily raise the sugar concentration by adding pure sucrose to the juice.

Procedure. The destemming, crushing, and juice separation in preparation for making film sherries are just as described earlier for dry white wines. The tough and pulpy character of the flesh of the Palomino makes it difficult to get a good yield of juice from this variety. Long contact of the juice and skins is to be avoided, nevertheless, since excess amounts of tannin materials will be extracted and result in a dark brown wine and coarse or astringent taste. One must remember that the film yeast flavor is delicate and does not show to advantage except in a clean and neutral base wine. The undue browning of the juice during the time required for draining or pressing may be combatted by adding SO_2 to the must immediately after destemming and crushing. If a slow juice separation—4–6 hrs—is to be used, one is well advised to add as much as 200 parts of SO_2 by weight to each million parts juice.

For the new wine to contain the desired 15–16 vol % alcohol, it is necessary to add some sugar to the juice or the partially fermented wine except in cases of unusually high sugar content in the grapes. In any event, it is necessary to have an accurate estimate of the sugar concentration in the juice before more sugar is added and before fermentation starts. It is better not to add the calculated amount of sucrose to the juice immediately but to make two or three smaller additions after the fermentation is actively under way. This technique is known as syruped fermentation and can yield wines of as much as 17 vol % alcohol under favorable conditions of yeast nutrients and temperature control.

In Spain, *yeso* (plaster or $CaSO_4$) is added to the grape juice during the crushing and pressing. This dissolves some of the solid cream of tartar which lowers the pH and slightly increases the bitterness through the introduction of the SO_4 as shown in the equation

$$CaSO_4 \text{ (s)} + KH(C_4H_4O_6) \text{ (s)} \rightarrow Ca(C_4H_4O_6) \text{ (s)} + H^+ + K^+ + SO_4^{2-}$$

The home winemaker may follow the Spanish practice by adding 5–6 grams $CaSO_4$ per gallon to his freshly pressed juice.

Fermentation. The juice should be inoculated with an actively fermenting culture of one of the selected strains of S. *fermentati*. Such yeasts may be obtained from winery supply houses as pure cultures which must be multiplied to obtain enough inoculum for the large fermentation or as dried yeast which may be added directly to the juice. Alternatively, the juice may be fermented dry with one of the standard, non-film-forming yeasts, and then the dry white wine may be inoculated later with some of the film yeast in the film form. Although no carefully controlled experiments have been conducted to detect differences between wines prepared by these two techniques, it seems likely that there would be differences, and that those wines fermented with the selected film-forming yeast would have more flor sherry character. The film-forming yeasts selected for sherry production, in addition to probable better aroma and flavor production, are more alcohol tolerant. This is important in the production of film sherries because it is necessary to ferment all of the sugar of the must before satisfactory film formation can ensue. Further, there is always the danger of bacteria's use of traces of residual sugar in wines of low-to-moderate alcohol content (9–15 vol % alcohol), producing bad-smelling and tasting compounds which spoil the wine.

As the new wine nears the completion of its primary alcoholic fermentation, it should be analyzed carefully for both alcohol and sugar. If fermentation slows too much, the wine may be gently aerated by racking (siphoning, pouring, or pumping) it from one container to another, and, if the temperature has dropped to 60°F or below, it should be warmed to about 70°F and held until all the sugar has been converted to alcohol.

Racking. After fermentation and 2 or 3 day's settling of most of the pulp and yeast cells, the clearer wine (the upper 95–98% of the wine in the fermentation vessel) should be carefully siphoned or pumped into a clean container. This transfer operation of the new wine should be repeated after an additional week or 10 days; then the wine should be reasonably clear although not brilliantly so. Further rackings (two to four in the first year of aging) should be performed as sediment deposits in the bottom of the container. By the end of the first year the wine should be quite clear to brilliant. If the wine has been fermented with a selected film-forming yeast, the film should begin to form within a few days to 2 weeks of the disappearance of the sugar. Do not be concerned about loss of the film by racking operations as it will reform or, if desired, bits of film yeast may be transferred to the surface of the wine after racking. This is done by collecting some of the yeast on a stick immersed into the wine and through the film layer before the racking operation and floating the yeast off the stick onto the surface of the racked wine by repetition of the immersion. The wine should be stored in containers that are not completely filled and there should be air available to the surface provided there is some film yeast growth on the wine's surface. If no film forms within a few weeks, the wine should be kept in full containers as for dry white wines.

If the primary fermentation of the must has been conducted with a standard wine yeast rather than with one of the selected film-formers, the racking and clarifying operations should be conducted just as for a dry white wine until the wine is reasonably clear, and it is desirable to establish the film yeast on it. The wine is put into a bottle or other container large enough so that the wine only fills it two-thirds to three-fourths full. Bits of film yeast from another container having a good film growth on the wine's surface are transferred to the new wine by a stick. The stick should be soaked in very hot water or steamed for some time before use so it is sterile. The container should be covered or stoppered so that some air can enter but dust and insects will be excluded. Absorbent cotton plugs in small openings and layers of filter paper or multiple layers of cheese cloth for larger openings are satisfactory.

The new sherry should be smelled and tasted at frequent intervals (biweekly to monthly) to verify that sherry rather than wine vinegar is being produced. Normally, the development of film character in the wine is a slow process (not much being apparent in less than six months), but the odor of wine vinegar (the ester, ethyl acetate) will be readily apparent on smelling and tasting if the sherry is contaminated with vinegar bacteria and is acetifying. In the unfortunate event that vinegar is being made and if it is detected in the very early stages, the process can sometimes be reversed by increasing the alcohol content to 16% and establishing a vigorous film growth. Usually, however, the home winemaker must be content to let the vinegar process run its course— use the product for dressing salads and making pickles—and very thoroughly clean and sterilize all of the equipment contaminated with vinegar bacteria before using them for wine again.

Near the end of the first year of the new film sherry, it will be apparent on sensory examination which of the wines are clear and are developing a film character. An alcohol analysis at this time will verify that they contain a safe quantity (15.5–16 vol %). Such wines may be considered to have finished their nursery or *criadera* stage and be ready to blend into the youngest stage of the *solera* (collection of barrels containing film sherries of differing ages fractionally blended). Wines not suitable to enter the *solera* at one year's age may be held as individual, unblended lots for an additional year to see if they develop satisfactorily or they may be diverted to other use (dry white table wine or vinegar).

Solera. The establishment of a *solera* by the home winemaker is a time-consuming and difficult project. Once it has been established, however, the pleasure of drinking one's own film sherry is generously adequate compensation. The *solera*, basically, is a collection of barrels arranged in stages such that each successive stage contains wines of increasingly complex blend and greater average age. The establishment of a four-stage, film-sherry *solera*—a feasible project for the home winegrower—would be done as follows: In the first year 80 gallons of film yeast-fermented Palomino would be made, and, when clear, divided equally between two 50-gallon oak barrels. The film yeast would be established on the surface of the 40 gallons of wine in each of the barrels. In the second vintage season the process is repeated, providing two more 50-gallon oak barrels of film sherry. In the spring after the second vintage, approximately 10 gallons of wine are carefully siphoned from beneath the film growth from each of the barrels of the older or first stage. This wine represents the first product of the *solera*. Eleven gallons of wine are then siphoned from each of the barrels of the second or younger stage. These 22 gallons of wine are blended together, and then 11 gallons are put into each of the first-stage barrels, care being taken to introduce it slowly and beneath the film so that there is minimum disturbance of the film growth. This sequence of operations is repeated in the third and fourth years with the result that, at the end of five years, a four-stage *solera* containing two barrels per stage has been started. It is to be noted that at this point (end of the fifth year) the wine of the first stage will contain some of each of the four earlier vintages while later stages will each contain less complex blends of successively younger wines. In successive years it is necessary to produce only enough wine to replace that taken from stage one plus that lost by evaporation and spillage. The average age of the wine in the first stage increases until after about 15 years it reaches an equilibrium average age of about seven years.

While the *solera* can be operated according to a strictly programmed sequence of times and amounts, it is easier for the home winegrower to make withdrawals from the first stage as needed for drinking—say four times per year—being careful not to withdraw more than one-fourth to one-third of the contents of the first stage during any one year. Each time a withdrawal is made, it is replaced by an equal amount of wine from the next younger stage—the fourth stage being replenished from the nursery or *criadera* wines as they become clear and recognizable as film sherry material. Since there is a considerable amount of evaporation from the barrels, it is necessary to have some 25–35% more *criadera* wine to enter into stage four of the *solera* than drinking wine is removed from stage one. The eight-barrel, four-stage *solera* as described will yield some 20–25 gallons of film sherry per year. This is equivalent to eight to nine cases of sherry and should be adequate for the requirements of the average family. The size of the *solera* may be adjusted as desired, of course, subject to the legal limit of 200 U.S. gallons per year. Using barrels smaller than 25–30 gallons is not recommended because the amount of evaporation becomes excessive and the smaller barrels are more subject to cracking staves and leakage from the drying of the upper staves. Also, the smaller barrels tend to make overly woody sherry.

Ideally, the *solera* should be housed in a cellar or room of constant temperature of 65°–68°F and of constant relative humidity in the 30–40% range. The room should be isolated from vibrations, and there should be adequate lighting when the wines are being transferred from stage to stage. Temperature control is important as it influences both the rate of yeast growth and flavor production and the rate of evaporation. The relative humidity value is important because it controls the concentration of alcohol in the wine: both water and ethyl alcohol, the two principal components of film sherry, diffuse through the oak staves of the barrels at rates determined by the state of the oak, the temperature, and their effective molecular weights. At the outer surface of the staves, the net rate of evaporation into the atmosphere is governed by the temperature, the specific heats of evaporation, and, most importantly, by the concentrations of alcohol vapor and water vapor in the atmosphere. A high relative humidity (high concentration of water in the atmosphere) effectively impedes the net loss of water from the wood surface through condensation of water molecules onto the surface layer nearly as fast as water molecules leave by evaporation. As a result there is a preferential evaporation of alcohol since the alcohol concentration in the atmosphere is relatively low. Thus, the alcohol concentration of the sherries will decrease and possibly will drop to a concentration suitable to grow vinegar bacteria. A low relative humidity, conversely, effectively encourages a preferential loss of water from the sherry resulting in alcohol concentration increases. This tendency, while preferable to lowering alcohol concentration, does have the disadvantages that the alcohol concentration may rise so high that film growth becomes impossible, and, at lower relative humidities, the upper (unwet) staves of the barrels tend to dry out. This latter effect becomes troublesome when more wine is added to the barrel and leaks between staves appear. Ideally, the relative humidity should be adjusted to the value which, for the individual cellar, barrels, and wines, maintains the alcohol concentration between 15.5 and 16.5 vol %. It may take several years of small adjustments and many alcohol analyses to arrive at the ideal relative humidity value for the particular cellar.

The home winemaker is encouraged to keep an accurate cellar log of the movements of wines through his *solera*. This will not only permit calculation of the losses by evaporation but will permit accounting for the relatively large amount of wine on hand in case of inquiry by federal authorities.

Film yeast wines do not benefit from bottle aging. Indeed, they are at their best when drunk from the last stage of the *solera* and deteriorate in flavor with time in bottles.

Suggested Reading

1. Amerine, M. A., Singleton, V. L., "Wine—An Introduction for Americans,"
   University of California Press, Berkeley, 1972.
2. Amerine, M. A., Joslyn, M. A., "Table Wines," University of California
   Press, Berkeley, 1970.
3. Fornachon, J. C. M., "Studies on the Sherry Flor," Australian Wine Board,
   Adelaide, 1953.

RECEIVED August 20, 1973.

INDEX

A

Acid(s)
of native American grapes 94
production, effects of climatic
conditions 96
ratios, Brix 106
for several fruits, organic 20
in wine, organic 16
Acidity, analysis of volatile 136
Acidity, titratable 18, 163
Acidulating agents 155
Additives, prohibited 141
Aerobic bacterial spoilage 213
Aging 227
changes in wine color during .. 82
Alcohol, Tobacco, and Firearms,
(BATF), U.S. Bureau of 106
Amelioration of musts 106
American
grapes
acids of native 94
pectin content of 99
sugars in native 90
grape varieties, winemaking from
native 88
hybrids 15
and hybrid varieties, anthocyanin
pigments of native 104
oak 261
Amino acids on fertilization,
dependence of 2
Analyses, wine 226
Analysis 134
of diglucoside anthocyan
pigments 139
methods required in winery
operations 135, 142
acetaldehyde 143
calcium 145
citric acid 143
color 145
copper 144
fixed acid 142
glycerol 144
histamine 144
hydroxymethylfurfural 143
iron 144
lactic acid 143
malic acid 142
pH 143
potassium 144
simultaneous determination of
acids 143

Analysis (Continued)
methods required in winery
operations
sorbic acid and sorbates 144
sorbitol and mannitol 144
tannin 145
tartaric acid 143
total acid 142
of metal contaminants in wine .. 138
for pesticides and fungicides ... 141
for sulfur dioxide in wines 137
of volatile acidity 136
wine 226
Analytical fractionation of phenolic
substances of grapes and wine 184
Anthocyanin(s) 51, 291
classification of wines by 69
in the different species of *Vitis*,
distribution of 57
diglucosides 53
for grapes and other fruits 30
pigments, analysis of diglucoside 139
pigments of native American and
hybrid varieties 104
Aroma 292

B

Bacteria 282
freeze-dried 162
lactic acid 151
malo–lactic 294
Bacterial inoculation, procedure for 159
Bacterial spoilage, aerobic 213
Bacteriological stability 154
Balling, degrees 285
Barrels 255, 286
Berry, total extractable phenols per 30
Biological–technological sequence.. 1
Bisulfite ion in solution, sulfur
dioxide as 233
Botrytis cinerea 12, 57
Brandy
distillate, fusel oil content of .. 234
distillation, single and split
columns for 237
distilling wines into 232
Brix 285
/acid ratios 106
values for eastern grape musts .. 91
Browning 281
oxidative 119
Bulking agents 108

C

Cap 294
Carbonyl reactions as a flavonoid
assay 200
Cell yield 165
Chardonnay 291
Chemical composition of fruits ... 12
Chromatography, paper 295
Clarification of wine, fining and .. 121
Cleaning, equipment 286
Climatic conditions on acid produc-
tion, effect of 96
Color
during aging, changes in wine .. 82
chemistry of red wine 50
determination of wine 71
Colorimetric analysis of phenols .. 186
Commercially important grape spe-
cies of the United States 18
Condensed tannins 63
Contaminants in wine, analysis of
metal 138
Control of microbial disorders ... 131
Cooperage surface-to-volume ratio 266
Cooperage, wood 254
Copper cloudiness, iron and 125
Coumarins 195
Criadera stage 302
Crushing–maceration–pressing
operations 279

D

Decarboxylation 151
Defects, oxidation 127
Diacetyl 155
Diethylpyrocarbonate (DEPC) .. 297
Diglucosides, anthocyanin 53
Diglucosides character, transmission
of the 69
Disorders of wine, non-microbial .. 122
Disorders of wine, microbial ...127, 131
Distillation single and split columns
for brandy 237
Distillation, wines for 233
Distilling wines into brandy 232
Double salt treatment 154

E

Enzyme(s)
cofactor nicotinamideadenine
dinucleotide (NAD) 171
in grapes 41
nomenclature 178
Enzymology of malo–lactic
fermentation 171
Equipment, buying foreign 218
Equipment cleaning 286
Esters, excess formation of 282
European oak 262
Exact threshold values 271
Extractives, oak 275
Ethyl acetate 302

F

Federal regulations 288
Fermentation
care of wine, post- 120
control of malo–lactic 156
enzymology of malo–lactic 171
inhibition of malo–lactic 157
malo–lactic 107, 109, 151, 295
stimulation of malo–lactic 158
temperature of 290
Fertilization, dependence of free
amino acids on 2
Film sherry 299
Filtration, sterile 157
Fining agents, use of 122
Flavan molecules, polymerization of 65
Flavonoids 193, 268
Flavonoid assay, carbonyl reactions
as a 200
Flavor complexity 154
Flavor intensity 271
Flor sherry 299
Folin-Ciocalteu (F-C) reagent ... 186
Formaldehyde precipitation of
phenols 201
Fox grapes 89
Freeze-dried bacteria 162
Fructose ratios at maturity, glucose 14
Fruits
anthocyanins for grapes and other 30
chemical composition of 12
organic acids for several 20
as raw materials, grapes and other 11
Fungicides, analysis for pesticides
and 141
Fusel oil content of brandy distillate 235

G

Glazed crocks 293
Glucose/fructose ratios at maturity 14
Grape(s)
acids of native American 94
enzymes in 41
fox 89
maturation, development of
pigments during 76
musts, Brix values for eastern .. 91
and other fruits, anthocyanins for 30
and other fruits as raw materials 11
pectin content of American 99
skin 159
waxes 6
sugars in native American 90
varieties, protein pherograms of
different 4
varieties, winemaking from native
American 88
and wine, analytical fractionation
of phenolic substances of .. 184
Growth rate 165
oak 264

H

History of winemaking in the
United States 88
Home winemaking 278
Humidity, relative 304
Hydrocarbons in wine 7
Hydrogen sulfide 294
Hydrolyzable tannins 63
Hydrometer 285
Hybrids, American 15
Hybrid varieties, anthocyanin pigments of native American and 104

I

Inorganic constituents in wine 32
Instability, tartrate 122
Internal reflux ratio 248
Iron clouds 17
Iron and copper cloudiness 125

L

Lactic acid 296
bacteria 151
in fermenting must 17
Lactic bacteria, malo– 294
Lactic fermentation,
malo– 107, 109, 151, 295
Lactobacillus 153
casei 167
plantarum 177
Leucoanthocyanins, principal
source of 31
Leuconostoc 153
oenos 153
ML 34 178

M

Maceration–pressing operations,
crushing– 279
Malate carboxy lyase 164
Malic acid 167, 295
Malic acid ratio, tartaric acid-to- .. 98
Malic and tartaric acids in must
samples 97
Malo–lactic bacteria 294
nutritional requirements of 161
Malo–lactic fermentation 107, 109, 151, 295
control of 156
enzymology of 171
inhibition of 157
stimulation of 158
Maturation, development of pigments during grape 76
Maturation, wooden containers in
wine 254
Maturity, glucose/fructose ratios at 14
McCabe-Thiele method 248
Metal contaminants in wine,
analysis of 138
Methyl anthranilate in *V. labrusca* 102
Microbial disorders, control of ... 131
Microbial disorders of wine 127
non- 122

Micronutrients 297
Molar absorptivities 187
Molybdophosphoric anions 199
Must(s)
amelioration of 106
Brix values for eastern grape ... 91
lactic acid bacteria in fermenting 17
samples, malic and tartaric
acids in 97

N

Nicotinamideadenine dinucleotide
(NAD), enzyme cofactor 171
Nitrogenous substances in wine .. 20
Nonflavonoids 268
Nursery stage 302
Nutritional requirements of malo–
lactic bacteria 161

O

Oak
American 261
extractives 275
flavor 291
growth rate 264
species 261
Organic acids for several fruits ... 20
Organic acids in wine 16
Organoleptic character of wine .. 50
Oxaloacetic acid decarboxylase
activity 177
Oxidation defects 127
Oxidation, wine 278
Oxidative browning 119

P

Palomino 299
Paper chromatography 161, 295
Pasteurization 157, 297
Pectin content of American grapes 99
Pectins in wine 37
Pectolytic enzymes 109
Pediococcus 153
Pesticides and fungicides, analysis
for 141
pH 163
Phenol(s)
colorimetric analysis of 186
content of wine, total 185
formaldehyde precipitation of .. 201
per berry, total extractable 30
formaldehyde resins 201
Phenolic
compounds during vinification,
evolution of 79
substances of grapes and wine,
analytical fractionation of .. 184
substances in wine 27
Pherograms of different grape
varieties, protein 4
Pigments, analysis of diglucoside
anthocyan 139
Pigments during grape maturation,
development of 76

Plaster 300
Polymerization of flavan molecules 65
Post-fermentation care of wine ... 120
Pressing operations, crushing–
    maceration– 279
Protein
    cloudiness 125
    pherograms of different grape
        varieties 4
    stability 23
Punching down 294
Pyruvic acid 167

Q

Q. robur 262
Q. sessilis 262
Quality control check points 220
Quality control and evaluation,
    wine 212
Quercus alba 258

R

Racking 290, 301
Raw materials, grapes and other
    fruits as 11
Rays of wood 257
Reagent, Folin-Ciocalteu (F-C) .. 186
Red dry table wines 291
Reducing agent, addition of sulfur
    dioxide as 280
Reflux ratio, internal 248
Relative humidity 304

S

Saccharomyces cerevisiae 235
Saccharomyces feremntati 299
Sauternes 298
Sauvignon Blanc 291
Schizosaccharomyces 154
Sensory properties 14
Semillon 291
Sherry, film or flor 299
Single and split columns for brandy
    distillation 237
Skin waxes, grape 6
Solera 302
Sorbic acid 297
Species of the United States, com-
    mercially important grape ... 18
Split columns for brandy distilla-
    tion, single and 237
Spoilage, aerobic bacterial 213
Stability, bacteriological 154
Stabilization, wine 116
Standards, analysis methods re-
    quired by legal or regulatory 135
Starter cultures 158
Stem 285
Sterile filtration 157
Storage area 286
Succinic acid 296
Sugars in native American grapes 90
Sugars in wine 13

Sulfur dioxide
    as bisulfite ion in solution 233
    as reducing agent, addition of .. 280
    in wines, analysis for 137
Surface-to-volume ratio, cooperage 266
Sweetening 297

T

Tannins 62, 254
    astringent 287
    bitter 287
    condensed 63
    hydrolyzable 63
Tartaric acid 296
    -to-malic acid ratio 98
Tartrate instability 122
Tartaric acids in must samples,
    malic and 97
Technological sequence, biological– 1
Temperature of fermentation 290
Threshold values, exact 271
Titratable acidity 18, 163
Total phenol content of wine 185
Tungstophosphoric anions 199
Tyloses 258

U

United States
    Bureau of Alcohol, Tobacco and
        Firearms (BATF) 106
    commercially important grape
        species of the 18
    history of winemaking in the .. 88

V

Vinegar, wine 302
Vinification, evolution of phenolic
    compounds during 79
Vanillin 287
Vitamins in wine 33
Vitis
    distribution of anthocyanins in the
        different species of 57
    labrusca 89
    rotundifolia 15
Volatile compounds in wine 39

W

Waxes, grape skin 6
White dry wines 289
White Riesling 291
Wine(s)
    aging of 227
    analyses 226
        methods for 134
    analytical fractionation of pheno-
        lic substances of grapes and 184
    by their anthocyanins,
        classifications of 69
    color 291
        during aging, changes in 82
        chemistry of red 50
        determination of 71
    for distillation 233

Wine(s) *(Continued)*
fining and clarification of 121
hydrocarbons in 76
inorganic constituents in 32
maturation, wooden containers in 254
microbial disorders or 127
   non- 122
nitrogenous substances in 20
organic acids in 16
organoleptic character of 50
oxidation 278
pectins in 37
phenolic substances in 27
post-fermentation care of 120
quality control and evaluation .. 212
red dry table 291
stabilization 116
sugars in 13
tannins in 62
total phenol content of 185
vinegar 302

Wine(s) *(Continued)*
volatile compounds in 39
white dry 289
yeasts 283
Winemaking
home 278
from native American grape
   varieties 88
practices, modification of 105
Winery size 215
Wood cooperage 254
Wood, rays of 257
Wooden containers in wine
   maturation 254

Y

Yeasts 282
classification of 283
wine 283
Y(glucose) 164
Yeso 300

The text of this book is set in 10 point Caledonia with two points of leading. The chapter numerals are set in 30 point Garamond; the chapter titles are set in 18 point Garamond Bold.

The book is printed offset on Danforth 550 Machine Blue White text, 50-pound. The cover is Joanna Book Binding blue linen.

Jacket design by Gerald Quinn.
Editing and production by Spencer Lockson.

The book was composed by the Mills-Frizell-Evans Co., Baltimore, Md., printed by The Maple Press Co., York, Pa., and bound by Complete Books Co., Philadelphia, Pa.